MEN OF METALS

An exciting career among the pathfinders of modern metallurgy

Samuel L. Hoyt

American Society for Metals
Metals Park, Ohio 44073

Nothing contained in this book is to be construed as a grant of any right of manufacture, sale, or use in connection with any method, process, apparatus, product, or composition, whether or not covered by letters patent or registered trademark, nor as a defense against liability for the infringement of letters patent or registered trademark.

Library of Congress Cataloging in Publication Data

Hoyt, Samuel Leslie, 1888–1977
Men of Metals.

1. Hoyt, Samuel Leslie, 1888–1977
2. Metallurgists—United States—Biography.
3. Metallurgy—History—20th century. I. Title.
TN140.H754A34 669'.0092'4 [B] 78-31521
ISBN 0-87170-059-X

PRINTED IN THE UNITED STATES OF AMERICA

Contents

Chapter 1

Student Days in Germany, 1911-13

Meeting leading German metallurgists; learning German; German pedagogy; travel experiences.

I was graduated from the University of Minnesota in 1909 with the degree of Engineer of Mines and spent two years at Columbia in postgraduate studies. It was there that I decided to switch from mining to metallurgy. I was on my way in 1911 to Germany and France to continue in metallurgy and to write my Ph. D. dissertation for Columbia. Upon discussing the possibilities with Dr. William Campbell, my metallurgy professor, he advised, "If I were in your position I would spend one year in Germany with either Tammann or Guertler and one year in France with Le Chatelier. I would also take some work in physical chemistry." He was the one who, with my father's assistance, started me on a career that was to introduce me to many famous Men of Metals, in this country and abroad.

Actually, fate played a major role in my choice of metallurgy by guiding me to Dr. Campbell. I went to Columbia to major in geology and ore dressing, and after selecting the courses I wished to take I presented my list to Professor Walker, head of the department. He approved but added a course in metallurgy under Campbell. Unknown to him, he had, by a stroke of his pen, determined my future career. The lectures in metallurgy were so fascinating that I decided to switch my allegiance — but not until I had checked the World Almanac and found that metallurgy was a large and important industry.

William Campbell was a quiet Scot and a student of Roberts-Austen, a major figure in English metallurgy. He had come to Columbia to study with Henry Marion Howe (of whom, more later) and had remained to be his assistant. I have a photograph that shows Howe and Campbell along with Irving Langmuir, Bradley Stoughton and others. Later, Campbell was Howe Professor of Metallurgy and enjoyed a long and distinguished career as teacher and consultant. Although he was reserved in manner, he was a friendly person and was always glad to greet former students.

Columbia attracted young men of exceptional talent and promise, and I should like to mention a few before boarding the ship for Germany.

Harvey Mudd was studying to carry on his father's extensive mining interests. He was a serious, soft-spoken young man with a delicious sense of humor. With superior intelligence and judgment he was easily marked as a success and lived up to that estimate in later years. After Harvey's premature death, in 1955, Mrs. Mudd founded the Harvey Mudd College in Claremont, California, as a fitting memorial.

There were several others who became outstanding in later years but who were just offstage as Men of Metals. I will mention just three. The first, Wellington Koo, would surely be a Man of Metals had he chosen a metallurgical instead of a diplomatic career. I knew Koo in a class in mineralogy as a quiet and distinguished student. I figured that he would enter mining or metallurgy in some capacity, but he heeded the call of his country instead.

The second was as close to metallurgy as ore is to metal, which is not close at all. He was Perry G. Harrison, who mined a lot of iron ore in Minnesota and became vice-president of Oglebay Norton Co. He was a brilliant student with great promise of a successful career, which was one reason why I introduced him to a very special little lady who became Mrs. Harrison.

The third was Henry Allen, who worked on nickel steels in our laboratory. He was a quiet worker who seemed to know what he was doing — which is more than can be said for most of us. He fulfilled that early promise and later became chief metallurgist of Henry Disston and secretary of the famous Franklin Institute in Philadelphia.

We now board the *Oscar II* for the adventure in Europe. Some wonder why I did not include England, where, insofar as we knew, there were more outstanding metallurgists than in any other country. I was following Dr. Campbell's advice, and, in addition, the opportunity to learn two important languages was highly appealing. Furthermore, my father had always wanted me to study in Germany, and I had great respect for his judgment and wishes. In any event, the die was cast.

The trip over was fabulous, with excellent companionship and food — so pleasant that I have always preferred to cross by ship rather than by air. After a stop at Christiania (now Oslo), I went directly from Copenhagen to Jena, where I took German in the summer session at that famous university. Almost immediately I had an opportunity to try my high-school German. After signing in with the registrar and paying my fees, I inquired about a place to live. I was referred to a lady standing nearby who had a pension, my first "pension mama." Frau Demuth knew no English, so while walking up to her place I tried what I called "conversational German." I asked a few questions, but only after figuring out each time what the answer might be so I could understand it. Not once did I receive an answer in my expected terms, which demonstrated that school-taught German did not suffice for conducting a conversation.

I was immediately immersed in German, both printed and spoken — at the university, living with a German family, and at social and cultural events. The latter were usually held at the town square just inside the old wall gate. At the university I first became truly interested in that incomparable literary team, Goethe and Schiller. My lecturer was Professor Schmidt, a Goethe scholar. Schiller's famous summer house was nearby and afforded a convenient opportunity for a visit and a lecture. Goethe used to come over from Weimar for greatly appreciated intellectual companionship, and we were shown the stone bench where they had sat and the outdoor fireplace. On a visit to Weimar, I visited Goethe's equally famous Gartenhaus, on the bank of the Ilm. At a much later date I revisited both places under greatly altered circumstances, as we shall see.

After the summer session closed, I continued my studies of German with a private lesson each evening with Dr. Demuth and in preparation each morning for the evening session. In addition, I had a rare privilege of learning and practicing German the easy way. Fräulein Johanna Sondhaus, a local schoolteacher, had the delightful habit of coming by each afternoon to go with me on visits to local points of interest. These included the old town, the seven hills and the battlefield at Auerstadt, where Napoleon defeated the Prussians in 1806. The tactics that had been successful for Frederick the Great were no match for the unorthodox tactics, movements, and artillery of Napoleon.

The time came at last for the decision: Tammann or Guertler. Actually, it was not made on a personal basis. I had become indoctrinated in the attractions of Berlin, and I was aware of the outstanding reputation of the Royal Institute of Technology at Charlottenburg. Furthermore, Walther Nernst had the chair of physical chemistry at the University of Berlin. I recalled the admonition to study physical chemistry, so, taking everything into account, I left Jena for Berlin.

During my first weeks in Berlin I enjoyed an interlude of the kind that was frequently available. It was a refresher course on the microscope that was given that fall in Munich by the top optical physicists of the Carl Zeiss Works. So it was down to Munich, with side trips to Innsbruck and Neuschwanstein, the fairy palace of King Ludwig of Bavaria.

This seminar cleared up the corrections of apochromatic objectives for chromatic and spherical aberrations for me. I understood for the first time, for example, why the use of a yellow filter was needed with the ordinary, or achromatic, objectives. The more highly corrected apochromatic objectives permitted the use of white light, or even of blue light to secure higher useful magnifications. Since then I have also understood numerical aperture and the corresponding resolving power of an objective or the limitation of its useful magnification. A given optical system can reveal particles, or reproduce them in the image, only down to a definite and limiting size. During the 1920s there was a flurry of excitement (and publicity) about supermagnifications, but that died out quietly as the concept of empty magnification became broadly known.

Neuschwanstein has the reputation of being the most beautiful and picturesque of all royal palaces. Those who have seen the moving picture about the Brothers Grimm will appreciate that. The interior, though unfinished, is also beautifully decorated and is strongly reminiscent of Richard Wagner. King Ludwig was Wagner's patron when he was in difficulty with the law during the revolution in 1848.

Innsbruck was a peaceful and quaint medieval town in glorious mountain surroundings. It was typical of the times when people sat leisurely in Konditoreis while sipping hot chocolate and reading the news of the day.

Back in Berlin, I tackled the institute to gain admission to classes and laboratory. My experience with the German educational system had started. I had brought official documents (registrar's records) to show the courses in which I had credits at both Minnesota and Columbia — they were not enough. So I sent for my diploma from the University of Minnesota — it was not enough. Following instructions, I then sent for my high-school diploma to show that I had adequate credentials for entering Minnesota. It was explained to me that some of our high schools did not prepare adequately for admission to a university.

My first contact with the faculty was Geheimrat (Privy Counselor) Mathesius, head of the metallurgical department. He came from an old and distinguished family and was a blast-furnace man from the Ruhr district in Westphalia. I noted that he was distinguished in appearance and dignified in manner, which befitted the top chair in Germany, as he engaged me in a general conversation and inquired into my plans. After approving the courses that I wished to take, he turned me over to Dr. H. Hanemann,

Dozent on the faculty and in charge of the metallographic laboratory. It was there that I completed the laboratory work on my dissertation on the constitution of the ternary alloys of copper, tin, and zinc. It was eventually published in the Journal of the Institute of Metals for 1913.

My two years' association with Dr. Hanemann showed him to be quietly reserved with a sound and logical approach to metallurgy. He was Professor Hanemann later, but we all called him "Doctor" in those days. He gave me assistance as I needed it but left me largely on my own.

A highlight of my experience in the metallographic laboratory was provided by one who could never qualify as a *Man* of Metals but she would most assuredly be on any list of *Women* of Metals — Fräulein Angelica Schrader. She had completely mastered specimen preparation and photomicrography, the best evidence of which was the "Atlas Metallographicus" of Hanemann and Schrader. I owe her a personal debt of gratitude for helping me to become skilled in the art. Her work, by the example it set of her beautiful photomicrographs, has contributed substantially to the high standard of the art today.

Fräulein Schrader, now Dr. Schrader, is one of a small group of her sex who have contributed handsomely to metallography. They are a fine group with excellent careers and should be an inspiration to young women who favor science. We will have more on Dr. Schrader and the ladies as we go on.

I followed Dr. Campbell's advice and registered for Dr. Guertler's lectures. His presentation was lucid and logical and treated the phase rule and physical chemistry deeply enough to give metallurgists an understanding of alloy systems and phase equilibria.

After the first lecture I went up to introduce myself and to convey greetings from Dr. Campbell. I was surprised to have him ask in perfect English, "Are there any questions? You may not have understood all that I said." He had been on the faculty at the Massachusetts Institute of Technology and his mother was an American, the daughter of General John Sedgwick, a Union soldier who fell at the battle of Spotsylvania Courthouse.

I learned a lesson in semantics in his first lecture that, in principle, has disturbed me ever since. Professor Campbell frequently used the term "mixed crystals" when discussing alloys such as copper-nickel, iron-nickel, and alloys of similar structure. Under the microscope they appeared as simple polygonal grains like pure copper or iron. I used to ponder the meaning of the term: Did it mean very tiny crystals of the two metals that were mixed together like sugar and salt crystals? Dr. Guertler discussed the same alloy type and referred to it as *Mischkristall* in German, and my eyes opened. It means *solution* — only crystallized or solid. In other words, the nickel, for example, actually dissolved in the copper, like sugar in coffee. This was a case of faulty and inexcusable translation from the German in which the term originated. Dr. Guertler was the only German with whom I conversed in my mother tongue.

Many visits to Dr. Guertler's home followed my introduction as I was helping on some of his work. I prepared English translations of abstracts for his International Journal of Metallography, which he could have done as well as I did. I recall one German phrase, with surprise and embarrassment, which I mistreated badly. I have done my share of translating, and that experience taught me a lesson, to wit: The translator should know the subject matter in both languages and then render it in English as he would have phrased it were he the author.

Dr. Guertler was a genial host as well as being an informal teacher at his home. He was a student of Tammann at Göttingen, which was always a mark of distinction. In addition to being a teacher, Guertler was also researcher, editor, consultant, compiler, and writer. His monumental "Metallographie" is eloquent testimony to his zeal in the last category. Incidentally, I had arranged to translate it but World War I intervened.

I attended several lectures of Professor Nernst at the University of Berlin. Though they were few in number, I was glad I could see the famous man in action. Due to the conditions (please note that word) in force in the registrar's office at the university, I had to drop out. I called on Professor Nernst in his office to discuss my situation and committed a faux pas of some magnitude. I referred to the *Umständen* involved, whereas I should have used *Zuständen* — which is similar to using "red tape" instead of "conditions." As he reacted and said something about *Umständen,* I became aware that I had said something I shouldn't have.

We had some young men with us who gave promise of success in later life, and I should like to mention several. Guy Wainwright was a welcome addition to the foreign contingent as a friendly and communicative individual. His father was president and principal owner of the Diamond Chain Co. in Indianapolis, so Guy was there to learn more of the characteristics and the sometimes devious ways of steel. He took over the business later as a sound businessman. In the 1940s, Art Focke, then his chief metallurgist, had arranged for me to give a talk before the local chapter of the American Society for Metals. Upon visiting the plant, Guy and I had a good session on our Berlin days, which had been heightened by his presence.

We also had a famous Croatian student, Franz Hanamann. His name caused some confusion until we violated a rule of German by accenting the two similar names on the second syllable of each, as "nay" for Dr. Hanemann and "nah" for Franz. With Alexander Just, Franz was co-inventor of the tungsten lamp filament as a significant improvement over the carbon filament. He was working on a Dr. Ing. degree, and had returned recently from the United States, where he had sold his patent rights to General Electric. He was living comfortably on the proceeds, I can add. He was then nitriding steel with ammonia but missed the lower temperature for nitriding steel that Fry used at Krupp for his Nitralloy process.

Franz was a genial person whom we all liked, especially for his humor and willingness to help others. He had planned to come to the United States for the 1912 meeting of the International Association for Testing Materials in New York City but changed his mind quickly. The reason blared forth in the headlines one morning: TITANIC SUNK. After World War I, he went back to his native country and became president of the University of Zagreb.

One day a young man presented himself at the laboratory and announced he might be joining us. His attitude seemed to say that he might, just possibly, if he liked what he saw. We noted particularly a quiet air of self-assurance and a straw hat, obviously too small, worn on the side of his head. That was the young man who would be the first to unravel the mystery of the age-hardening process and later become president of the International Nickel Co. — Paul Dwyer Merica. After graduating from Wisconsin and teaching for two years in China, he was at the University of Berlin working for a Ph.D.

Paul must have liked what he saw, because he joined our laboratory group. His work on the magnetic effects of stressing steel kept him cooped up in a small testing laboratory just off the main laboratories. I was preparing alloys in the furnace room adjacent to it and I distinctly recall his not unpleasant tenor singing "Your little hands are cold" from *La Bohème.* Also, and long before students swallowed goldfish, Paul put on a yogurt-eating contest at his place. I declined politely, as I could not eat even one cupful. He was a great addition to our group, loved by all, and an interesting companion. Which reminds me that Paul and I decided one time to stay up all night and then to take in an airplane race the next day. It tickled Paul to note the incognito name of one participant on the program, which was followed by an asterisk. The footnote gave his real name as Count So-and-So. I also recall that Paul could beat me at tennis, and I regretted that we could not race in shells, as that was my specialty.

I must also mention another man who was at work in the furnace room. He was E. H. Schulz, who was working on stresses in quenched steel. We will meet him again later and under unusual circumstances.

The metallurgical world is now well acquainted with William Kroll of titanium fame. I knew him then as Guillaume, his Luxembourg name, although he was not in our group at the laboratory. Geheimrat Mathesius conducted his students on a tour through the Rhine-Westphalia steel districts, and I was permitted to join. We were split into pairs for traveling, which is responsible for my becoming well acquainted with William. It happened that I was given a very uncongenial partner, which he observed. After a day or so, he arranged a swap in partners and I was with William for the remainder of the trip. (Thanks, Guillaume.)

I recall him as a very friendly individual with a keen and alert mind. I wish I could say about him, as I can about Paul Merica, that I was sure he would some day climb the heights in his profession. At that early date I could know nothing of his persistency, skills, and willingness to think ahead of his times.

One day in the fall of 1911, Dr. Hanemann called me into his office. Dr. Walther Mathesius, son of the Geheimrat, wished to talk with me because he was leaving soon for the United States. Dr. Hanemann briefed me quickly on Walther and referred to him as *ein sehr begabter Mensch*, or a very talented person. Many years later, as Walther, Mrs. Mathesius and I were at lunch in Chicago and reminiscing on his successful career with U. S. Steel, I mentioned how Dr. Hanemann had referred to him. That really pleased him, and especially, I believe, because there had been a certain rivalry between them. It also spoke nicely for Dr. Hanemann's objectivity. That meeting in 1911 started a friendship that was maintained up to his death. Back in Charlottenburg, he queried me on conditions in the United States and the new way of life he was about to enter. He also seemed to admire my German, because he vowed to do equally well in English.

In those days in Charlottenburg, metallurgists were making slow progress in understanding metals and alloys, with the emphasis on their physical makeup or constitution, phase changes, microstructure, and properties. Then one day Dr. Hanemann told us about a new technique that was to open for examining solid-state structure, or the arrangement of the atoms on the crystal lattice. Max von Laue had just announced that crystalline solids produce diffraction patterns by X-ray bombardment. The atomic planes of the crystal act on the X-ray beam like diffraction gratings do on light. His name is perpetuated in X-ray crystallography by Laue diagrams. Actually, Laue was trying to determine the wavelength of X-rays and hoped that a crystal would serve as a grating.

The name Adolf Martens was well known because Floris Osmond had coined the name *martensite* for the structure of quenched steel. I was surprised to learn that he was still alive and holding classes on testing materials. It didn't take long for Paul Merica and me to join.

Geheimrat Martens started his life's work as an engineer for the state railroad and was concerned with construction, materials, and acceptance tests. At the age of thirty he was called to the institute in Charlottenburg as assistant in, and then director of, the department of testing materials. As this activity expanded and his ability became recognized, he established the Materialprüfungsamt (Materials Testing Bureau, or MPA) in Gross Lichterfelde, a suburb of Berlin.

Martens contributed abundantly to the improvement of testing equipment and methods, especially increasing the accuracy and the reliability of the tests. Furthermore, he did not test just for the sake of testing, or as a check on the specification requirements, but insisted that the test results bear a rational relationship to the service requirements. That surely is reasonable, and I have often wondered why we were so unnecessarily tardy in adopting tests for the very real hazard of notch brittleness. The Geheimrat's goal was even more fundamental than that, however, because he sought to improve the reliability and performance of German manufactures and thus to make "Made in Germany" a mark of excellence.

Henry Clifton Sorby was the first to develop the microscopic examination of metals, but Martens was also a pioneer. When Martens started his own work in 1874, he was unaware of Sorby's publication and was thrown on his own resources to develop equipment and methods. Later, he worked with Carl Zeiss to construct the excellent Martens-Zeiss microscope. Curiously, neither Sorby nor Martens adopted metallography as a major activity.

When Paul and I attended his lectures, the Geheimrat was quiet, dignified, and even reserved in manner, yet we were all aware that a vast store of knowledge and experience lay behind his words. After explaining the various tests he would expound on the significance of the results and the true meaning of accuracy. In a few lectures he broadened our understanding of materials and their applications to give a deeper conception of the field of our profession. Paul and I were among the last to attend his lectures, as he died in 1914.

After the lectures we would proceed to the laboratory, where we either witnessed tests, or performed the tests ourselves on his machines, with his test bars with spherical seats, and with his precise mirror extensometer. I still have my own test sheets as souvenirs.

On the last day we visited the micro-laboratory that was in the care of his early *Mechaniker*, or laboratory assistant. There the *Mechaniker* described the technique and equipment carefully and with obvious pride.

One day Dr. Hanemann showed me a large book, Part II of "Martens's Handbook of Materials Engineering" (to paraphrase the title) by E. Heyn, and advised me to take his lectures. "He is the best man for the application of metallographic principles to materials technology and I believe you will enjoy him." So I signed up.

Emil Heyn (or, to give his full name — Geheimer Regierungsrat Professor Dr. Ing.e.h. Emil Heyn) was head of the metallographic department of the MPA and also director under Geheimrat Martens as well as a professor at Charlottenburg. He was a student of Ledebur and joined Martens while he was still at the institute, about 1898. His metallographic studies developed into what became his life work, metallurgical science and its application in practice. For example, in 1898, he developed deep etching and macroetching with copper ammonium chloride, both still valuable techniques for investigations of steel. In 1900, he examined the effect of reducing gases on copper that contains copper oxide, or hydrogen embrittlement of copper. In the same year, he studied hydrogen embrittlement of steel and recovery from the effect.

In 1904, Heyn contributed to the correct understanding of the iron-carbon alloys of high carbon content. That was the so-called double diagram which shows the iron-graphite stable alloys and the iron – iron carbide metastable alloys. Charpy, in France, had reached the same conclusion, so it is the Heyn-Charpy diagram. That distinction persists today. His work on internal stress and "season cracking" was equally pioneering and guided future work in the field.

Heyn's lectures reflected his considerable experience, while his presentation, combined with a fine sense of humor, made them enjoyable as well as instructive. It was through him I first became interested in the notch effect and the notch-brittle behavior of ductile steel. He was one of a limited few who recognized that a normally ductile steel *could* fracture in a brittle manner, just as cast iron breaks, if a notch is present. He also made the point in his book. I cannot resist noting that this highly treacherous behavior was not recognized widely, nor did it become the object of intensive investigations of brittle fracture until a lapse of about forty-five years after Charpy, Heyn, Izod, and others demonstrated the notch brittleness of steel.

Heyn was well-known abroad but not as well-known in this country, judging from the reactions when I mention his name. His lectures were extraordinarily popular at the institute and attracted so many students that they had to be held in the main lecture room. He was completely at home as a lecturer and was welcomed at international gatherings, where his leadership made him a popular chairman. His command of English, French, Italian, Swedish, and Russian was welcomed for use in interpreting.

Just prior to his passing in 1922, at the age of 54, Heyn was chosen to head the newly established Kaiser Wilhelm Institut für Metallforschung, the nonferrous research institute. Upon his death, Dr. O. Bauer, his associate for nearly twenty years and one of my professors, carried on until the institute was transferred to Stuttgart in 1934.

Heyn was not the only metallurgist who was highly regarded abroad and less well-known here. Another was Dr. Carl Benedicks, student of Henri Le Chatelier and Floris Osmond at the Sorbonne in Paris, and professor at the technical university in Stockholm as well as director of the Swedish Metallografiska Institutet.

I was aware of the famous Swedish steel (my excellent razor blades, for example) and so I decided to go up for a visit. Dr. Guertler told me to include Dr. Benedicks and promised to advise him of my coming. Dr. Guertler went on to say that Benedicks was truly gifted in devising advanced experiments and in implementing them with home-made equipment. I received a warm welcome (maybe because I came from Minnesota), which was followed by a visit through his laboratories and workshop. Dr. Benedicks was a true gadgeteer. I valued that visit highly because in those days we had to make many things ourselves, and I was due to take over a laboratory myself when I returned home.

Earlier in this discourse I mentioned Professor G. Tammann of Göttingen. Although I did not study with him, I did make his acquaintance when I was on a trip through the Rhineland. After I had visited the medieval town of Goslar, which included the Barbarossa palace, and the famous School of Mines at Clausthal in the Harz Mountains, I went to Göttingen.

At Göttingen I made my way to Tammann's laboratory, which was in his house, or vice versa. I found him a genial and cultured gentleman who received me cordially although I was an unknown from across the ocean. I knew by then his high position in metallurgical science and his many contributions to constitution diagrams and phase equilibria. Tammann was a dedicated leader in these fields and contributed to them as no one else had. He published many papers and wrote several books that publicized his work and marked him as the leader in that phase of metallography. (I trust the reader will not assume that I equate metallic microscopy with metallography.)

If Tammann observed some aspect of metal behavior that he could not explain, he frequently did something about it. For one, he observed that metals could be plastically deformed, that cold worked metal became stronger, and that strength and ductility reverted closely to their initial values after annealing at an elevated temperature. Ewing and Rosenhain had published the pioneer work on the mechanics of plastic deformation in 1900, and Tammann went on from there. Although it was before X-ray methods were available and long before the role of dislocations became (reasonably) clear, Tammann contributed valuable concepts to the field. Without going into details, a unique contribution at that time was a *Zwischensubstanz*, or intergranular material, that he felt played a dominant role in the over-all behavior. To demonstrate the presence of this material, he treated cast cadmium with an aqueous solution of iodine. The metal dissolved and left a transparent and insoluble network. I am unaware of any follow-up of this finding although this broad field has inspired intensive studies right up to the present.

An outstanding contribution was his determination of resistance limits in solid-solution alloys of a noble metal with a less noble metal — for example, gold with copper. In ammonium sulfide solution the copper is dissolved when present in the alloy above a certain percentage, the resistance limit. When the copper content is below that limit, it is not dissolved. The explanation of this effect required assumptions of the spatial distribution of the less noble atoms in the parent lattice. By means of excellent X-ray diffraction work, Ed Bain cleared this point up later and opened the field of order-disorder atomic arrangements.

After a visit to his laboratories I returned to the office to say goodbye. He then presented some of his reprints to me, souvenirs of a stimulating visit. I will add one more comment. In a biographical sketch, Masing, his successor and former student, quoted Tammann as telling doctoral candidates when preparing for the examination, "The presentation must be brief, clear, and complete" — a good admonition to all who write reports.

Leaving Göttingen, I was on my way to Paris via the Rhine, the Vosges Mountains, and the Black Forest. I enjoyed everything along the Rhine except the wine. (My German friends were horrified at the thought of a wineless Rhine trip, even when I explained I didn't drink wine or anything stronger.)

My first stop on the Rhine was a visit to Marksburg, a castle that had been restored. Louis XIV had destroyed all those German castles, except Cochem on the Moselle, so this was a rare opportunity to look back into the Middle Ages. However, duty called and I had to curtail that trip in order to visit a lead refinery in the vicinity. It was unique in that silver was still being recovered from lead by additions of zinc. This refinery was a museum piece and I felt I had to see it. I don't recall just how I gained entrance, but it must have been by some unorthodox method — as I learned when the astonished manager caught up with me and escorted me to the nearest exit.

I was to join Dr. Hanemann in the Black Forest about a week later, so that interim gave me an opportunity to go into the Vosges Mountains in France and brush up my French. I went via Strasbourg and Colmar in Alsace to Gérardmer, a small mountain town and outpost of the French army. I stayed in the principal hotel amid glorious scenery and spent my time in studies of the language. I developed a real interest in French upon discovering the presence of the charming daughter of the proprietor. I recalled Johanna Sondhaus of Jena and pictured equally delightful lessons in conversational French along the beautiful paths of the region. But the results were completely negative — apparently, French was to come the hard way.

After that brief visit in Gérardmer, I crossed the Rhine at Colmar to Freiburg and went up to a quiet little *Gasthof,* or inn, in Buchenbach. I found it, as usual, clean and neat, with good food. Speaking of food, I made my discovery of the famous black honey of the Black Forest — there is no connection, because the honey is really black from local flowers. A continental breakfast is never better than with hard rolls and that honey.

In the mornings I continued writing my dissertation and I spent the afternoons with Dr. Hanemann discussing what I had written and how best to write it. Dr. and Mrs. Hanemann and I also enjoyed walks through the region.

As soon as that week had come to an end I bade goodbye to the Hanemanns and said to myself, "Paris, here I come." I was off at last for a year with Professor Le Chatelier, or so I supposed.

Paris — A Strange Interlude

After some sightseeing and house-hunting in Paris I settled in Pension Verdin on Avenue Victor Hugo for the serious business of learning French. I wanted to be able to converse in French before approaching Le Chatelier at the Sorbonne.

I took French lessons each morning with the daughter of a professor at the university and studied most of the remaining time. I made good progress inasmuch as I had studied French at home. I also had a windfall. There was an American girl at the pension, a Miss Claire and the sister of Mrs. Kresge of the Kresge stores. She told me that she planned to go to Germany and would help me in French if I would help her in German. So we made a deal.

Miss Claire was hailed as "the second Melba" by the opera star herself, and so was anticipating an operatic career — but not in France. She was in demand in Parisian social circles and was well-known locally for her pleasing personality and beautiful voice. She helped me greatly with French but she had a tale of woe. When I asked her why she wanted an engagement in Germany instead of France, she said she had offers in opera in France but couldn't go along with the unwritten part of the contract, or the extracurricular activities to which she would have to agree. In those days the royal courts of Germany had their own opera companies and she wished to join one.

One time she demonstrated her considerable fluency in French. Billy Papke and Georges Charpentier were scheduled to fight for the middleweight championship and I mentioned that I would like to see it. (I had recently knocked down Count de Grassi at Columbia, who was being hailed as the White Hope, so I was interested.) Miss Claire insisted on negotiating for my ticket. To my surprise, she talked a blue streak at the window and I wondered what was going on. Papke won, but was overweight, and the next day I read the strange headline, "*Qu'il Perd, Gagne,*" or "he who loses, wins." Charpentier became the champion.

To get on with my French I talked to salespeople, conductors, and all sorts of other people to gain experience. However, I decided to use a policeman whom I passed on my way to the Arc de Triomphe as a yardstick of my progress. Then came the day when our conversation was fluent both ways and without errors. I had passed my final.

With Osmond, Professor Le Chatelier had established the Sorbonne as a top-ranking school for postgraduate studies in metallurgy, so it was with high hopes for a good year that I approached the Sorbonne that morning. Later in life I should be able to say, "Yes, I studied at the Sorbonne with Le Chatelier."

I explained my mission to an assistant and then was ushered into the office of Professor Le Chatelier. He arose as he saw me and grasped my hand in a friendly greeting. After I had explained what I had done at Columbia and Charlottenburg and told him that I wished to complete work on my dissertation in his laboratory, he felled me with words of doom "*Il n'y a pas de place.*" He simply had no place for me. He followed up by saying that if Professor Howe had written him last spring he would have reserved a place for me.

So that bubble burst in my face. Later in our conversation he said, "Let's see what other opportunities there may be here that will be agreeable and also meet your needs." He finally suggested that I study with Professor Léon Guillet at the Conservatoire des Arts et Métiers. That sounded agreeable because Guillet was well-known for his studies of alloy steels and as the developer of the famous French automobile steels when he was a consultant for De Dion-Bouton. He handed me a card of introduction as I took my leave.

I did have the privilege of visiting Le Chatelier's laboratories with an assistant who took me in his charge. A highlight was the laboratory of Osmond, who had died recently. I had also looked forward to working with *him,* to learn more about the metallography of steel. Osmond was the one to bring order to the microconstituents of heat treated steel and had named martensite, sorbite, and troostite. Those names persisted in our nomenclature for many years until Ed Bain came along. I also saw the original inverted-stage metallurgical microscope of Le Chatelier, the design that is universally used. I was also told of the Le Chatelier platinum and platinum-rhodium thermocouple that made it possible, for the first time, to measure high temperatures by the thermoelectric method.

The meeting with Le Chatelier, a famous and gracious man, is one of my fond memories. He was a kindly and friendly personality with individuality and strength of character.

Here came the incident that sent me back to Charlottenburg for my second year. Upon arriving at the Conservatoire I learned that Guillet was not there, but his assistant took me in hand and showed me the laboratories. He knew my plan and suggested that I call on the professor at his apartment, in view of the early opening of the school year. I decided to do just that, but somehow I got a hunch that there would be protocol to follow. So the next day I asked my French teacher for advice. Yes, there was protocol but she would handle it by composing a very proper letter for me to sign. She also suggested that I procure a letter from the United States consul.

After I was fully armed I went to Guillet's apartment and was admitted by his butler. He asked me to be seated in the antechamber while he took my documents in to the professor. As Guillet came out to meet me, I noted that he was a large man of rather stiff and imposing bearing and very formal in his approach to my interview. He greeted me at once with "My office is at the Conservatoire. I can be consulted there during my office hours." I was not of a mind to be stopped so quickly and I mentioned that I had to make my plans and time was limited, but he stopped me by repeating his first words.

I had met many Europeans by then and had found them to be uniformly friendly and politely hospitable, so Guillet's reception came as a rude shock. I decided immediately that I could not enjoy a year with him.

I left the apartment saying to myself, "Au revoir, Paris. Berlin, here I come."

Berlin Again

Back in Dr. Hanemann's laboratory, I completed the laboratory work and wrote my dissertation for submission to Columbia University and as a paper for the Institute of Metals. Although the experimental technique was not as elegant as it is today, I was able to establish the ternary diagram of the copper-tin-zinc system from pure copper to about 50% copper. I established the freezing points by plotting cooling curves of the melt through solidification. One day I allowed one of my ternary alloys to cool much further. I expected the plot to show one heat effect at about the temperature shown on the copper-tin diagram for the eutectoid. To my surprise, another one showed up nearly 100 °C higher as well.

I discussed this discovery with Dr. Guertler and he told me to look for a corresponding transformation in the copper-tin binary alloys that must have been missed by previous workers. By highly sensitive thermal analysis I located a new line or transformation and a new phase in that system and, of course, I was thrilled upon making a contribution of my own to the constitution of this complex system. This was all published in my Institute of Metals paper of 1913. I reasoned that the transformation must be a eutectoid transformation even though the microscope did not show a typical eutectoid structure. A eutectic that is close to a pure metal shows no eutectic structure: I found no eutectoid structure: Ergo, this eutectoid must lie close to the composition of the tin-rich phase. Later work by others confirmed this.

I can claim the new line in the diagram as my contribution, but the explanation I gave was only my best interpretation of my data and observations. I was fortunate that later work showed it to be correct. The literature is unclear or even incorrect on priority — which, after all, is really important only to the innovator.

In a laboratory that was full of activity, interesting items would come to our attention. I have mentioned E. H. Schulz and his work on quenched steel, and Franz Hanamann and his work on nitriding. I should also mention Dr. Hanemann's special melting and casting techniques for high-carbon melts of iron and carbon to study the system as close to the pure carbide as possible. As an example of a different experience, Dr. Hanemann showed me a microsection of an early Celtic dagger that contained martensite. That is the constituent of heated and quenched steel, and its presence demonstrated that the art of hardening steel by heat treatment was very old. More recently, when reading Homer's *Odyssey* I ran across much earlier practice, to quote:

> "And when a smith dips an axe or adze in chill
> water with great hissing, when he would temper it,
> for hereby anon comes the strength of iron."

The next and the last event of these two years abroad was the fall meeting of the Institute of Metals in Ghent, Belgium. I had become a member and I was scheduled to present my paper. Of course I was thrilled, but I was also apprehensive because I had spoken only once in public and that was a talk before a Sunday-school class on the power of music. I need not have been worried, because the English were kind and sympathetic. Even when adverse comments are offered in discussion, the reasons are stated as honest convictions. Furthermore, the discussions are well thought out and ably presented so they add constructively to the paper and to the literature. The English maintain these discussions on a high level. They are a delight to read. It was a treat for me, a young man from the United States, to attend this meeting and to observe the English in action and how they handle protocol.

The mayor of Ghent and local dignitaries attended the opening session and offered the official welcome, to which our president, Professor Huntington, responded in equally fitting words. The sessions were also formal, in that they followed a fixed routine, giving authors and discussants ample opportunity to speak. An informal and, in places, amusing account was published shortly thereafter in the Iron Monger. It was very well done and gave a good review of all that took place, and included personalities. I was pictured as "Glad Eye," with a crew cut.

In addition to Dr. Guertler, who read a paper, I met a number of the English participants and their ladies — between sessions, at meals, and at garden parties. The English are masters of the art of pleasant camaraderie and informal conversation, putting everyone at ease. They are good listeners as well as talkers. I should like to mention a few.

Sir Gerard Muntz, Bart., exemplified to me the aristocracy of English industry. His Muntz metal was the brass antifouling sheathing for ships. Sir Gerard took an active interest in the institute and in the sessions; in conversation he was friendly, with quiet and informal dignity.

Dr. Walter Rosenhain, of the National Physical Laboratory in Teddington, was well-known from the fifteen years of research back of him. The amorphous-phase hypothesis was being discussed very actively in those days and was brought up again by his presentation of a paper. This and similar attempts to account for various ways in which metals behave under heat and stress continued for several decades. It is clear now that they represented the application of logical reasoning to inadequate facts that gave conflicting and inaccurate conclusions.

Rosenhain rates in the top bracket of Men of Metals. He was prolific in research and resourceful in devising experiments and technique. His papers and lectures were classics of the art of scientific discourse. It was a treat, and a lesson, to hear him discuss papers of other authors and handle discussions of his papers. Whether on the offense or the defense, he was always on the firing line with facts, persuasive argumentation, and logic, confounding the adversary. He was fluent in French and German, an asset at international meetings at which he was frequently in the chair.

He went to Cambridge on an 1851 Exhibition Scholarship and remained in England. While he was at Cambridge, he and Ewing made their pioneer disclosure of the crystallographic mechanism of plastic deformation.

Dr. Rosenhain became head of the metallurgical division of the National Physical Laboratory in 1906, and remained there until he retired in 1931 to enter consulting practice.

Professor H. C. H. Carpenter (later Sir Harold) was the scholarly type and was very active in the institute. (The H. C. is for Henry Cort, an ancestor who introduced the rolling mill for wrought iron plate.) When I met him at Ghent, he was just transferring from the University of Manchester to assume the chair of metallurgy at the Royal School of Mines in London. He filled the post of Percy and Roberts-Austen so creditably that he was later knighted — a considerable honor for one of our inconspicuous trade. He told me he was going to the States that fall in preparation for his new duties. I suggested that he should by all means include Minneapolis and the Minnesota School of Mines at the University. I also invited him to be our house guest during his stay, since I was to be on the faculty at that time.

Carpenter had a highly productive career. With Edwards, he had worked on the copper-zinc system, which I had also worked on, so we enjoyed discussing points of mutual interest. Our discussion of the exceptional character of the intermediate beta phase was really interesting, because neither of us understood it.

In Ghent, I also met Professor Thomas Turner, the cast iron authority of Birmingham. His daughter was the wife of G. Shaw-Scott, the secretary of the institute. I can easily remember the name of his university. He told me he had just returned from the United States and had visited a southern city that, to his surprise, was called "Birm-ing-ham," whereas his home town was called "Birmingum." That reminds me of an Englishman who was in New York on business. He had telephoned his man that he would be at the "Berkley" and would await a call there for an appointment. However he pronounced it "Barclay," as they do in London, and numerous calls to the real Barclay failed to reach him.

Cast iron was well-known and in those days was used primarily because it was easily cast into shape and was machinable. Little attention was paid to its properties. Professor Turner had a more enlightened view and he studied cast iron as an engineering material. Hardness was an important property, and he devised a special method to determine it. With a sharp diamond point he made a series of scratches under decreasing loads. He would then determine the narrowest scratch by feeling them with a quill point. The corresponding load was a measure of the relative hardness.

The pleasant interlude at Ghent came to an end, and I was soon on my way across the Channel for a brief visit to England. I engaged an Englishman in conversation as I wished to learn more about our cousins. I had read "England and the English," companion to "Germany and the Germans" which I thought was well written, so I wished to do some checking. The background was as follows.

War clouds were on the horizon during my two years in Europe, not to mention the revolution that was brewing in Hungary. Germany seemed to feel that she was hemmed in by Russia, France, and England, and not allowed to develop as a world power. France was resentful of the defeat of 1871 and the loss of Alsace and Lorraine, and knew that Russia and England would help in case of trouble. The countries I visited were building their war potential, as was made clear in the press. I recall the bold headline in a Paris newspaper, "The Aerial Supremacy of France," followed by an account of her new air force. During my second year in Germany, that challenge was met by a "voluntary" subscription levied on all Germans, including the Kaiser, to build up the German air force.

The situation became especially critical when the Kaiser sent warships to Agadir to emphasize Germany's interest in Africa. The French ambassador to Berlin held prolonged discussions with the German foreign minister over that incident, and each morning we would get the papers to learn the latest news. One or the other had to back down, and France finally won that round. The morning of that announcement, the *Mechaniker*, or technical assistant, in our laboratory told me, "Bismarck advised against laying claim to any part of North Africa. The Kaiser should not have acted so hastily." His comment was about as satisfactory as anything I got from the papers.

Once, while those negotiations were in progress, I was a guest in the home of a well-to-do German family just outside Berlin. The lady of the house personified German culture, with the bearing and composure so well-portrayed on the screen by Dame May Whitty. The inevitable discussion of the war situation came up, although the company was musically oriented, and she lost some of her composure as she pictured the possibility of Germany being destroyed by the French air force. She was really disturbed.

The Englishman on the Channel boat was calmer about the situation, and philosophically contrasted peacetime life with what it would doubtless change to if war came. He commented that Englishmen had a contented and peaceful existence, inherited from the past as a mold into which they fitted comfortably. If war came, he assured me, everyone would be in training, on the battle line, or helping on the home front. How prophetic!

Upon closing this account I should like to set down a few experiences that I, as a young American, was able to have, and that I recall as being interesting and educational.

Speaking of being an American brings to mind an incident at Pension von Förster at 182 Uhlandstrasse. My "pension mama" announced one day that three "fellow Americans" were to be with us for a few days and she knew I would be happy to meet some "countrymen." When the day arrived I was introduced to three brothers from a country in Central America. Their English was as negligible as my Spanish, and we conversed in their nearly negligible German. Then came a post-mortem. It dawned on me that we had practically preempted the name "American." I asked Frau von Förster, "What should we call ourselves: '*Vereinigten Staatler*'? or 'United Staters'?" Neither sounds acceptable. I had been forewarned at my pension in Jena. A young German there told me that his mother had lived in America and was very fond of Americans. In my innocence I inquired where she had been and he replied, also innocently, "In Brazil." There is a good question — what should we call ourselves?

In those days, Berlin was beautiful, clean, and interesting, and we foreigners took full advantage of its educational, cultural, and scenic assets. I have already commented on the first, but we also had the best in stage and opera. In addition to excellent German plays we also had Shakespeare in German. It is said that Shakespeare is exceptionally well-done in German because of translation by the best German writers and the use of leading actors even in the minor roles. In opera, we had local talent, and guest stars, and even Caruso, whom I had heard frequently when at Columbia. We also had Arthur Nikisch and the famous Berlin Philharmonic Orchestra, with soloists like Eugène Ysaÿe, the Belgian violinist, and Ossip Gabrilowitsch, the pianist who was breaking in as a conductor. Gabrilowitsch was especially interesting to us because he had married Clara Clemens, Mark Twain's daughter. I had the pleasure of meeting her at a small musicale. One time, I had a seat at a Nikisch concert back of the podium at the rear of the orchestra. That is an excellent vantage point because one can observe how the conductor manages and controls his orchestra.

The surrounding country provides beautiful lakes, like Wannsee and Bukow, forests, palaces, quaint little towns, and Potsdam with its replica of Versailles and the music room where Frederick the Great played his flute. These could all be reached by local streetcar or by state railway.

A feature of our attendance in Dr. Hanemann's laboratory was excursions, or *Ausflugs*, into the surrounding country. The scheme was to ride to one town, then walk cross-country to another, with a stop at a *Gasthof* for lunch, and then ride back home in the evening. Dr. Hanemann would discuss German history and literature as extracurricular education. Those informal talks gave us a better understanding of the country in which we were living.

One time I asked Dr. Hanemann which of the German kings or emperors was the greatest. He thought a moment and then said, Otto the Bear. I had expected Carl the Great (Charlemagne, as we call him), Frederick Barbarossa, Frederick II (Holy Roman Emperor), Frederick the Great, or maybe the Great Elector who led his country to a rapid recovery after the Thirty Years War. One time he and Paul Merica entertained us with a discussion of philosophy. Paul was well versed in the subject because he had to pass an examination in it for his Ph.D., as I did not for mine.

Looking back, I am greatly impressed by the willingness of a busy professor to use his time in this way for the benefit of his students. His talks were objective discussions of men and events. He realized that his country, like all others, included the entire spectrum, from bad to good, but felt that Germany was not properly understood in other countries. He pointed out that Germany was not just a nation of poets and farmers, but had made great progress in the arts and sciences and had also become a leading industrial country. There was one other point he made clear. Whereas Germany had been the battlefield of European armies for centuries, she would try to avoid that in the future by being strong as a military power.

At this time I should like to make note of a few unusual observations. On our tour through the steel plants of Westphalia we were shown the Harmet process in operation. The modern generation has probably never heard of it, but for a short time it promised to increase the yield from steel ingots. A plunger or ram pushed the ingot up in the mold to keep pace with the shrinkage and so to eliminate the pipe cavity. It accomplished that satisfactorily, but the segregation of harmful impurities remained, hence its demise.

It was in France, however, that I saw sights which remain fixed in my memory and may be gone forever otherwise. One time I witnessed a street fight between two men, broken up, as is the tradition, by nearby women. I wonder if the women are still as competent. The modern equivalent is, possibly, the verbal eruption that occurs after an auto collision on a Paris street. I also witnessed a strange fight on a wharf at Boulogne. The technique was a jump in the air to be followed by planting a foot squarely in the face of the opponent.

Paris surprised me with compressed-air buses on the streets and trackless-trolley cars in a suburb, just as we see them now. The most delightful sight was the bird charmer at the Place de la Concorde. Surrounded by a hundred or more birds, he would call out a name. Immediately, one little bird would respond by alighting on his arm for the reward of grain. Other birds would then do the same thing at calls of their names.

I feel sure that two other sights will never be seen again. One was a Kaiser Parade of the local garrison in Berlin, and the other was the appearance of brilliantly uniformed Russian nobility at their Paris embassy. They always preened themselves on state or religious holidays.

That experience of two years abroad impressed me so favorably that I have always encouraged young people to travel and study in foreign countries. Outside the United Kingdom one will acquire a new language, although there are some who would even include England. If one is so-minded he will also become acquainted with the people and their country. Of course, friendships with people in one's own field will be especially valuable.

We can now note a proliferation of foreign assignments of one kind or another, both governmental and private. I believe I sense a start, long overdue, of using people who qualify in the language of, and familiarity with, the host country. Too frequently in the past, we have used personnel who knew little or nothing of either language or country. That was a grave mistake and might have been avoided. Yet with a dearth of fully qualified prospects, there are many times when unqualified personnel must be used. I noted both aspects of this situation as far back as 1911. Surely we like to have foreigners come here who speak English and know something about us, especially our good side, and they will have profited much more from their sojourn than they would have if every local contact had to use the foreigner's language.

My two years in Europe came to an end, whereupon I was on my way to Minneapolis and the University of Minnesota. I had been engaged to start a Department of Metallography at the School of Mines and I arrived on the job in September, 1913.

Chapter 2

Teaching Years at Minnesota, 1913-19

Establishing a department; studies of steelmaking for NAS-NRC; experiences in teaching and in the field.

IN August, 1913, I was on the ocean and on my way to my alma mater, the University of Minnesota, where I was to be assistant professor of metallography in the School of Mines and head of the department. My time in Berlin had come to an end, also my formal education. I was about to enter the adult world!

After a brief interlude of a pleasure trip through the Bavarian and Swiss Alps, a few days in Paris at my Pension Verdin, the meeting of the Institute of Metals in Ghent, and some sightseeing in London, I began preparations for my first job — teaching at $187.50 per month! I was quite aware of my evolution from student to teacher, from one who attended organized instruction to one who had to organize instruction. The thought was humbling, and the prospect was challenging. I was armed on the ship home with copious notes from Columbia and Charlottenburg, but my principal companion on the voyage home was Desch's "Metallography." Dr. Cecil H. Desch was professor of metallurgy at Glasgow; his book was the standard text of that period for English-speaking students. He was known then, making a name as a sound student of physical metallurgy with marked ability to express himself forcefully and clearly.

In those days physical metallurgy was relatively simple in content, yet the mechanics of preparing to teach it gave me plenty to think about. However, the broad outline of an introductory course was not too difficult.

Then, as now, knowledge of alloy constitution, or the constitution diagram, was a prerequisite for understanding alloys. That gave me my start of the phase rule of Willard Gibbs.

Although Gibbs was not a metallurgist, we should pay tribute to him for his basic contribution to our science. When J. J. Thompson advocated Gibbs for the Davy Medal in 1890 for his paper "On the Equilibrium of Heterogeneous Substances," he commented, "It will have great influence on chemistry." Thompson might have added "and metallurgy," because the phase rule has been an indispensable guide to understanding our metals and their alloys of different compositions and at different temperatures.

These equilibria had been discussed by H. W. Bakhuis Roozeboom and illustrated in his classical treatise, "The Heterogeneous Equilibria." They were represented by constitution diagrams, which, like maps, gave the alloy compositions and temperatures over which the individual phases were in stable equilibrium. Roozeboom's contribution ended the old or empirical treatment of alloys and gave us the modern or scientific method by using one of the great generalizations of our science. Where there had been uncoordinated chaos, the great Dutch scientist brought order and understanding with the most necessary single contribution to physical metallurgy.

After laying the basis for understanding metallic alloys, I would then proceed to thermal methods for establishing alloy diagrams and for the examination of metals and alloys by use of the metallographic microscope. That was about as far as I got on board ship, but it was a good start on blocking out the early part of a course on metallography. All I would have to do additionally would be to fill in the details and to work up the remainder of the course. That would take care of the introductory course; I would have time, later on, to work up a course on the important metals and alloys.

I arrived in Minneapolis fairly well prepared for my new job. However, I had no office, no classroom, and no laboratory! In my absence the School of Mines had burnt down. I found that the faculty was all housed in the new Mechanical Engineering Building and that I had a desk in the dean's outer office along with Miss Finnegan, his efficient secretary.

Dean W. R. Appleby greeted me warmly and quickly made me feel that I was one of the family. The dean was ideally suited to run our School of Mines, although that was better appreciated by graduates than by undergraduates. His long-time associate, Professor E. H. Comstock, has written an excellent biographical appreciation to which I can subscribe most heartily. After graduating from Williams College and doing post-graduate work at the Columbia School of Mines, he went west to Minnesota to establish the new School of Mines and remained at its head for his entire career. During his tenure he played a leading role in handling the state's interests in the iron-ore operations and in promoting the development of the taconite ore bodies. Incidentally, he preferred being dean of the faculty to being president of the university. During my six years on the faculty I enjoyed and appreciated the strong and sympathetic support I received from him.

I didn't know it when I arrived, but the dean and I had to find someone for me to teach. I had no ready-made classes, this in spite of the great play that the dean of chemistry had made to have me attached to his department. He had insisted that metallurgy was a branch of chemistry and that his students should be trained in it. I got a student finally — Arvid Nissen. He was working on argentiferous galena ores for an M.Sc. degree in geology. His special interest was the mode of occurrence of the silver.

Arvid was the keen, clear-cut type of young man that any teacher likes to have as a student. With a teacher/student ratio of one, we got along right well. There was no problem in registering him in my lecture course in metallography, but I had to have a laboratory with facilities for melting synthetic materials and for microscopic examinations. I was able to solve the problem of a laboratory with the assistance of Dr. R. O. Beard, assistant dean of the medical school. He had married the mother of my good friend of Columbia days, Perry Harrison. In a conversation with Dr. Beard, when I mentioned my lack of laboratory space, he offered me a room in the Medical Building.

Arvid's work consisted, in part, of preparing synthetic melts of lead and silver sulfides. The other phase was the microscopic examination of those synthetic specimens and of natural ores. For this work I had inherited a Leitz metallographic microscope from Ed Newton of the Mines Experiment Station. It was in the attic of the Chemistry Building. Arvid got his M.Sc. degree, which was followed by the publication of his dissertation.

As it turned out, my lectures in Metallography determined Arvid's career. He became fascinated with metallurgy, as I had by Campbell's lectures, and switched over completely from geology. After graduation, he worked for the Inland Steel Co. at their Indiana Harbor Works and the Illinois Steel Co. at their Gary Works. Later, he joined Edgewood Arsenal, where he invented and developed the mechanism that made mortar fire possible — a timely and useful contribution. Finally, he joined the Pennsylvania Forge Corp. as chief metallurgist and had an active and interesting career.

For a while, my assigned duties were few and far between. That gave me an opportunity I learned to appreciate; I had time to study the literature and became aware of so much that I didn't know. I had heard that teaching is a valuable educational process; I soon discovered the truth of that statement. Another valuable lesson was the need to choose words carefully and to say what I had to say in simple, understandable terms *à la* Desch, Rosenhain, Jeffries, Mehl and those others whose talks and writings are so pleasurable. On the applied side, I was grateful for the basic training I had received on the constitution and behavior of metals and alloys. I had concentrated on nonferrous metallurgy, yet I experienced little difficulty in handling problems of ferrous metallurgy.

I have noted that the old School of Mines had burnt down. That gave me the chance to plan my department in the new building. My office was really a seminar room with a large table for class work, and display cabinets along the walls. The metallographic laboratory was large enough to accommodate several large microscopes; this was fortunate because I wanted my students to have adequate facilities to gain a working knowledge of metallic microscopy. In addition, I had a small laboratory for mechanical testing and a furnace room in the basement.

My secretary, Miss Carrie Green, whom I had found in a business school as a stenotypist, developed into a Woman of Metals. Her mechanical shorthand proved to be exceptionally useful in my work. Many metallurgical terms were new to her, yet she never failed in transcribing my dictation. Her method was to take them down phonetically and then find the word in the dictionary. Carrie had taught biology in high school in North Dakota, with microscopic work in the laboratory. I encouraged her to try similar work on metals, with the result that she developed into a metallographist *à la* Fräulein Schrader. She was the thread of gold that ran through several administrations of the department. The students liked her and learned good laboratory technique with her friendly assistance. Carrie had a fine career for the rest of her life.

A few words may be in order about my students and teaching them the science and technology of metals and alloys. I believe there is a discernible relation between the two. For several years my classes were composed of graduate students and faculty members who wished to acquire an understanding of our most important technical materials. Those classes were a delight for a professor.

This situation reflected the relatively minor status of physical metallurgy as a subject for undergraduate curricula. In other words, my classes consisted of more mature people, rather than undergraduates who had elected metallurgy. In this way my students differed from those in other departments of engineering and technology. In those days, only a small number of schools taught metallurgy for its own sake in preparation for a career. That situation existed in spite of the economic significance of our basic industry and the painful lack of well-trained professional metallurgists. In industry, we had physicists, chemists, and engineers who had been "requisitioned" for duty as metallurgists and had to prepare themselves on their own to handle their jobs. I believe there was no other basic profession in which that abnormal situation existed.

My classes had to have laboratory experience to round out their training in metallurgy. Though I was largely on my own, the dean provided as much as his funds made possible for the metallographic and testing laboratories. I was also able to get my pride and joy, the excellent Weston potentiometer for pyrometry and thermal analysis. When I needed furnaces for melting or for determinations of melting points and phase transformations in the solid state (critical points) I had to make them myself.

Pyrometry was in its infancy here at that time. An incident may serve to bring that out. A local truck manufacturer was experiencing rear-axle failures. The president, Mr. Briggs, asked me to determine the cause so he could place a claim for damages against the Timken Detroit Axle Co. I found that faulty heat treatment and sharp corners of keyways were to blame.

The next summer, I visited the plant in Detroit to check on their heat treatment and with the thought that they just might want me to help them out of their difficulties. When I inquired into their use of pyrometers for temperature control, I was assured they had a better method. The furnace operator, an old hand, could judge temperatures better by eye! That ended that, because they didn't realize they needed help. I got no consulting job, but I did learn that some people who were handling and treating steel were poorly informed on metallurgy. I recalled losing steering arms and front wheels off our little Oldsmobile back in 1902 and 1903 for the same reason.

Handling thermocouples for accurate temperature measurements and preparing microsections are part of the complement of a trained metallurgist. I taught the former by having the students calibrate a Hoskins thermocouple, starting with the individual wires. Then I had them weld the hot junction, using an ordinary lead pencil as an electrode. The calibration was a good test of skill and gave valuable training in technique. They found that establishing a cold junction of 0 °C has its problems, depending on the precision that is required.

I had a similar philosophy or pedagogical technique when teaching metallographic practice. The student would prepare a microsection, etch it for structure and then sketch it freehand under the microscope. A beginner would usually draw scratches, etching defects, and so on, providing a prelude to a discussion of how to do the work correctly.

We had some men on the campus who became prominent in later years. The first was a visitor from abroad, Professor Carpenter, whom I had met at Ghent. He was touring the United States prior to assuming the top chair in metallurgy in the United Kingdom at the Royal School of Mines, London, and came to Minneapolis to look into our iron-ore work. He was our house guest on beautiful Lake Harriet.

Professor (later, Sir Harold) Carpenter spent most of his time at the School of Mines and the Mines Experiment Station on various iron-ore problems. (The U. S. Bureau of Mines also had a station on the campus, which made the university its headquarters for that field.) Finally, the time came for me to take him to the railroad station to continue on his way. At the station he stepped to the ticket window and asked for a "shayd-ule." I noticed the puzzled look of the ticket agent and I had him give the professor a "timetable." I also learned how to request a timetable when in England.

One of my students was Professor Burt Newkirk of the mechanical engineering faculty and later an expert with General Electric on balancing equipment.

Dr. Francis C. Frary was professor of chemistry and, although he was not one of my students, we became well acquainted. Even in those days he was recognized as one who would surely make his mark — which, of course, he did. Though electrochemistry was his main field, he had the ability and energy to work in other fields as well, including metallurgy. As one of his contributions he invented a new way to harden lead — Frary metal. I was discussing it with him one day in his laboratory when he wished to know if a certain circuit was live. He simply put his finger in the socket and found that it was. A youngster of mine did the same thing at home, but out of curiosity. He got a shock and exclaimed, "It bit me!" Francis's quiet demeanor would hardly suggest it, but as an undergraduate he was a strong man on the gym squad. I recall him in those days as pleasantly communicative and ever ready to discuss matters of mutual interest.

At the time of World War I, when Francis left Minnesota, the chemistry department had to take on three new men to handle his work. One of them came over to take my course in physical metallurgy. After a tour of duty in the Army, Francis became a Man of Metals as longtime Director of Research for the Aluminum Co. of America.

In this connection, I should also mention H. E. Bakken, who took my classes and later was one of Francis's principal assistants at Alcoa.

When I arrived at the School of Mines, we had a young instructor of mathematics on the faculty who was destined to make a practically priceless contribution to our steel industry and hence to our economy. I became acquainted with Ed Davis shortly after my arrival. Although he was teaching mathematics, he was a graduate electrical engineer from Purdue with a hobby, or personal enterprise — a magnetic log washer for concentrating iron ore. The magnetic field increased the apparent weight of magnetic iron minerals and increased the efficiency of the process.

Ed's work at the Mines Experiment Station was the start of an extraordinarily fruitful career which culminated several decades later when the low-grade taconite ores of the Eastern Mesabi Range became a lifesaving source of iron ore. The range is estimated to contain nearly two billion tons of ore. So, "Mr. Taconite."

Over that intervening period, Ed experienced a long, demanding, and, at times, even a frustrating dedication to the central idea of converting "waste" taconite into a priceless asset. In our field there must be but few achievements of equal brilliance and economic significance. Although the continuing support of Dean Appleby and the loyal assistance of his staff were needed to continue the fight in spite of losing battles, it was Ed who led to the victorious outcome of the war. One battle was that of overcoming the resistance of blast-furnace operators to using concentrated and pelletized ore, although it was a far superior charge.

I recall Ed as a sociable and companionable person who established friendly relations with associates and students. One incident stands out in my memory as I think back on those early days. His family and mine lived a few doors apart, and we both had a problem that is now solved by calling in a baby-sitter. At that time, baby-sitters had not been invented, but Ed was equal to the situation and came up with a solution. He wired up each nursery with transmitter and receiver, permitting either of us to listen in and check if all was well at the other end of the line. The system worked very well.

Our new building, Appleby Hall, was a delight to work in and is still in use. It was my home-away-from-home for several years. In addition to teaching, I was called in on consulting jobs. These brought steel sharply to my attention, with its diversity of problems of production and use. My work had been in nonferrous metallurgy, raising a question. In those days the gap between ferrous and nonferrous metallurgy was wide, so wide that I wondered if my training was adequate for steel. Although the heat treatment and structural changes of steel were unique, I found that my training in constitution and in the basic principles of metal behavior enabled me to bridge the gap. I also learned that my outside practice provided worthwhile examples for classroom instruction. They pepped up and illuminated academic discussions, and at the same time illustrated the use of abstract reasoning in solving practical problems.

I have mentioned the case of the truck axles. Another case arose from an epidemic failure of fishplates or splice bars on the Soo Line Railroad. After studying typical examples, my principal finding was that the steel had a poor structure for that application. My observations of the behavior of the rail joint as a locomotive passed over it showed that the fishplates failed from notch fatigue. If I could improve the structure, and do it cheaply, I might improve the service performance. After some laboratory testing with production in mind, I proposed oil quenching after the last pass in hot rolling. The Illinois Steel Co. installed oil quenching at their Joliet plant and proceeded to supply satisfactory fishplates, which stopped the severe losses of my client. Always on the alert for more instructional material (or was it for enhanced revenue?) I thought that the Soo Line might ask me to work on other problems. Instead, I learned that one swallow does not a summer make.

One time I was asked to conduct an unusual extracurricular school. Mr. L. B. Douglas, manager of the Illinois Steel Co. warehouse, explained that his men, and especially the executives of the companies to which he supplied steel, would benefit if they had a more adequate understanding of steels and their applications. Would I give a series of lectures to carry out his idea?

I thought he had an enlightened viewpoint, and I agreed to handle the course. His customers would gain a better comprehension of steel and this would lead to improved communication with the steel salesmen and greater assurance that the steels were correctly selected and processed for their applications. Finally, I thought I might dispose of the too generally accepted concept that "steel is steel."

Those classes were well attended, although I confess I had my problems as instructor. I went all-out to fulfill the purpose of Mr. Douglas — "to teach customers of the company the various kinds of steel best adapted to their needs." Apparently, the course was satisfactory; a representative of Illinois Steel stated, "The success of the scheme is attested by the interest shown and regular attendance." Bringing seller and buyer together in this way in an open forum should be an inspiration to others to do likewise.

I was not at all complacent about the book learning of my student days, and I felt the need to know more about the industry into which my students would go. I also wished to get practical experience for my own sake and to become acquainted with people in the industry. Therefore, I spent several summers in steel mills. In 1914, I tried to get a summer job at the South Chicago plant of the Illinois Steel Co. but Mr. Field, general superintendent, was not hiring anyone during the then current economic depression. He first rejected the idea of my just visiting but relented when I explained that I and my students would benefit by my acquiring practical experience. He turned me over to Mr. de Pierrefeu ("Piff"), chief metallurgist, who ran the laboratory. He had my Soo Line report on splice bars and wished to discuss it and my suggestion for quenching from the rolling heat. (On August 4, I saw big black headlines announcing World War I. Piff returned to France as an interpreter and lost his life in an auto accident.)

The laboratory was one in which Professor Albert Sauveur had worked as a young man. He was one of the first to photograph opaque objects with X-rays. By then he was on the Harvard faculty and had attained considerable stature as a metallurgist. Our laboratory took pride in that association — especially Mr. Post, who was Sauveur's successor in charge of metallography. I was surprised that Sauveur had had a poorly equipped laboratory, but that was par for the course in those days.

A major attraction at the South Works was their electric-furnace shop and its relatively new practice. Mr. Pritz, the superintendent, permitted me to study his practice as much as I wished to. I also spent time at the Bessemer and open-hearth shops, where I learned something of the art of steelmaking as well as the difficulties of making it work.

Mr. Field impressed me greatly. We would discuss steel in his office or chat on the front steps of his home. To him I was a *rara avis*, a college professor doing time around a steel mill in the summer. As one who could pass authority to subordinates and make them take it, he was a shining example. The Gayley dry blast for improving blast-furnace efficiency was a hot subject in those days, especially so because Mr. Gayley was a member of the board of U. S. Steel. The South Works, the "University of South Chicago," was elected to try it out but had abandoned it. At a meeting of the board, Mr. Field was asked why. He replied that he didn't know, which shocked the board, but that he would call in his blast-furnace superintendent to explain. That man was Mr. Herman Brassert, a top authority and eminent consultant in later years. As I recall the reason, the saving in operation did not overcome the cost of dehumidification. The principle was sound, but was too expensive in those days.

I took advantage of being in Chicago to visit other steel mills. One that I recall with pleasure was the Columbia Tool Steel Co. My desire to visit it came from my father. I recalled hearing him tell about Mushet steel as an improvement on carbon tool steel. Then later, he told about high speed steel, a still greater improvement. His favorite tool steel salesman, Mr. Scheid, had introduced those new steels to his plant. This induced me to go to Chicago Heights, where I met Mr. Claridge, the elder, and president of the company. The next generations of those families have carried on the tradition of making and selling fine steels.

Steelmaking has its spectacular moments, and that was especially true of the crucible process at Columbia. Heats of 100 pounds were melted in crucibles in a standard pit furnace of the day. The quality of the finished product depended essentially on the purity and make-up of the charge. According to a Sheffield saying — "If you put the Devil in the pot, you will get the Devil out of the pot." The act of pulling the pots and pouring the ingots was so exciting that it would have been worthy of being recorded by motion pictures in color, although I fear that the opportunity is lost forever. Lifting the crucible up out of the furnace called for a strong man.

In addition to the melter, another man of considerable skill was the tilter of tool steel bars. After breaking down the ingot into billets and converting them into oversize bars, the final size and shape were produced by "tilting." The tilter sat on a swing chair while he passed the bar back and forth under a broad hammer. The work was done at a low red heat and was done so skillfully that the finished bar had the surface appearance of a bar that had been drawn through a die. Fracture tests revealed a very fine and uniform structure.

To round out my summer in Chicago, I visited the Edgar Allen Manganese Steel Co. (burglarproof safes), the Railway Steel Spring Co. (locomotive tires), and the Gary Works of U. S. Steel.

In the meantime, J. Hunter Nead appeared on the scene as chief metallurgist at Minneapolis Steel and Machinery Co. He had been working at the Watertown Arsenal on ordnance metallurgy with Professor Fay of the Massachusetts Institute of Technology. He came to my laboratory from time to time with problems he had on shell production for an ordnance contract. As I recall, he was especially concerned with steel analysis and heat treatment to avoid a ferrite network structure. We became acquainted on a first-name basis before he left for a wider field.

During the next summer, I returned to the South Works to extend my studies of electric-furnace practice with Mr. Pritz. In the meantime I had studied Rodenhauser's book, which described very similar practice. In those days, deoxidation was completed by manganese and silicon additions. The use of aluminum was regarded almost with horror and certainly as an ill-advised crutch. I also learned, as a minor mystery, that Bessemer steel was about fifteen Brinell hardness numbers harder than electric-furnace steel and that open-hearth steel was softer than either. We didn't know much about nitrogen in those days. Their strongest steel was a silico-manganese steel heat treated to 260,000-psi tensile strength and 7% elongation.

Since my visit the previous summer, Walther Mathesius had transferred from Worcester, Massachusetts, to the South Works. He was back to his first love, the blast furnace, and was serving under Brassert. We had a reunion, our first meeting since early in 1912. He was speaking excellent English, as he had vowed he would. He was then starting an outstanding career.

I tried to visit the Gary plant of U. S. Steel but was refused admission according to orders from the New York office. I did have a good session with Arvid Nissen at the local YMCA. He was enthusiastic about the experience he was getting as delay clerk at the open-hearth shop. The next day was

Sunday, which gave Arvid and me the opportunity to witness steel rolls being poured at the Steel Foundry Co. The president, Mr. Corlis, expounded on the difference between furnace and ladle additions. The former practice was said to give a chill depth of 2¾ in., as compared with ¼ in. for the latter — with no explanation of the cause.

I closed my visit to Gary at the American Sheet and Tin Plate Co. of U. S. Steel by witnessing their production of tin plate and galvanized sheet. I guess they hadn't heard from New York.

After Gary I traveled to Middletown, Ohio, to visit the research laboratory of the American Rolling Mill Co., then a unique institution in the industry. My special interest was Armco ingot iron, described for me by Mr. Aupperle, chief metallurgist. It is a specially purified, very-low-carbon ingot iron made in the open hearth. It was converted into sheet and had the virtue of superior resistance to atmospheric corrosion. (A polemic raged in those days over calling the product of the open-hearth steel furnace "iron." Professor Sauveur, a consultant of American Rolling Mill, hit the nail on the head by pointing out that Armco ingot iron was technically pure iron with practically no carbon, and was iron regardless of how or where it was made.) A large test rack outdoors exposed samples of various irons and steels to the atmosphere to compare their corrosion resistance.

I went next to Pittsburgh, where I visited Mr. D. M. Buck, chief metallurgist of American Sheet and Tin Plate Co. I knew he was advocating copper-bearing steel with about 0.25% copper, which improved the resistance to atmospheric corrosion. He, too, had test racks, and I was not surprised to see that specimens of his steel performed very well in the Pittsburgh atmosphere. He told me that copper neutralized the effect of sulfur in Armco ingot iron and contributed importantly to its corrosion resistance.

It was at this time that I conceived the idea of running some independent tests of samples to be supplied by the producers. I will return to the results of those tests later, as they were not ready to be reported until the end of my stay at Minnesota.

In the meantime, a highly controversial and even vitriolic polemic raged on the relative merits of Armco ingot iron and copper-bearing steel. Allerton S. Cushman, a consultant in Washington, D. C., was a violent and eloquent supporter of the claims for ingot iron and, I understood, a very formidable opponent in open discussion. Mr. Buck had nobody of that caliber to support him. Both consumers and users must have been in serious doubt when faced with a decision of which to buy. With all of that fire and smoke, I thought that it was high time for someone (that is, me) who was not directly involved to run competitive tests and let the chips fall where they may.

In Pittsburgh, I also met Frank Speller, chief metallurgist of the National Tube Co. He became a leading authority on corrosion and was then concerned with soil and water corrosion. A current question at that time was whether to use steel pipe or wrought iron pipe for plumbing. Professor Howe had examined their relative merits by studying specimens of each kind after considerable service and had found no significant difference. Speller had also looked into this situation and had decided that the resistance of boiler tubes would be increased by working the surface an extra amount when the skelp was hot rolled. His process was called "Spellerizing," although its formal name was "roll knobbling."

I witnessed the production of lap and butt welded pipe at the huge National Works. That spectacular sight impressed me greatly as an example of modern production.

The home office of the Carnegie Steel Co., the major division of U. S. Steel, was in Pittsburgh, and there I met Mr. T. S. Unger, their chief metallurgist. The effects of impurities in mild steel was a live subject on which he had collected considerable material. After giving a brief rundown of his activities, to illustrate for me what a man in his position finds to do, he showed me his display on phosphorus. That impurity, though well above the specified limit, did not induce brittleness when the carbon was low. Apparently that was forgotten when, many years later, high-strengh low-alloy (HSLA) steels were being introduced that contained phosphorus as a beneficial alloying element. I knew about Stead's brittleness, caused by high phosphorus, and noted that Unger had restricted his tests to cold deformation of the finished product.

I had tried to secure summer jobs previously, but the depression of those times made that impossible. The situation had improved by 1916, with full-scale production for World War I, and I received a very favorable opportunity at the Midvale Steel Co. in Philadelphia. Mr. Radcliff ("Mike") Furness, superintendent of melting, forging, etc., wrote that if a college professor preferred a steel mill to the outdoors for a vacation, he was welcome to a place on his payroll. Of course it was a humble assignment, but I was happy to be accommodated because Midvale was recognized as a high-grade mill.

On my first day, Mike Furness took me around his precincts to acquaint me with their various shops and to let it be known, I suspect, that he sanctioned my presence. He also assured me that they had no secrets as far as I was concerned. I took advantage of his generous offer to contribute to my education by spending Saturday afternoons on their reports and records. As a major supplier, Midvale was in full production on heavy naval ordnance, armor-piercing projectiles, armor plate, high speed steel, and various items of alloy steels. It was an ideal situation for a young metallurgist's holiday.

Midvale used acid open-hearth steel for gun forgings, to secure adequate bursting strength as judged by radial tests. Basic open-hearth steel was used for armor plate, because it gave the desired fibrous structure for better resistance to penetration. Armor-piercing projectiles were made of crucible steel, because top quality was needed. The production of all three required great skill, which cannot be acquired overnight. That was the reason for our slow start in ordnance production after a period before World War I during which our facilities had been dismantled. A small but highly vocal and organized group had been denouncing warmongers. That had had a disastrous effect on our munitions industry, because, I was told, they went all out by dismantling their plants to disprove the allegation. For example, the art of producing armor-piercing projectiles has been called the highest metallurgical art — which I can believe, having studied it at Midvale. Equipment and qualified personnel cannot be provided overnight, and the same situation is true for armor plate. Even at the time of World War I, two of our major suppliers failed to pass the Navy tests.

Mr. Furness was justifiably proud of the steel they made. One time, as he was explaining their open-hearth practice, he commented, "So you visited the mills in Pittsburgh. Well, if that is steel, this isn't." He was an able and dedicated production metallurgist who appreciated the value of quality in all of its ramifications.

After our tour of the shops, Mr. Furness took me to the "microlab" to turn me over to Mr. A. H. Miller, who was in charge. He had excellent personal qualifications, both social and professional. His formal education in this country was supplemented by a year at the Sorbonne under Le Chatelier, which aroused my envy. After introductions, he explained the work of his department and gave me a desk for the summer. That desk was an excellent choice for me, as will become clear later.

I had two principal assignments — studies of high speed tool steel and of foreign inclusions ("dirt") in forgings. Foreign inclusions in steel can come from simple entrapment of slag or refractories, or as the result of reactions within the steel, especially from reactions between alloying addition agents and impurity elements, principally oxygen and sulfur. I was to concentrate on the latter types of foreign inclusions.

My samples were obtained by adding a deoxidizer (manganese, silicon, or aluminum) to a spoonful of molten open-hearth steel that was in the prerefined or "wild" state. At that stage, the dissolved oxygen in the steel was reacting with the carbon, producing numerous gas bubbles and a display of sparks from the surface. The deoxidizer quieted the metal and produced oxide inclusions for later examination. We then poured a small ingot and had it forged to a small square bar. To secure specimens of sulfide inclusions I added flour sulfur to the spoon, and the effect surprised me. The evolution of sparks, or the wild condition, was suppressed immediately, in the same way that the strong deoxidizer had acted.

The work on foreign inclusions became quite fascinating, and when I was writing my report I had a vision of a highly interesting and fruitful field for further inquiry. I had pressing duties to meet and no opportunity presented itself to complete an organized program, even though I later received a Carnegie Fellowship from the British Iron and Steel Institute.

My work on foreign inclusions led to examinations of coupons from gun forgings, armor-piercing projectiles, and other forgings, and to comparisons of basic and acid open-hearth steels and crucible steel. One time, I examined a large number of coupons from ordnance forgings of which some had passed the tests and others had failed. I got the idea that if the first test fails, another one has a good chance of passing, unless there is something really wrong with the steel.

I had heard of *Kruppsche Krankheit* (Krupp sickness) of the Krupp ordnance steel. At Midvale, I became better acquainted with "temper brittleness," which is our term. That steel was an extra-deep-hardening steel because it contained more hardening elements (chromium and nickel) than our current 4300 steels have. It must be quenched from the tempering temperature to avoid notch brittleness, the so-called sickness. The notched-bar test showed the difference very nicely between slow cooling and fast quenching, but the cause was not known. The palliative effect of a molybdenum addition *was* known, however, and the French specified it for their ordnance steels.

Speaking of molybdenum reminds me that the men in the shops had a hard time pronouncing the name. Being practical people they solved the problem by calling it "molydiddelum." It may be noted that they preserved the chemical symbol, Mo.

Midvale had a heat treatment that gave the optimum mechanical properties. They called it "fibering." The test coupon was first heated to a temperature well above the regular quenching temperature and then quenched. Then it would be quenched again, but this time from the correct temperature, and drawn or tempered just below the critical point. This treatment refined the grain size and produced the best structure for the final heat treatment. It was designed to eliminate the "butterfly" or "cloud effect" that was due to a heterogeneous distribution of the fine carbides. The latter were too fine to be resolved under the microscope, but the effect was visible.

Coming to high speed steel, at that time we had only a hazy or incomplete understanding of its ability to machine steel when its cutting edge became heated to or just below a dull-red temperature. Nor was the cause of its ultimate failure understood, although a simple carbon tool steel would lose its cutting edge at once by tempering and becoming soft. Enough of theory — I was asked to learn more about the effect of heating the steel to just below the melting point. The idea was that the higher the quenching temperature the greater the cutting efficiency.

I learned something from machinability tests of different steels, but not as much as I learned from a microscopic examination of tools that had been tested to destruction. The conclusion seemed to be obvious that the cutting edge failed by the softening process of tempering the quenched structure. Upon etching the end of the failed tool, a much darker structure contrasted with the light color of the body of the tool. I toyed with the idea of using tempering tests as screening tests in experimental work prior to the conventional lathe tests of tool life.

In my report to Midvale I wrote: "During the cutting operation, the effect is that of drawing the temper of the quenched steel and, of course, depends upon both time and temperature. The value of high speed steel here lies in the ability of the martensite to resist the action of temperatures, even up to a red heat, for an appreciable length of time." Note — I proposed to work along these lines to check steel compositions and heat treatment without the necessity of running off preliminary cutting tests.

One of the bright spots of my summer was in the microlab — sitting right across the desk from me, in fact. It was a large double desk, and Francis B. ("Frank") Foley was on the other side. Frank was a jack-of-all-trades and master of them too. He was one of Miller's assistants and was an enthusiastic and hard worker who came in close contact with shop operations and laboratory investigations. He was very friendly and helpful, and although so much of his experience was gained in his daily work he knew what he was doing and why. I would have been dull indeed if I had not recognized a particularly clear and effective thinker. After I returned to Minneapolis, with my work coming along nicely, I approached my dean with the suggestion that Frank would be a fine addition to the staff. He was.

I should like to mention two other men with whom I established an association. Dick Patch, a former organic chemist on the faculty at Harvard, worked on special steel assignments for Miller. I never did learn how he became a metallurgist. A few years later he moved over to E. F. Houghton & Co., as a vice-president. He knew his way around Midvale and helped me to get my work done. Doc Kelly, chief chemist, was especially helpful to me on the project on foreign inclusions. Like Dr. Patch, he too left Midvale to go with F. G. Budd.

On my last day at Midvale, Mike Furness asked me to come to his office for a review of my summer's experience and to give an oral report on my projects. I assured him that I deeply appreciated the opportunity he had given me and that I would be a better teacher as a consequence. His parting words were full of significance. He told me that anything I would say about Midvale Steel would be good. How truly he predicted!

That fall, school opened as usual, and shortly afterward Frank joined our team. He quickly became popular with the students, because they respected his know-how and he was always ready to talk to them on even terms. We didn't know it then, but Frank was off to an interesting and productive career.

One day a sprightly young man came to my office at the school of mines and announced himself as engineer of tests of the Northern Pacific Railway. I thought to myself — maybe here is another railroad job. Instead, he advised that he had no funds but did have metallurgical problems. Would I help him on that basis? So we worked together on several of his problems, with the dean's blessing.

A little later he told me he was worried about the quality of their main locomotive side rods, which were made in the NP shop in Brainerd, Minnesota. His coupon tests failed too often to meet the requirement for ductility. The failure of a side rod can be disastrous. About that time, I was once on a train bound for Chicago that came to an abrupt stop alongside the Mississippi River. I went up front to learn what had happened and found that a side rod had broken. After some fixing we went on with only one cylinder. I was told then that the failure of a side rod can cause a derailment if the broken part flails around.

Frank was as keen as I to go to Brainerd to observe their practice. Briefly stated, we found that one end of the bloom was forged and then put in the hot furnace to heat the shank or central section, which was forged next. After forging the second end, the rod was finish machined. That was all.

The shop had been built on soft ground, which accounted for a relatively high forging temperature. The first end had been reheated (that is, badly overheated) with no subsequent work, which was the reason it failed in the test. Frank proposed annealing the side rod to restore the ductility. Fortunately, they had a suitable furnace in which we treated it and saved the rod. Everybody was happy except us. We wished they would anneal all of their side rods.

In 1917, the United States entered World War I. The School of Mines and the Department of Metallography became involved in the war effort on a large-scale study of steelmaking. This program was initiated by some groundbreaking work on our manganese situation by Ed Newton of the staff of the Mines Experiment Station. For present purposes, all I need to say is simply that Ed was responsible for bringing the seriousness of the situation to the attention of government officials and others who were concerned with our increased steel production. Basically, it was due to the lack of domestic manganese ores and the harassment of ocean shipping by German U-boats. By developing the facts, Ed succeeded in inducing the U. S. Bureau of Mines to set up a section of the War Minerals Investigation to deal with the problem. It was the Manganese Section, under C. M. Weld.

A twofold attack on the problem of conserving manganese consisted of a study of blast-furnace practice and a study of the use of ferromanganese and other alloys in steelmaking practice. The former had the objective of reducing the manganese losses during the production of ferromanganese and in blast-furnace practice as the steel mills, and the latter the conservation of manganese in steelmaking and the substitution of alternative alloys. P. H. Royster, of the Bureau of Mines, was put in charge of a task force on blast-furnace practice and I was asked to head a task force on steelmaking.

Royster, Foley, and I initiated the field work by consulting with experts on our problems and locating suitable plants for our operations. One of the first persons our team called on for advice was Bradley Stoughton, able secretary of the American Institute of Mining Engineers. He was a good contact for our mission because he was heavily involved in numerous non-Society problems of mining and metallurgy. I started to explain the work we were doing for Weld's Manganese Section and had stressed our need for cooperation with selected steel producers, when he stopped me with, "Cooperation, cooperation! Everyone wants to cooperate these days but on what and how? I just received a telegram from Washington saying that they will cooperate. I haven't the slightest idea of what is meant." After he got that off his chest he proceeded to give us the help and advice we needed. I was also pleased that cooperation was the order of the day, without which we would have been lost.

On another of these exploratory visits I called on Professor Comfort A. Adams, chairman of the Engineering Division of the National Research Council. Some may be surprised that he had so great an interest in the fine points of steelmaking that I wished to discuss it with him. He was professor of electrical engineering at Harvard at the time and consultant to the General Electric Co. After we had closed our discussions of steelmaking, he lit another cigar and surprised me with an account of the virtues of the Single Tax.

Our interest in manganese conservation led us to Niagara Falls to look into the production of substitute steelmaking alloys and into possible processes for utilizing domestic low-grade manganese ores. Our principal contact was Dr. F. M. Becket of the Electrometallurgical Division of the Union Carbide and Carbon Corp. He had developed special electric-furnace practice for the silicon reduction of the ores instead of conventional carbon reduction. In that way he was able to produce low-carbon alloys. Under his guidance, the division had become the largest producer of these ferroalloys in the country. However, our hopes for securing useful data and information were not realized, as Dr. Becket explained that he could not issue any information on his processes or permit us to visit the plant. He did inform us on the availability of ferroalloys, which added to that phase of our program.

Our quest for information did not end in Niagara Falls. We found a ferroalloy plant in Chattanooga, Tennessee, where we were given helpful details of alloy compositions, recoveries, and power and materials consumption. That gave us a more complete picture of both the technology and economics of alloy production. Chattanooga is a beautiful locale as well as the site of a famous Civil War battle. We were able to visit Lookout Mountain and Missionary Ridge, where Grant, Sherman, Thomas, and Hooker got the Union Army out of a most difficult dilemma.

After this preliminary survey we returned to Minneapolis to line up our field programs and to organize our teams to do the work. My principal assistant was Ralph Dowdell, a student and a candidate for the degree of Met. E. I greatly appreciated his willingness to work, which characterizes those who wish to succeed, and his enthusiastic desire to add to his knowledge of steel. He was an excellent assistant in the field and in the laboratory, and in all his work he demonstrated the characteristics that later made him a leader in his profession. Other members of my team, with one exception, were my students, whom I selected because they were willing and enthusiastic workers.

The exception was Howard Scott, a young man in the Metallurgical Division of the National Bureau of Standards under Dr. George K. Burgess. "Scotty" was a conscientious operator with a sense of humor which lightened those scorching-hot days in the mills and on travel. The reason for his presence on our team was my need of a good man to take temperature measurements of the heats of steel we studied. I had asked Dr. Burgess, a top authority in that field, to recommend someone for that assignment.

To complete our team, we had two Bureau of Mines men back at the School of Mines for any laboratory work that might become necessary. They were Alex Field, a physical chemist, and C. E. Plummer, an analytical chemist. They both enjoyed good reputations. At Minnesota, Plummer analyzed many of the slag and metal samples we had taken in the field. His forte was getting results quickly and accurately.

Our field work came in the summer of 1918, one of the hottest of that period. An open-hearth floor or electric-furnace shop was a particularly uncomfortable place to work in, but we all realized that we had a job to do. To maintain our schedule we had to travel by night and always in upper berths in those non-airconditioned days. We welcomed those occasions when we would be at one place for one or more nights, so we could freshen up with a bath. Our eating facilities were useful principally for supplying sustenance rather than pleasure. That was especially true of lunch at some neighborhood restaurant or lunch counter. Ralph would call for the ketchup and then plaster his plate with it while commenting, "That should kill the taste."

One of the practices we wished to study was the use of molten spiegel for the manganese addition. It appeared to have advantages of better mixing with the molten steel and better utilization of the manganese. To study that practice, we visited the Lackawanna Steel Co. at Buffalo, New York, and Dr. George Waterhouse, chief metallurgist. He was a graduate of Sheffield University and a former student of Professor J. O. Arnold, a well-known (and controversial) metallurgist of the era. He came to the United States on an 1851 Exhibition Scholarship to study with Howe and Stoughton at Columbia. (That scholarship has produced a number of leading metallurgists, such as John Johnston, Matthew Hunter, and Walter Rosenhain.) After Lackawanna, Dr. Waterhouse joined the faculty of MIT and remained active as teacher, administrator, consultant, and government advisor.

We found George to be cooperative and helpful, with an intelligent interest in our work. He was heavily involved with the molten-spiegel practice and with a duplex process operated in conjunction with a 300-ton tilting open-hearth furnace. Their practice was new to me, as it was to most steel men. That 300-ton furnace was a monster, including the gear for tilting it. About 100 tons were tapped at a time; the spiegel from a nearby cupola was then added to the ladle. As I recall, the steel was of good quality, but the process has given way to conventional stationary open-hearth practice with various improvements in technique and furnace design.

I was convinced that the addition of deoxidizer and recarburizer in the molten condition had merit for conserving manganese and for improving the steel quality and uniformity. So we decided to look further. We located similar practice at the United Alloy Steel Co. at Massillon, Ohio. There we met Martin Schmid, chief metallurgist, who explained that they melted ferromanganese in an electric furnace and added it to a ladle of electric-furnace steel. It sounded promising. Then he took us to the melt shop to witness the practice. To our horror, and his, the ferromanganese had been largely converted into slag and the demonstration had to be called off.

By adding some conventional open-hearth and electric-furnace practices to the unconventional methods I have mentioned, the sum total of our visits gave us a good sample of the use of manganese in current steelmaking practice. Our routine procedure consisted of taking slag and metal samples of the steel bath at predetermined periods during refining, finishing, and teeming. At the same time, Scotty took temperature measurements. These samples and temperature measurements were accumulated back home for processing. We also had mill records of each heat that gave us the furnace charges and recoveries along with observations on the heat performance. The immediate objective was a determination of the efficiency of the use of manganese, which would indicate the best practice for its conservation. We had no part in setting the amounts that should be specified for heat analyses. A secondary objective was a more adequate understanding of the physical chemistry of the various steelmaking practices.

Uppermost in my mind was the possibility of using silico-manganese in steelmaking practice and also of utilizing low-grade domestic manganese ores. This stemmed from some experience I had had with Professor Thomas Bains, Jr., of the School of Mines faculty, with high-silica, low-grade manganese ore of the Cayuna Range. That ore was attractive as a possible source of manganese, but no economical method was known for getting rid of its silica. I don't recall all the details at this late date, but Tom and I had developed a process for reducing the ore completely instead of getting rid of the silica. We simply mixed coarsely ground ore with the correct amount of carbon for reduction and heated it to a surprisingly low temperature and produced a silicon-manganese-iron alloy. We took our process beyond the laboratory stage by making 500 pounds of the alloy in a small open-hearth furnace in a local foundry. The analysis came 58.3% iron, 22.2% manganese, and 16.75% silicon. We tried to patent our process but the Patent Office refused to see novelty in it. I never heard of anyone doing it, before or since.

As an aftermath, I attempted to resurrect the process during World War II, when we had a manganese problem again. One time, when I got as far as Minneapolis, I decided to go up to Crosby on the Cayuna Range to visit my friend of Columbia days, Perry Harrison. Perry was president of the Evergreen Mining Co. and was closely acquainted with the ores of the range. I told him about our work in 1916 and then inquired if that high-silica, low-grade manganese ore was still there. Perry just shook his head and said he wished he knew ore that ran that high in manganese.

My interest in silico-manganese having been awakened, it took active shape as a search for users. I wished to learn, if possible, if silico-manganese was as advantageous as my reading had led me to believe. I found that the McConway and Torley Foundry in Pittsburgh had used it in their acid open-hearth practice and that they were willing to let me review their records. I was particularly interested in its effectiveness or efficiency as a deoxidizer. Fortunately, there were periods when they could not secure silico-manganese and had to use ferromanganese plus ferrosilicon. I examined their records, which covered three years. On the average, the single alloy gave a 12% higher recovery of manganese, principally because of the protective effect of the more powerful deoxidizer, silicon. Furthermore, we held that silico-manganese would be more apt to produce deoxidation products that would liquate out of the bath and produce cleaner steel. In my AIME paper "The Use of Manganese Alloys in Steel Making," I reproduced excerpts of those records to illustrate or compare single-alloy and two-alloy additions.

Those results confirmed what I had read and been told. My next step was to find an integrated mill that used basic open-hearth practice and would be agreeable to running some tests. Silicon was a *bête noire* in many ways in those days — steel mills wanted no part of silico-manganese, except one. My inquiries took me to Tom Girdler, who later became president of Republic Steel, to learn if Jones and Laughlin would cooperate. Instead of sizing up my request personally, he told me to contact Herbert Graham, the superintendent of an old, old, basic open-hearth shop at their Aliquippa Works in Pittsburgh. When I told Herbert what I wanted done, his response was typical. "Sure, I'll try it." Working with Herbert was refreshing, because he appealed to me as being both progressive and aggressive. Most open-hearth superintendents would worry about losing a heat, but he was willing to take a chance. His later career confirmed his early promise, as is well known.

I still had to find some silico-manganese for the trials. It was in short supply because it was imported, but inquiries led us to the Wheeling Mold and Foundry Co. in Wheeling, West Virginia. This time, travel was not by a dusty train, but by a riverboat from Pittsburgh down the Ohio River. We had few respites like that one during our hot summer. The management was cooperative and agreed to supply us with enough silico-manganese for two heats of steel — mission accomplished. The Bureau of Mines agreed to purchase it and ship it to Jones and Laughlin in Pittsburgh.

All of this took time, delaying the trials until after school had opened and, ultimately, even after Armistice Day. By then the pressure was off the problem of manganese. In spite of that, I pushed to have the trial heats run, if only to demonstrate, as I hoped, that silico-manganese could be used in basic practice and that it was a superior alloy for deoxidation.

I was not able to witness the test heats, but I arranged with the Bureau of Mines to have Frank Foley observe them and report the results. His reports verified, or at least supported, the points I have made here on silico-manganese practice. Insofar as I know, that was the first time silico-manganese was used in this country in connection with the basic open-hearth practice. Since then, that alloy has been used very effectively, especially for high-manganese steels with 1 to 2% manganese. It is added to the furnace at a critical stage of the refining period as a block. Its use in that way reduces the amount of ladle additions and produces better steel as compared with adding ferromanganese solely to the ladle.

Mr. Weld called a meeting of the Manganese Section to discuss the field work and to consider possible future work or, more realistically, to close the program. I presented a preliminary and summary report in addition to the progress reports I had submitted as the work progressed. At that time I had additional evidence on microstructures. Howe attended this meeting — one of the few opportunities I had for a person-to-person discussion with him. He was universally recognized as dean of United States metallurgy, with vision to see deeply and broadly. The career of Henry Marion Howe is so well known that I need not recount it here and will only say that one young metallurgist will always recall a brief association with keen satisfaction. When I was a student at Columbia I missed the privilege of studying under him because he had left with the exodus of top faculty members a few years before.

I presented photomicrographs of inclusions that were produced by silicon deoxidation. I commented that they had a "doughnut" structure or appearance. The name came from their spherical shape and dark rim. Howe was especially interested in them. They were essentially silica and were described later by Merrill Scheil and me when examined microscopically by reflected polarized light. At that meeting, Howe asked pertinent questions and illuminated various findings of our field studies with penetrating discussions.

A couple of incidents will show that some of my time was spent, by accident, on something other than steelmaking.

On one of my visits to the Bureau of Standards I had an interesting session with Paul Merica. After completing his work in Berlin for the Ph. D. degree, he had taken a position at the University of Illinois, in the materials testing laboratory. We had considerable correspondence in those days, some of which related to an opening at the Bureau of Standards in Burgess's division. Paul asked me for a reference, putting me in a dilemma. If I wrote how I really felt about Paul, I could easily appear to one who did not know him, to be overdoing it for a friend. Yet I did want to get across a true picture of his ability and to establish that Paul would be an exceptionally fine addition to the staff. He remained with the bureau for several years until he left to join the International Nickel Co.

On this visit Paul had an exciting research to discuss. He had been working on Duralumin, Wilm's hardenable aluminum alloy that Zeppelin used for his dirigibles. The heat treatment, which increases the hardness and strength, differed radically from known hardening treatments and was a prominent metallurgical mystery of the day.

Paul first showed me photomicrographs of his alloys. This came as a surprise, because aluminum had defied attempts to prepare microsections properly. As a second surprise, those micrographs revealed no difference between the hard and soft conditions. Paul's understanding of constitution diagrams guided him to precipitation hardening by an alloy compound as the mechanism of the hardening treatment. He reasoned that the first step of quenching from an elevated temperature retained the alloying addition in solution in the aluminum and gave the soft condition. Reheating the quenched condition to a low temperature must produce a submicroscopic, and therefore undetectable, precipitate of the alloy compound. That precipitate, or dispersion, produced the hard condition. That was a concept of the first magnitude and has led the way to numerous other alloys that are strengthened by the same mechanism. This work was published in the classical paper of Merica, Waltenberg, and Scott. I was told a few years later by Dr. Rosenhain of England and Professor Guertler of Germany that they, too, had arrived at the same precipitation-hardening mechanism to account for Wilm's hardenable alloy.

Another experience, of an entirely different kind, I recalled vividly a few years later. Once, when I was in Chicago overnight and ready for a diversion, I asked the hotel clerk where I might go. I figured later he had not sized me up correctly, but he suggested a café and nightclub on the South Side. I found the place easily enough, but once inside I wondered what I had let myself in for. A lot of people of both sexes were sitting at tables, drinking and listening to some weird music that rankled my more classical ears. Just as I had decided I was in the wrong place, my eye caught two men looking my way. One was looking directly at me and saying something to the other that I felt instinctively meant no good for me. I left at once and was glad to be breathing fresh air. Back at the hotel I asked the clerk, "Who was that man out there who gave me the ugly look, the one with a big scar on his face?" "Oh, that was Scarface Al Capone. He is a big shot in these parts."

During this field work I had numerous discussions of steelmaking with metallurgists, foremen or superintendents, and melters. Slag and metal compositions and temperatures were recognized as vital parameters; indeed, that was the reason for our numerous samples and temperature measurements. The whole process of melting, refining, and deoxidation was simply an art that we planned to examine more closely by applying the principles of physical chemistry to our test results. We all held that a better understanding of the reactions would guide to more uniform production and might lead to useful innovations in steelmaking practice. I also had Alex Field in mind as one who could take on the physico-chemical treatment of the results. This phase of our program began shortly before Armistice Day, 1918.

We got a good start on the analyses of the slag and metal samples; but with the loss of interest in the work as the manganese situation reverted to normal and with my resignation from Minnesota, our work in this field was not completed. The chemical analyses were expertly handled by Plummer, but determinations of oxygen posed some serious problems. The standard method was Ledebur's hydrogen reduction of the oxide and the recovery of the water that resulted. Dr. J. R. Cain of the Bureau of Standards was to carry out these determinations on cylindrical samples that we prepared for him. The need to heat the samples to above the melting point was objectionable, leading me to suggest dissolving the sample in a tin-antimony alloy that would lower the reduction temperature to 1100 °C. I had recalled from the German literature that Wüst has used that technique for carbon determinations by combustion.

The Engineering Division of the National Research Council took an interest in our work at about this time. There was a desire on the part of some of the leaders to promote a continuation of our work. Professor Howe outlined a research program on open-hearth practice and asked me to comment on it. My response was encouraging when I discussed his program and gave him my ideas. In conclusion, I wrote that there was "quite a possibility that the making of steel can be put on a more scientific basis." Howe was also joined at that time by Comfort Adams, Joseph E. Johnson, Jr., and G. H. Clevenger of the Engineering Division of the National Research Council.

I seemed to be the one in the middle, having just come off a large-scale field study with a large collection of samples and temperature measurements. That was the situation when the Engineering Division called a meeting to which all interested parties were invited.

We had a well-attended meeting of the project leaders of the war-minerals investigations, as well as representatives of the steel industry, at which each of us presented a summary report. After I had given my report, accenting future prospects, the quiet that followed was ominous. The manganese program and its adjunct, or a scientific study of steelmaking, were dead. During the war, cooperation was the key word, but afterwards, the steel companies went on their own. I had already joined the General Electric Co. and was glad I had started on a new career.

History shows that the physical chemistry of steelmaking was too important a subject to lie dormant. A decade or so later, a large-scale effort started that, in one form or another, has continued. Similar programs are carried out in foreign countries.

Incidentally, I have never regretted this wartime experience nor other contacts with steelmaking practice. Though I was never one to participate in scientific studies of steelmaking, the knowledge I gained has been very helpful in my later career. I can recommend to metallurgists in the consuming and using industries that they acquire a working knowledge of steelmaking. That knowledge is especially valuable for steel selection and procurement for applications that emphasize economics and safety. I can also recommend that the actual scientific work in that field be left to the likes of Herty, Chipman, Fitterer, and Darkin.

I had the good fortune to meet and work with a number of men who were destined to contribute substantially to metallurgy while it was in a formative period. I should like to digress here with a few highlights of their careers.

I lost Frank Foley when he joined Royster's team, although we did work together on a part-time basis and Frank helped on my work on the use of silico-manganese in basic open-hearth practice. He was with the Bureau of Mines directly after the war and later had the distinction of assisting Henry Marion Howe on a research program. Still later, he returned to Midvale Steel as director of research, where he finished his active career. Upon retiring from Midvale he became a consultant with the International Nickel Co. in New York, commuting from his home in Philadelphia. Frank had a fine talent as a pen-and-ink artist. He used it when I knew him at Midvale to sketch split ingots to show shrinkage cavities, but developed his talent still further when at Minnesota by sketching interesting buildings and landscapes.

When Bradley Stoughton ran AIME he was also an innovator, a hazardous role to play in a venerable and established organization, but his ideas for progress were sound. That included the insertion of Metallurgical in the name. We were all shocked when he resigned. Not long after, he succeeded Professor Joseph W. Richards as head of the metallurgical department at Lehigh University. He was associated with that department, officially and unofficially, for the rest of his career.

Bradley was a warm and human person, as is reflected in the biography by Gilbert E. Doan, "Bradley Stoughton — Mankind Was My Business." That was brought out no more clearly than by his major contribution to procuring the eight-hour day in the steel industry. The superintendent of Open Hearth Shop No. 3 at South Works had indoctrinated me with forceful arguments in favor of it as compared with the twelve-hour shifts his men had to work, but he had no opportunity to advocate the change where it would count. After retirement, Bradley continued to be active, professionally and socially.

Dr. F. M. Becket of Union Carbide had a long and successful career, making important contributions to the production of the widely used ferritic and martensitic stainless steels. He also solved the pernicious problem of intergranular attack of the austenitic chromium-nickel stainless steels by adding columbium (niobium in Europe) to combine with the carbon in innocuous form. That alloy became type 347 and solved a serious problem of the production of stainless steel equipment, as I know from personal experience.

Ralph Dowdell remained at Minnesota for most of his career. After Oscar Harder succeeded me as head of the department, he stayed on as an assistant and then went to the National Bureau of Standards briefly, where he helped Herbert French in a large-scale investigation of the heat treatment of steel. When Oscar left for Battelle Memorial Institute, Ralph returned to Minnesota as head of the department. He remained in that position until he retired.

Howard Scott made good progress at the bureau, and was a co-author of the classic paper by Merica, Waltenberg, and Scott on the mechanism of precipitation hardening. He was finally picked off by Westinghouse, where he became head of metallurgy and ceramics at the research laboratory and closed out a fine career.

Alex Field left the Bureau of Mines shortly after the war to join the National Carbon Division of Union Carbide and later went on to fame with the Rustless Iron Co. during the depression years of the 1930s. Both he and Rustless were absorbed into Armco Steel, where Alex closed his career.

Herbert Graham remained with Jones and Laughlin as their top technical man for the rest of his active employment. It was entirely appropriate that the new Jones and Laughlin research laboratory was named in his honor. A special contribution to the consuming and using industries was his work with the intermediate manganese steels. They are the Jalloy steels, which take advantage of the beneficial and economic characteristics of that steel alloy. In his later capacity as director of research and technology, he demonstrated the advantage of using a production man in research, if he has the scientific and technical capacity. For one thing, the people in the shops are aware that he knows their side too, and must talk sense.

I have referred to atmospheric corrosion tests of common sheet steels. Just before I left Minnesota those tests were completed and the results were published in 1919 in *Chemical and Metallurgical Engineering*.

My appetite for studying corrosion was whetted by my summer experiences, by a situation that developed in the Minnesota State Highway Department on the switching of some identification tags, and by tales of the wrought iron chain bridge at Newburyport, Massachusetts. I also knew about the iron pillar at New Delhi, India, which raised a question of the influence of a natural protective surface on corrosion resistance. I was able to secure part of a link from that century-old chain and found that the surface was well preserved. To check the corrosion resistance of the iron proper, I cut off a small sample to expose a freshly cut surface. It proceeded to rust quickly when exposed to the atmosphere. So much for the claim that had been made that those links were highly pure iron and therefore corrosion resistant.

For my own corrosion tests, I secured from the suppliers samples of various sheet steel products that ranged from Bessemer steel to Armco ingot iron and copper-bearing steel. Specimens of each were placed on racks on the roof of the School of Mines. After three years' exposure, the results were consistent and appeared to be reliable and worthy of being published. I summed up the test results by commenting, "The general conclusion which, it is believed, may be drawn from this exposure test is that the so-called copper-bearing steel, in which the copper content is about 0.20-0.25%, offers the greatest resistance to corrosion of the common sheeting materials." It is only fair to say that the HSLA steels of the 1930s and 1940s are even better.

My tests brought out one point for the first time. I used two series of tests that ran parallel and directly compared removing the rust surface after each exposure and then re-exposing the specimens with allowing the rust surface to remain intact. Some protection was provided by the surface film, possibly most of it, but removing the surface and again exposing bare metal did not change the order of excellence of the test materials.

Atmospheric corrosion tests have been run in great profusion since then under the aegis of the American Society for Testing and Materials. My results have been overtaken, although, if I may say so, they have not been disproved or controverted. While confirming the effect of copper, the ASTM tests also brought out the benefit of chromium and phosphorus additions.

Turning to a sideline of my teaching experience, the geology department had become interested in the microscopic examination of opaque minerals, yet had neither instructor nor facilities for teaching it. I agreed to take it on as a temporary solution until they could take over. I was not too poor a selection for the course, because I had studied geology with Professor J. F. Kemp, and petrology and mineralogy with Professor C. T. Berkey while I was a postgraduate student at Columbia. This was interesting work and could be a full-time job for a teacher provided he was well-qualified on the phase rule and constitution diagrams. I ran into that quite early, because a geologist and a metallographist, viewing the same microstructure, will (or did) interpret it in two different ways.

I was given a specimen of copper ore that had an easily identified Widmanstätten structure. It would be clear to a metallurgist what the origin of that structure must have been, yet the geologist differed radically and said that the matrix mineral had been fractured along cleavage planes and had then been mineralized by penetrating waters. Thus there were two completely different concepts, mutually exclusive, and one had to prevail. Oscar Harder told me of a similar experience that he had had with this same structure. He used X-ray-diffraction patterns of a quenched specimen and a slowly cooled specimen to demonstrate the difference and hence the origin of the duplex structure. When he tried to persuade the geologists, he had no more success than I.

I had another case which must have been a classic. One day, a geology student, Tom Broderick, came over and said he wished to have a research project in opaque mineralogy — would I suggest a subject? I got a textbook of mineralogy and studied various possibilities. I finally selected for study a field that was unexplored at the time. Between the lead sulfide, galena, and the antimony sulfide, stibnite, my textbook listed a whole series of minerals that looked suspicious. Each was identified by composition, structure, and properties that made it appear as a separate mineral. I just didn't believe it.

I had Tom prepare a series of synthetic melts that covered the entire range. The microscopic examination revealed that the lead-antimony sulfide, Jamesonite, was the only true mineral in the series. All the others were mixtures in various proportions! It was obvious then that a new approach to mineralogy was in order. I had no difficulty in selecting one series as a bellwether; there must be others that one could spot if he were conversant with constitution diagrams. I resigned at this time and lost track of the work. I plan someday to check on the present treatment of this phase of mineralogy. Tom Broderick continued in geology and became chief geologist of Calumet and Hecla.

Just before I resigned, someone from the department of surgery came over with some surgical implants and asked me to cooperate on a research program. Actually, I accomplished very little in the short time available. A major problem was the detection of corrosive attack by body fluids. Speedy answers were desirable, but checks by weight loss or other conventional methods did not appear to be desirable. I polished the implants by metallographic technique and returned them to surgery for their tests. If the microstructures were developed, chemical attack would be shown. I felt that was a sensitive method for the purpose. Incidentally, I have read recently that this method was used by the Biological Materials Department of Northwestern University.

In my consulting practice at Minnesota and in teaching, I often recalled Heyn's lectures on the significance of notch brittleness of ductile steels and the importance of notched-bar testing. At this time, there was no general recognition of the effect nor of the need to use the test to control steels and their heat treatment. This negative attitude prevailed even though the facts were available in the literature and could be checked in the laboratory. I decided to inform my students by lectures in which I cited well-authenticated data on a variety of steels and by laboratory demonstrations. Some of this material is reproduced in my book, "Principles of Metallography."

I have always regarded Georges Charpy and Professor E. Ludwig as the two prime movers in this field in the early 1900s. Charpy's name is well-known now and will remain in our test vocabulary as long as Charpy tests are run. He was truly a Man of Metals as a practicing metallurgist with broad interests and noteworthy accomplishments. It is not so well-known that he was the first to demonstrate the growth of excessively large grains in a metal after a small amount of cold work followed by annealing at an elevated temperature. This effect is infrequently mentioned, yet is important if coarse grains are to be avoided in a part. A similar grain-coarsening can occur during hot working if a final reduction is too small.

Even before 1919, brittle fracture of steel and its relevance to engineering practice had been so clearly demonstrated that I felt I should spread the gospel. To that end I had two papers published — one by AIME and one by ASST (American Society for Steel Treating, now ASM). I would like to say here that I could do a better job now, but in youthful enthusiasm for a worthy cause, I felt that the mere disclosure of that treacherous behavior would instantly create interest in it. Insofar as I know, nothing resulted from either of those two papers or from my account in "Principles of Metallography," although it was used as a textbook for about three decades. Incidentally, Charpy wrote a discussion of my AIME paper, starting some interesting correspondence.

In spite of their lack of acceptance in 1920, I held to the faith thereafter that brittle fracture and notched-bar testing were too important to be so generally ignored by our metallurgists and engineers. Even though I was occupied with other matters, attempts on my part to bring it out in the open became a continuing enterprise. I don't mean to imply that I was alone in recognizing brittle fracture, yet others of the same persuasion were few and far between. They had no more effect or influence than I had.

Howard J. Stagg, Jr., was possibly the first metallurgist in the United States to use a notched-bar test to guide the heat treatment of steel. He discussed his work as chief metallurgist with the Halcomb Steel Co., in Syracuse, in an early talk before the Steel Treating Research Club of Detroit, a forerunner of the American Society for Metals. His test results were illuminating and relatively new at the time, but would be regarded as conventional today. I was interested in his comment, as a disclaimer, that his results cannot be "translated into terms of the Fremont, Charpy, or other machines."

Howard was a popular speaker in those days and for years thereafter because of his accumulated knowledge of steel and his interesting presentation of his subject. One talk that drew especial audience interest was "Why

Did the Damn Thing Fail?" That has always been a fascinating subject. Howard was ahead of his time by three decades.

It was about this time that a great light opened my eyes to the correct understanding of the Charpy, or notched-bar, test. Then, as now, it was called an (or even *the*) impact test. According to general usage, it was supposed to indicate behavior of a component or structure under impact or shock loading regardless of the geometry of the part or its stress field. The notch was simply ignored under this concept. I realized then that the notch, and notch behavior, were the features of the test and that the correlation with service was with parts that had a notch of one kind or another, or a sharp change in cross section. I included data in "Principles of Metallography" (1920) that clearly brought out the vast difference between simple impact behavior and notch behavior. Charpy was not confused on this point. He simply used a pendulum type of testing machine to secure a simple and direct measure of the energy absorption in breaking his test bar.

I left Minnesota at the end of the school year of 1919 to join the General Electric Co. I had promised Dean Appleby that I would write up my lecture notes and they appeared as "Principles of Metallography" and "Metals and Common Alloys," published by McGraw-Hill. They were used as texts for many years until McGraw-Hill wrote in 1947 that they were out of print.

Teaching is a great teacher. That holds for both organizing the subject matter and for its presentation. Stated more directly, I became better informed on physical metallurgy and I learned to use langauge that my students would understand. That latter statement may sound cryptic, but I learned that good English, as contrasted to everyday English, was a *must*, or I would get a blank expression on even my better students. I would advise my friends that my competence at the close of my teaching days was not measured on an absolute scale; I deal here with a before-and-after comparison. I came to regard the habitual use of good English by the faculty as a part, even a necessary part, of the equipment of university professors. It does not mean that they are fuddy-duddies but simply that they have mastered one of the tools of the profession.

I found life on the campus of a large university to be a delightful experience. We were all of the same social stratum and had common interests; the work was congenial and moderately well paid. Our Campus Club gave us a convenient meeting place for lunch and an opportunity to discuss, and occasionally to come up with a solution to, world affairs. Most of us had just ordinary intelligence, but we did have our geniuses too. I recall Professor Richard Scammon of the medical school in particular. He would join a group of similarly equipped professors after lunch for a game of words. One of them would name a word (usually one that I had never heard before) and would then get its meaning from one of the others. The latter, in turn, would give a word, and so on. The Scammons were both highly intelligent, as could also be said of their young son. Whenever young Richard was around, he was in evidence. He has grown up now, was recently Director of the Census, and has lived up to his early promise.

Toward the end of my regime in 1919, Dean Appleby went on a scouting trip and returned with Oscar Harder to be my successor. The dean picked a good one. Oscar came from Oklahoma and did his graduate work at the University of Illinois for the Ph.D. and was at the Mellon Institute in Pittsburgh when the dean found him. With a good command of physical metallurgy and the ability to meet all comers on a free-and-easy basis, he made a fine record at Minnesota. A major contribution that I recall was his pioneering work in dental metallurgy in cooperation with the College of Dentistry. Oscar remained at the School of Mines for about ten years, or until he left to join Battelle Memorial Institute in Columbus, Ohio, as assistant director.

As for me, I read Jack London's "The Call of the Wild" with sympathetic understanding. I got the notion that I wanted to have a more intimate and lasting association with the world out yonder. A good opportunity came in 1919 when I left for Cleveland to join General Electric.

Au revoir, academia.

Chapter 3

The General Electric Company, 1919-30

Nela Park and the Research Laboratory under Dr. W. R. Whitney; running the weekly colloquia; lamp metallurgy; extended tour of Europe.

IN the spring of 1919, I got a letter from Colonel F. M. Dorsey (ret.), manager of the new Experimental Engineering Laboratory (EEL) at Nela Park, in East Cleveland, Ohio. He was looking for a man to take charge of the metallurgical department, and Dr. Zay Jeffries had recommended me for the position — would I be interested? A visit to Nela Park convinced me that I should make the change.

Nela Park was the home office and headquarters of the National Electric Lamp Association, a group of lamp companies in the Midwest. Nela Park was owned by the General Electric Co. and contained one of its two lamp divisions, the other being the Edison Lamp Division in Harrison, New Jersey. Nela Park, a large executive and technical center, had facilities to handle all phases of the lamp business from research through lamp manufacture.

Colonel Dorsey's brilliant record as head of a chemical-warfare unit at Nela Park during World War I had attracted the attention of Messrs. Terry and Tremain, the able managers of NELA. This had led them to establish the EEL for chemical engineering, metallurgy, and lamp manufacture. The colonel was a sound visionary and an aggressive leader. Our work at Nela Park requires some introductory comments.

Zay Jeffries was primarily a consultant with the Cleveland Wire Works, but he had also become an important person at Nela Park, where he consulted on a more informal basis with management, the research laboratory, and various technical groups. As though that were insufficient to consume his energies, he was also director of research of the Aluminum Casting Co. in Cleveland.

It was the manager of the Cleveland Wire Works, Burnie Lee Benbow ("Bennie" for short), who asked Zay to consult with him on tungsten. That came about primarily because tungsten metallurgy was new and radically different from conventional metallurgy. The Cleveland Wire Works needed help to bring order out of chaos, and the assignment fell to Zay. A few words of explanation may be in order.

Although Edison started the incandescent-lamp business, the newer tungsten filament had largely replaced his carbon filament. That substitution had resulted from the work of Just and Hanamann of Vienna. The latter was the Franz Hanamann of my Charlottenburg days.

Those first tungsten filaments were very efficient in terms of lumens of light per watt, but they were weak and fragile. I recall when they were first used on the Lake Harriet streetcar line in Minneapolis. They gave a new and brighter light but one never knew when a lamp would fail. Coolidge's ductile tungsten put an end to that, but it also introduced a new set of production problems. I have written the story of Coolidge's brilliant achievement, so I need not discuss that here. It appeared in *Metals and Alloys.*

In production, however, a major defect soon developed. Although Coolidge's process worked satisfactorily in that it produced a ductile tungsten wire, when operated in a lamp the filament was prone to fail by "offsetting." That resulted from the growth of coarse grains that would partly slip past each other and produce a hot spot that was followed directly by a premature burn-out. The offsetting was eliminated by doping the metal with fine thoria particles, which greatly retarded grain growth and gave a normal life.

About this time, Dr. Langmuir, at GE's Schenectady research laboratory, had developed his gas-filled lamp with a coil filament. It was a significant improvement over the vacuum-type lamp with a straight filament, but had the disadvantage that the coil drooped after too short a time. That destroyed the advantage of the coil, because the filament temperature fell off and with it the efficiency of light production. What to do?

The answer came from an unexpected source. Nela Park had a man who was almost a free-lance researcher, Dr. Aladar Pacz, pronounced "Packs." The objective was a tungsten filament that would have sufficient hot strength to prevent sagging. His 218th trial was a success, and the process was adopted for production. The wire was christened 218.

The 218 wire was made by adding sodium and potassium silicates during the chemical processing. Although the additives are lost during the production of the wire, the filament recrystallizes into extra-long grains whose boundaries run lengthwise. That structure provides adequate sag resistance, even for double-coil filaments. Oddly enough, the addition of barium oxide also produces extra-large grains, but with transverse grain boundaries that ruin the hot strength.

Tungsten behaves in reverse when compared to our common metals. Special treatments had to be developed to process it from ore to lamp filament, because simple or conventional processes were inadequate. As a metal it is a maverick, a paradox. Copper, for example, is soft and ductile when it has a fine and uniform grain structure. Tungsten is soft (200-plus Bhn) but exceedingly brittle when in the same structural condition. Furthermore, copper loses ductility when drawn or cold worked to a fibrous structure. Tungsten acquires ductility when cold worked to a fibrous structure, as in Coolidge's process.

Before consistent filament performance could be assured, the Cleveland Wire Works had many production problems to solve; Bennie realized that he needed expert assistance and called in Zay. He could not have picked a better man — Zay had the competence to work effectively and the willingness to think. Some chill cast aluminum alloys were in trouble from shrink cavities. Zay studied the problem and recommended using a copper chill to induce prior solidification at that location. "Put a chill in a chill casting? That's ridiculous!" was the reaction at the foundry. Zay insisted, with good reasoning, that it be tried. It was, and it worked.

I will add my own tribute, that putting together all the parts of that metallurgical puzzle of filament performance was a major achievement. All who are acquainted with Zay's publications of that period will recognize that his interest in physical metallurgy was intensified by his work for Bennie on the maverick, tungsten. They will also see the application of his philosophy — that to recognize the unusual, one must know the usual.

Bennie had an outstanding career himself as the longtime manager of the Cleveland Wire Works. He demonstrated his ingenuity very early by developing equipment and practice for drawing spool lengths of wire. That substantially reduced the costs of both wiredrawing and lamp manufacture. He went on from there in his own independent way to develop his department into a model division of the company. He never hesitated to innovate even though he would be going counter to current thought. His mass-production methods for producing metal powder are one good example, even if "it can't be done." He demonstrated that a good reason is no match for good reasoning. Bennie richly deserved the tribute paid to him by Jeffries and Archer in their dedication of "The Science of Metals" — "To Burnie Lee Benbow, who typifies the progressive executive, this book is dedicated."

Members of Bennie's staff who worked with Zay were Pete Tarasov, metallographist, whose son, L. P. Tarasov, is well-known for his work on abrasives and grinding; Bill Sykes, who conducted more basic studies on tungsten and its alloys and on Carboloy; and Ed Bain.

Edgar C. Bain will come up again, but it was at the Cleveland Wire Works under Zay Jeffries that he started his brilliant career. His investigation of the copper-gold alloys by X-ray diffraction was a classic contribution to our understanding of certain solid-solution alloys. Taken in conjunction with Tammann's work on resistance limits, the existence of order-disorder transformations became a reality. The nickel-chromium system was originally held to form a continuous series of solid solutions but that crystallize in two different structures. Ed showed by X-ray analysis that it was a two-phase system. Another study was on high speed steel and the cause of red hardness. Ed collaborated with Zay on these studies at the Cleveland Wire Works.

While I was being indoctrinated in tungsten metallurgy, our laboratory was being built. My first order of business was securing George Bierman, a graduate student of mine, as my principal assistant. Many will recognize the name, for he was the brother of Bernie Bierman, the longtime coach of Minnesota's Golden Gophers. I had additional assistants, as well as the assistance of other sections of EEL that were equipped to handle every phase of our field from ore processing through life testing of lamps.

We had an interesting experience with an electrolytic method for making Dumet, or "sealed-in wire." That is the short length of wire that is sealed in the stem of the lamp and connects the outer lead wire to the inner lead to the filament. It was being made by brazing a seamless copper tube over a core of the low-expansion alloy Invar. The copper provided the necessary electrical conductivity and the core gave the composite wire the same radial expansion as glass. That method of production had the appearance of a poor metallurgical process though it did give good wire.

Our objective was to develop a continuous process by depositing the copper electrolytically. There must be a good bond between copper and core to avoid a leakage of air into the lamp, which prompted me to attempt to weld the two together. A quick check of alloy constitution diagrams divulged nickel as being soluble in both the iron-nickel core and copper. From this basic concept came the scheme of the process.

After preliminary work on plating copper on rotating Invar rods in a cyanide bath, we devised equipment to use Invar wire for the starting material. The wire was first passed through a rotating end mill to clean the surface, then through a nickel-plating bath and finally through a hot copper cyanide bath. Instead of using a long plating bath we used a small compact tank and passed the wire through it as a helix. We circulated the bath rapidly through an outside filter, which made it possible to shorten the length of wire in the bath by the use of a high current density, and also maintain a clean bath. Hot swaging followed plating, which caused the nickel to diffuse into both sheath and core and weld the two together. This method of joining has more recently been developed as diffusion bonding. The final steps were wiredrawing to final size and borating the surface.

We had little trouble with copper plating, but the art of nickel plating in those days left a lot to be desired for one who was not heavily involved in it. Once we had a good deposit, we had little difficulty in making good wire. However, the postwar depression caused the EEL to close. All that I accomplished was a patent on the process.

Wrought iron has been used for a long time. In finding a new use for it I learned more about it. Dr. George Shaw, our chief chemist, was developing a new or improved process for treating tungsten ore. The process consisted of fluxing the ore at a high temperature in a rotating cyclinder that discharged the fusion continuously at the exit. Major Fogler, our chief mechanical engineer, complained to me one day that one of his cylinders had failed prematurely. When I found that a good cylinder was made of wrought iron whereas the failed one was made of steel, I told the major to specify wrought iron. When the replacement arrived I found that it, too, was steel. The supplier told me that he made no distinction between wrought iron and steel and supplied whatever he had on hand. I told him that when I asked for wrought iron, that was what I wanted. When the next one arrived I filed a flat spot on it and examined it under a magnifying glass. That inspection revealed the telltale slag stringers of wrought iron, and the cylinder was accepted for service.

During the work on tungsten ore, an amusing incident occurred that demonstrated Colonel Dorsey's active interest in our programs. I went to his office one time to discuss something and found him confronted with a mass of calculations and with a puzzled look on his face. Seeing me, and with a sheepish grin, he commented, "I've been working up a heat balance on Shaw's process. It comes out an even 100% and I can't find where I made my mistake!"

Our work in metallurgy covered tungsten, including particle size and distribution of the metal powder, Dumet wire, and lamp metallurgy, including the important filament microstructure, various causes of premature lamp failures, and appraisals of new processes.

Dr. Pacz dropped in once in a while to talk tungsten and to learn what was new. He was an outstanding individualist who was allowed to work as he chose. Terry and Tremaine were broad-minded that way and especially toward the one who had made the gas-filled lamp a commercial commodity. It was Pacz who invented the modified aluminum-silicon alloy Alpax. By the addition of a small amount of sodium, he had converted a poor alloy into a very good alloy. On one of his visits he told me what induced him to take up aluminum. He had brought over some articles of deep drawn aluminum that he had anodized (I assume) and had colored in various ways. I remarked that I was surprised he had done so much with aluminum, and that I had figured he was a tungsten man. His reply surprised me. Zay Jeffries was an aluminum man but had switched over to tungsten, so Pacz decided to include aluminum in his research. Thus it was that aluminum-silicon alloys became important.

The production of a white lamp, or an inside-frost lamp, was a major problem of the lamp industry. Surface coatings were unsatisfactory. Sandblasting the inside surface made an excellent lamp, but the bulb shattered on the slightest provocation. Orders were out to stop all work on sandblasting. The project appeared to be hopelessly bogged down.

Then, one day, a lamp-development engineer went to Mr. Enfield, manager of the lamp development laboratory, and said he had something to show. He placed three inside frosted bulbs on Mr. Enfield's desk and then put down three more which alternated with the first three. He flicked the first one over on the desk and it shattered into many pieces. He repeated with the second, and it simply bounced. The next two sequences repeated the performance — convincing evidence that the inside-frost lamp had at long last arrived. The secret of success was an acid treatment after sandblasting, to reduce the effect of the notches. After Enfield hit upon that remarkable and useful innovation, one wonders if he ever secured a patent on his invention. Our Patent Office found nothing novel or patentable in what he did, although we may presume that it did see something useful.

Colonel Dorsey had an administrative assistant, Lieutenant "Murt" Murtfeld. He was an able young man and was popular with all of us. One time I invited him to have dinner with me and also to take in a light opera. Prior to his tour of duty with Chemical Warfare he was a professional musician and had conducted Victor Herbert's operettas on tour. At the performance, knowing that I had dabbled in music, he asked if I could name the key in which the orchestra was playing. He had absolute pitch and knew at once, but I had to take my time. All I had to assist me was knowledge of the flute (my instrument) and limited knowledge of the strings and brasses. I also used an opera glass. After observing and listening I gave him my decision. Triumphantly he said, "That's correct." How did I do it? That may be an interesting question to answer and I'll tell later how I did.

I met another nonmetallurgical man, Billy Evans, the well-known baseball umpire. During the off season he had charge of our gym program, which he ran expertly. I tried boxing with him, but it was a mismatch since he was over twenty pounds heavier. Naturally, I discussed baseball with him. He was behind the plate during the season's first game in Washington at which President Taft threw out the ball. Walter Johnson pitched one of his copyright games that day. I once asked Billy about the 1919 World Series that resulted in the Black Sox scandal. Williams pitched the first game with Billy behind the plate, yet he never suspected foul play. He told me that Williams never had better control. He walked more men than usual, and at critical times, but he missed the plate by no more than an inch or less. On another occasion, Cicotte cut off a ball from the outfield that would have stopped a run at the plate. Billy wondered why the experienced Cicotte suffered such a lapse of judgment but did not suspect that it was intentional. He rated that Chicago team as the best in baseball to date.

One day at Nela Park I received a letter from Mr. Leverett Saltonstall, a Boston lawyer and later United States senator from Massachusetts. A large molasses tank in Boston had failed and discharged its contents into the streets and surrounding property, causing great damage. Mr. Saltonstall wanted me to be his expert witness in a damage suit that was coming up. Professor Sauveur was the expert for the other side. That was one of the famous cases of brittle fracture, although in those days a trial on the basis of its merits would hardly be possible in view of the lack of acceptance of notch brittleness.

Before taking up this request with Colonel Dorsey, I discussed it with Zay Jeffries to get his reaction. Zay told me that Sauveur had taken a very positive though dubious position and suggested that I decline. I took Zay's advice.

Nela Park, the "University of Light," enjoyed a good reputation and attracted visitors. Our most famous was Dr. Albert Einstein, who had come over from Germany with Mrs. Einstein for a visit. The research laboratory held a reception for them, as had happened, to his surprise, on other visits. He was interested to learn that I had studied at Charlottenburg.

In Cleveland, I began to get an insight into reasons for selecting steels for specific applications. At Midvale, I had simply accepted the idea that the steel used was the most suitable, and I attempted to go no further. One day I visited R. R. Abbott, chief metallurgist of the Peerless Motor Car Co., for a discussion of automobile steels. The question of nickel steel versus chromium-nickel steel came up because Peerless was using the former and Cadillac the latter, as I recall. Considerable testing had revealed no significant difference, so I asked Abbott why there was a difference in practice. The answer was simple — Peerless was accustomed to nickel steel and Cadillac to chromium-nickel steel. Some years later, I was observing a heat of nickel steel being made for a GE rotor. When I asked why nickel steel, the reason was equally simple — the supplier had greater confidence in meeting the requirements than he would have with an alternative steel.

One of our leading lights in Cleveland was Joe Emmons, chief metallurgist at Cleveland Twist Drill Co. He was always ready to talk about steel, twist drills, heat treatment, or about nearly any other subject. One time, I showed Joe a competitor's twist drill that was supposed to be a good one, and asked him what he thought of it. He smiled as he calmly stated, "It's not a twist drill." Joe always had copious test and performance data to back his opinions and always had sound answers.

In 1920, the American Society for Steel Treating (later to become the American Society for Metals) established its headquarters in Cleveland after the fusion of the Steel Treating Research Society, of Detroit, and the American Steel Treaters Society, of Chicago. William H. ("Bill") Eisenman was the secretary and driving force of the new society. Bill retained his energy and unexcelled leadership throughout his career. His life and a history of the society have been written up in *Metals Review* of ASM.

Bill was a great salesman for the society, in the best sense of the term. Evidence of his ability are the national and regional metal exhibitions and the establishment of numerous local chapters of the society with local programs. He was also skilled in selecting a staff. Ray T. Bayless came out of industry to become the longtime editor of *Transactions*. It was an arduous task, but Ray thrived on it as a labor of love. He must have stored away information like a library does books. However, *Transactions* consisted largely of technical and scientific papers. To round out coverage in the metals field with more general articles and news, Bill started a monthly, *Metal Progress*.

Ernest Edgar ("Ernie") Thum was brought in from *Chemical and Metallurgical Engineering* as editor of *Metal Progress*. Ernie soon made it the preeminent magazine in its field for style, composition, and editorial content. That was his life's work and remains a fitting monument to his efforts. He was also a reporter for special feature stories. In that connection he came to A. O. Smith Corp., when I was there, to secure material for an article. His writeup surprised me for the amount of pertinent material he had secured on a brief visit.

The depression that followed World War I caught up with us at EEL, and although we had become a going concern under Colonel Dorsey we were the newest division at Nela Park and we had to go. A management committee went over our work to save whatever was held to be of greatest value. As a result, a few of us were retained to complete promising projects. Although my diffusion-bonded, sealed-in wire process was abandoned, I had a program on single-crystal filaments in progress which might lead to an improvement over the 218 wire. I was asked to continue with it.

At about that time, Dr. Willis R. Whitney, the famous director of the research laboratory in Schenectady, came to Nela Park with the option of acquiring any personnel or equipment he wished to have. I was far enough along with my single-crystal work to put on a fairly good demonstration for him. In any event, he invited me to come with him. That closed one epoch but opened the door to another.

I came out of this experience at Nela Park with deep appreciation of the effort that went into the production and distribution of our electric lamps. I have done little more than hint at it here, since that would be a story by itself, although, in miniature, it is comparable to the prodigious effort of our automobile industry in its yearly output. Of course GE was not alone in this. GE owned the important patents, but by no means did it have a monopoly. Westinghouse and other licensees had access to all new developments under GE's Mazda Service and could use them as they chose to. One rule was applied to all, requiring that they maintain the prescribed quality. Great credit is due our lamp industry for reducing lamp prices as lamp quality was increased, to give the user more light per watt or per dollar. The savings to the public of the improved lamps over the carbon-filament lamps has been estimated to be an astronomical figure — a great lesson of the worth of science and ingenuity.

The time came at last for the move to Schenectady. I acquired a Dodge touring car and had decided to drive through. I also decided that I needed advice on cross-country driving, because the farthest I had driven previously was from Minneapolis to Lake Minnetonka, a beautiful lake fourteen miles west of Minneapolis, and that was in a second-hand Ohio roadster of dubious pedigree. I broached the subject at a final conference in Mr. Enfield's office. In those days (1921), the highways were country dirt-roads and our only maps were U. S. Geological Survey quadrangle maps. After discussing routes, where to get gasoline, stopping places, and so on, the question of how to make good time came up. One of the men spoke up with the voice of authority and, incidentally, brought out the difference between then and now. "That's a good question. I know there are some who hit it up at thirty-five miles but I'll take my good steady thirty miles and get there just as soon." I followed his advice since my dependable Dodge was capable of it. So it was on to Schenectady, the city whose name few people could pronounce if they saw it written or could spell if they heard it. Actually it is a Dutch-Americanism from an Indian name for "end of trail."

I went to Schenectady with the feeling that I was going to the seat of industrial research, the famous GE Research Laboratory under Dr. Whitney. He was a strong, virile person with a voracious curiosity about everything that came to his attention, especially nature, whether outdoors or in the laboratory. Above all, and as a true disciple of Francis Bacon, he was a great believer in experiment and the use of the scientific method. That was equally true of basic research, evidence of which was his early treatment of Langmuir.

I should like to quote what Dr. A. W. Hull wrote in this connection: "He engaged Irving Langmuir and gave him freedom to work on anything that interested him. After three years, Langmuir was still deep in a study of heat conduction in gases. With no practical applications in sight, the board of directors questioned the advisability of keeping him. Dr. Whitney replied unhesitatingly and emphatically: 'You leave Langmuir alone. He is doing good work.' " The results were soon to prove it.

This is my view also, but I believe that all Dr. Whitney needed to know for him to say, "You leave Langmuir alone," was that lamps, vacuum tubes, and X-ray tubes were responsive to temperature and gases at various pressures, which Langmuir was studying.

I should like to mention Dr. Whitney's human side and illustrate it with an incident in which I was involved. One day he called me down to his office and asked if I would do him a favor. Some years earlier, a paperboy who was always at the main gate had had Dr. Whitney as a steady customer. He was a bright young man and was putting himself through law school. Dr. Whitney became impressed by his attitude and now wished to help him. He had become a practicing attorney and needed an expert witness for a case in Albany. Of course I would help.

I arrived in Albany with my binocular microscope and was ushered into a back room of the courthouse for a conference with the attorney. I was to be a surprise witness. An uncle and aunt had willed their farm property to the plaintiffs, an Albany couple. That will was contested by a hired hand and a maid in the house. They claimed that the owners had made a later will giving them the property. It was purportedly made out on a small card during a meeting in the kitchen. Was it a forgery? That was the question I was to answer.

During the noon recess I examined the card closely and noted some pertinent points. I gave the attorney a list of questions to ask the farmhand dealing with the pen or pens that were used and the sequence several statements on the card had followed. Those questions were asked and the defendant gave very positive answers. He testified that the entire will had been written by the uncle with one pen and that nothing had been added subsequently with another. He also testified on the order in which the statements had been written, a point that had to be in his favor if the will were to be accepted as valid.

When it was my turn, I testified that two different pens had been used and that the part about willing the property to the defendants had been put in subsequently with the second pen. This refuted the farmhand's testimony and was not disputed or controverted by my cross-examination. My testimony showed positively that it was a false and forged document.

I felt that I had won the case for Dr. Whitney's friend, but I reckoned without the judge. Instead, in effect, he ignored my finding of forgery and ruled in favor of the defendants. I felt strongly that it was a miscarriage of justice. The plaintiffs lacked funds to appeal the case.

Dr. Whitney's career is too well known to be repeated here. While he was the heart of the laboratory, he also had outstanding members on his staff. Dr. William D. Coolidge was one of them. He had two achievements in our field to his credit, either of which would make a fine career for almost anyone. He was not a metallurgist, yet I could make a good case for his ductile tungsten as the No. 1 metallurgical contribution of the laboratory. One might say that powder metallurgy was incidental to attaining his main objective, but he started it on its way to being an important industry. Specifically, he put tungsten to work in lamps, X-ray tubes, electrical contacts, and spot-welding electrodes, and his ductile tungsten led indirectly to the development of Carboloy and cemented carbides. Dr. Coolidge never published his own account of that metallurgical contribution, so I took it upon myself to do so in an article in *Metals and Alloys* to make a permanent record of it for metallurgists.

Our metallurigcal group had several research men who, in addition to advancing the arts, also functioned as brains or experts in their fields. They worked with control laboratories and production people in various plants of the company. W. E. ("Bill") Ruder had an important specialty under his wing — the silicon steel of Sir Robert Hadfield. It became tremendously important for use in transformers and electric motors by virtue of its low hysteresis loss. It was apparently a simple steel product, because it was just a low-carbon steel with a small amount of silicon, yet its manufacture, processing, and economic application required continuing study and control. Bill discovered some aspects of silicon steel behavior that had both scientific and technical significance — the outstanding response to grain growth on annealing after cold work, the improved magnetic characteristics of coarse grains, and the benefits of preferred orientation. The principles he developed for securing optimum performance are widely used. This was not a case of ordering a steel to a GE specification, because Bill had to work closely with the supplier, the Allegheny Ludlum Steel Corp. Norman P. Goss made an excellent contribution to silicon steel production by developing a continuous-sheet process.

Truman S. Fuller, with hearty and contagious laugh, handled copper and nonferrous metallurgy, though not exclusively. In the early days, copper was troublesome on account of its tendency to become brittle. Trum found that the cause was the reduction of grain-boundary copper oxide by hydrogen or other reducing gas. Even very small amounts of gas produce the effect. He was also one of the first to look into hydrogen embrittlement of steel. His paper "Hydrogen, the Dr. Jekyll and Mr. Hyde of Metallurgy" is still good reading. It is a very useful gas when used properly, but its effects can also be disastrous. Trum was active with chromium steels at an early date. A major contribution was his Turbuck, a strong and corrosion-resistant steel for use as turbine blades. He also promoted applications of aluminum alloys for GE equipment. His all-around experience and sound judgment fitted him well for the position of chief engineer of the Works Laboratory at the close of his career at GE.

Floyd Kelley was our specialist on high-temperature processing in hydrogen. He covered brazing and especially copper brazing, chromizing, and aluminizing or calorizing. These special processes are widely used today, but they all require intelligent handling as highly specialized processes. Floyd pioneered these industries as he developed the necessary know-how or expertise. The Schenectady Works installed several hydrogen furnaces for its production, one of which was nicknamed "the Zeppelin" for its size and shape. A smaller one exploded from the leakage of air. It was in the turbine shop, but, considering the gas involved, the explosion did little damage. The water that was formed as the result of the high-temperature reaction was decomposed into hydrogen and oxygen, which absorbed heat and produced a blanket effect. Floyd also turned his attention to Carboloy (see Chapter 4) and to substitutes. Out of that came tantalum carbide with nickel as the binder. It was especially effective in machining steel because it minimized wear on the tool, or cratering. He was also responsible for attaching the carbide insert to the steel shank by copper brazing.

Gerald Brophy and I shared an office with my Zeiss microscope, which was sent on from Nela Park. Gerry was our expert on steel and its heat treatment. I recall one of his projects. In those days, as old-timers will remember, we had streetcars that ran through town. They could be heard from far away. Gerry was asked to take the noise out of the gears. He did just that by modernizing their manufacture and heat treatment but was told "it costs too much." Like many production people, they wanted something better but revolted at the idea of paying for the improvement. Actually, that pressure to make things both better and cheaper is consistent with the function of research and development. It is a relentless pressure we live with in a society of free competition.

General Electric was heavily involved in welding as a device for making two pieces of metal behave as if they were one, or at least that was the general idea. Bare-wire welding was old, but the two pieces that were welded together did not behave as though they were one. The weld was brittle. Improved electrodes were available but they left something to be desired. One day Carl Ipsen came up to the laboratory to see Gerry. Carl was in charge of the electric furnace and welding group and later became secretary of the Electric Furnace Manufacturers Association. A company out west had developed a paper-covered welding rod that produced welds in steel that were strong and ductile and of even better quality then the parent metal. Carl wanted the research laboratory to give him an electrode that would do about the same but not infringe the A. O. Smith patent.

Gerry and I used to discuss our problems, and this one was no exception. He was having difficulties. GE had machines for welding continuously by using a coil of wire, so the development of a bare steel wire was desirable. Oxygen was thought to be the undesired impurity in the weld. One day I recalled my war work on silico-manganese and the way silicon protects iron and manganese from oxidizing. I suggested that he try a low-carbon steel with high manganese and silicon. He had some made up in our furnace laboratory and found a notable improvement in the weld. A commercial heat was made by an outside mill, but, inexcusably, it was too diffiicult to roll; it had too much alloy and more than the laboratory heats had. That concept was not pursued further, but I have noted that it has been applied more recently.

Gerry left the research laboratory to go with the International Nickel Co., first at the Bayonne research laboratory and later to the Hartford office, in charge of technical service.

We had a furnace laboratory for melting experimental alloys. Jack, one of the men in charge, was not only a good furnace man but also a Yankee wit who would frequently amuse us with his dry comments. I recall one about a particularly garrulous person who spoke a lot while saying little. "Yeah," Jack told us, "he's a man of few words, but he uses them over and over."

LeRoy Wyman joined GE as metallurgist on tungsten and lamp metallurgy at our Edison Lamp Works. While there, he introduced Carboloy dies to this country for drawing tungsten wire. They replaced the carburized chromium alloy dies. He came up to Schenectady to run our metallographic laboratory. Roy finished his career at the National Bureau of Standards, where he became a leader in establishing metallographic standards, especially as longtime chariman of the ASTM Committee on Metallography.

Burt and Louise Newkirk had moved to Schenectady since Minnesota days. Burt had put in summers at the research laboratory on balancing turbine rotors and had finally made a permanent change. The Newkirks were good company and had a fine family by that time. I remember their youngster, Jackie, a sprightly young man and a chip off the old block. He decided on metallurgy as a career and has had industrial as well as teaching experience. So the Newkirks provided our profession with a young Man of Metals; Jackie became Dr. John B. Newkirk, professor at the University of Denver, and a leader in metal physics. Come to think of it, I guess he is not so young any more.

A feature of life at the research laboratory was the Saturday-morning colloquium at which a staff member or invited guest would discuss his work. After the presentation, we had a discussion period with questions and answers. The research laboratory was a strong magnet for ranking scientists, because men like Whitney, Langmuir, Coolidge, Hull, Dushman, and others were well known. The staff enjoyed those sessions and received the stimulus for self-improvement that a youthful ballplayer does (meaning me) when he watches the pros perform (meaning the likes of Nick Altrock, Perry Worden, and Gavvy Cravath).

Dr. Whitney asked me to take charge of the colloquia one year. I don't now recall all of our guests, but some of them were Sir J. J. Thompson, a leading English physicist; Harlow Shapley, the Harvard astronomer known for quick wit and eloquence; Professor E. G. Concklin, our leading evolutionist from Princeton; Dr. Burgess, the director of the National Bureau of Standards; Nils Bohr, the eminent Dane whose work on the hydrogen spectrum started productive work on the structure of the atom; Arthur Compton, of cosmic-ray fame and enthusiast for Planck's quantum theory; Professor Sommerfeld, who was one of a small group whose work established modern physics; Professor Harold Urey of Columbia, whose field was isotopes and heavy water; and C. H. ("Boss") Kettering, of General Motors. I also induced Larry Hawkins to give us his comments on the research laboratory. He was Dr. Whitney's right-hand man, handling everything that Dr. Whitney chose not to handle.

My year could have no finer finish than a talk by Dr. Whitney himself. That took a bit of arm twisting (figuratively, of course). As the day approached, I became increasingly conscious of my job of introducing the speaker. How do you introduce your boss to his staff? Even after I was actually on my feet I still didn't know, in spite of advice from others, but I got an inspiration. So I said, "I suppose you are wondering what I am going to say. Well, some people eat the frosting first and the cake last, while others finish with the frosting. I am in the latter category. So, fellow staff members and guests, I give you the frosting of this year's colloquia, Dr. Whitney." From his smile and typical shake of his head I felt I had not done too poorly.

We also had a social feature that followed the lecture — a private luncheon at the Mohawk Club with a small group from the laboratory and our guest. One little incident lingers in my memory, and it may indicate the tenor of our table talk. It was Langmuir, studiously pondering the question — which would chill more quickly in a refrigerator, hot water or cold water?

I was very fortunate to be given that assignment. It brought me much closer to Dr. Whitney than would have been possible otherwise. He took a keen interest in those meetings, and that led to numerous discussions of worthy prospects and their subjects. For example, while I knew Boss Kettering by name, Dr. Whitney knew him personally from the time during World War I when they worked closely together on a war project. It was clear that we would have an exceptionally capable and versatile experimenter who was also an effective speaker with a down-to-earth philosophy. He was a researcher's researcher, and that came from the highest authority — Dr. Whitney. Boss Kettering came up with a typical Ketteringism which has stuck in my craw, to wit, "Beware of logic. It is an organized way of going wrong with confidence."

I also found Dr. Langmuir ever ready to comment and to offer suggestions. He told me we should have someone on the quantum theory and that Arthur Compton of Chicago was the man. I was pleased to carry on because, for one reason, the Compton brothers had married the Hutchinson sisters of Minneapolis and Enid was a classmate and friend. Their father was on the Minnesota faculty. Compton spoke at our colloquium at the time of the controversy over the Compton effect. Duane, at Harvard, had not been able to confirm the change in wavelength of scattered X-rays. Compton mentioned that and, with a smile, said he was going on to Harvard to find out where Duane went wrong. That episode has been written up in *Science* (Vol. 138, p. 794).

A small group of us met with Dr. Whitney each week to discuss lamp problems, with Saul Dushman as the leader. Saul was one of Langmuir's protégés on the physics of lamps and electron emission in vacuum tubes. (Saul made the enormous extrapolation that electron emission would follow the same laws that govern atom evaporation, and found that to be correct.) He was a very good man to have available on the staff, which was especially true for many of us because he was always ready to expound scientific principles to others who were not as well grounded as he. He opened my eyes on more than one occasion and helped me to establish the laws that govern relationships between experimental parameters.

One day Langmuir brought in a paper by Mark, Polanyi, and Schmid on the plastic deformation of zinc single crystals. The work was well done, and some of the phenomena were so striking that we were all excited. Normally, brittle zinc stretched in tension by several hundred percent. Dr. Langmuir thought that we should do something similar. I was working on single-crystal tungsten for lamp filaments, which prompted Dr. Whitney to ask me to include it in my program. I will return to the study of zinc a little later.

Our principal interest in tungsten was in its behavior at filament operating temperatures that were well above the softening or annealing temperature. The crux of my investigation was the effect of structure on the relation between the load imposed by the filament and its useful life.

Back at Nela Park, I had satisfied myself that the load, or rate of loading, had a decidedly important effect on the behavior. My equipment had been a Bunsen burner and two pieces of copper wire. Gripping both ends of a wire in my hands I had heated the center to a low red temperature and then pulled gently on the wire. It had simply pulled apart with no evidence of prior deformation. I heated the second wire in the same way, but this time I jerked it sharply. Then the wire necked down and deformed in ductile fashion. That was a striking demonstration, but we were interested in the behavior of filaments during low strain rates.

The production of single-crystal tungsten wire was largely a matter of technique along with the use of the proper tungsten as starting material. I knew that large grains of 218 wire grew parallel to the fibers of the drawn wire and that the velocity of grain growth was very high at high temperatures. Briefly stated, success must depend on using a very high temperature, a rapid rate of heating to the high temperature, and the type of tungsten that was amenable to the process.

I used a device that heated the wire continuously as it passed between two contacts in a hydrogen atmosphere. The trick was to do this at a very high temperature, even above the operating temperature of the filament, and then to control the speed properly. That, in turn, required exceedingly pure and dry hydrogen. This is much easier to describe than to do, but I did produce single-crystal wire. When I compared its sag resistance with that of 218 wire, the single crystal was definitely superior. I still have my photograph that shows the two test coils after heating. The improvement, however, was not sufficient to justify a recommendation that the process be tried by our lamp divisions. A paper on this investigation would have made an interesting case history, especially the selection and preparation of the starting material and the critical conditions for conversion to the final structure as a long single-crystal wire. In addition, we also ran across a method for making a massive single crystal that should be valuable for use as large components (thrust nozzles) at ultrahigh temperatures to utilize its high strength.

Paralleling the work on tungsten at high temperatures I also ran stress-rupture studies at room temperature on wires of the soft metals, zinc, tin, and lead. The purpose was to develop the relationship between load and life. I chose constant load in preference to constant stress, as d'Andrade had used in England, because the filament is loaded by its free weight.

Coming to the tests, the wire was suspended freely with the test load hanging from the lower end. The elongation was recorded on a motion-picture film while a seconds clock recorded the time simultaneously. When I plotted the times to failure on a log scale, I found a straight-line relationship with the square root of the load. I believe that was the first time this had been observed. At first glance this might appear to be a simple viscous-type of flow, but one test that took me by surprise negated that as being universally true. It was not a fluke, and I had the photographic record by which I exhibited it to others. The wire stretched rapidly under high loads as though it were to fracture quickly. Instead, the elongation decelerated suddenly and the load remained suspended for a significant time interval beyond that point. Apparently, this effect was caused by work hardening, as my test of copper at Nela Park had demonstrated.

Most of the studies of single crystals were made on zinc and copper, using the methods of Czochralski and Bridgeman, respectively. There is no need to discuss those tests, in view of the large amount of literature on them. Just one observation surprised me. The original paper by Mark, Polanyi, and Schmid indicated by a schematic figure that the crystal elongated by the slipping of rigid blocks over each other, like a deck of cards when spread. We could find no deformation of that type; in fact, a close examination showed that it did not happen that way. The serrations on the edge, which I expected to find, were not there.

The work with copper single crystals was equally interesting and had no amusing sequel. Incidentally, the copper crystals also did not deform by the process that Mark, Polanyi, and Schmid suggested. This led me to wonder why they proposed it, and why other experimenters neglected to check the point. I gave several talks on single crystals in those days, including a demonstration of a copper crystal about three quarters of an inch in diameter and ten inches long. To show how strong I was, I would bend the bar in my hands into a horseshoe shape. Actually, that showed how weak the single crystal was; one cannot do that with an ordinary copper bar. Then, to complete the illusion, I would pass the bar around the audience with the invitation to straighten it. Of course, no one could.

That was a demonstration impressive enough to be reported in the press. One day I received from a vaudeville performer who wished to secure some of my bars, a letter asking how much they would cost. Seems he did a strong-man act. I referred the inquiry to our special products division and was told to quote him $400 per bar. That closed the episode.

Along with the studies of single crystals and filaments, I was able to include some work on hardness testing. At about this time, interest developed in the hardening effect of plastic deformation. The common ball hardness test was used on test metals after increasing amounts of cold work. I delved into this problem by using Hertz's relation between the load on the ball, P, and the diameter of the impression, d, which was made in the specimen. The relation is $P = ad^n$ in which a is a constant and the coefficient n gives the capacity of the metal to increase in hardness with continued cold work. The constant n of soft annealed metals is relatively high but it drops to about 2.0 as the metal is cold worked. Beyond that point, additional cold work has little effect on the hardness.

We got two and possibly three interesting results from these studies. Hadfield's manganese steel differs from copper and similar metals by retaining work-hardening capacity (high n) even after severe cold work, while using the load divided by the projected area of an impression in Carboloy produced an improved hardness test for super-hard materials. By using the same principles with large, calibrated, copper blocks and one-inch balls, I devised a shop method for weighing heavy loads, such as turbine rotors. Knowing the unit pressure to produce the impression, and the area of the impression, would give the load on each ball.

By the time I arrived at the laboratory, Dr. Wheeler P. Davey had taken over X-ray work from Dr. Hull, who was busy with his numerous vacuum tubes. With the advent of studies of textures, preferred orientations, and grain size, Dr. Davey included structure work along with more routine work. Wheeler joined the Penn State faculty a little later and finished his career there. As a sideline and labor of love he worked for ASTM on diffraction patterns and was active on the Joint Committee on Chemical Analysis Data by X-ray Methods. He also inaugurated the Priestley lectures at Penn State; they were given by invited lecturers while in residence. I was happy to give the series in 1928. It was a pleasurable experience that took me back to my Minnesota days.

Wheeler was possibly one of the first to use a plastic bearing in place of the conventional metal bearing. I had noticed that he always had a fan running in his office, so one day I asked him about it. Then he told me about his experiment with the new type of bearing material. This work with single crystals and my association with Wheeler proved to be a good introduction to my European visits in 1925.

Chris Dantsizen, our chemist, and I were sent over to Germany primarily to investigate a German process for making aluminum from clay. Though that took up most of our time, we were able to visit lamp factories and research laboratories. We found that the Paris office of the International General Electric Co. (IGE) had excellent arrangements for visiting firemen in the form of a "See Europe Tour." Those visits were made possible by contracts which Mr. Gerard Swope, then president of the International General Electric Co., had arranged with European lamp factories. In other words, we visited affiliated lamp factories in France, Belgium, Holland, Switzerland, Italy, Hungary, Austria, Germany, and England. Our function was to disseminate and collect information on the technology of lamp manufacture. We were also in Frankfurt am Main and Leipzig-Bitterfeld on the clay process. This tour was not all business; some pleasurable interludes popped up unexpectedly.

After our visit to the lamp factory in Winterthur, Switzerland, we started for Milan but were stranded in St. Moritz, that beautiful resort in the Ober-Engadin. This trip came in winter, and we were experiencing a tremendous snowstorm. That was anything but an unhappy experience — a lucky windfall. Back home it had its repercussions. Why could we get to St. Moritz but no further for a couple of days? That was easy to answer. On the way up we were either in a tunnel or on a bridge. We had to leave for Italy via the Bernina Pass, where the snow had drifted so badly that trains were not running.

Another interesting interlude was a weekend in Venice. After the ride from Milan, we arrived there as evening dusk settled on the canals. Our taxi to the Royal Danieli Hotel was a gondola. Instead of remaining on the Grand Canal, the gondolier took the straighter route via side canals. They were narrow and dimly lighted, giving an eerie aspect to our surroundings. We wondered if we were really safe. Even the musical intonation of our gondolier as he honked when approaching an intersection was ominous. We enjoyed our brief visit to that uniquely interesting city.

On a trip like ours, the philosophy is "part of a loaf is better than no loaf at all," so it was on to Budapest. That city had changed since I was there in 1911, largely because Hungary had been on the losing side of a war. I recalled the words of a lieutenant of the Hungarian army on that first visit as he confided that they were preparing to revolt and become independent of Austria. History decreed Hungary's independence, but in 1925 the country was sadly depleted in land and resources. The low economic condition of Hungary was exemplified by our breakfast on the train from Trieste. It was the poorest I have ever had.

We visited the Budapest Lamp Works and were ushered into the office of Mr. Aeschinger, the president. He had a large map on the wall, and before discussing lamps he told us about the old Hungary as compared to the Hungary left by the Versailles Treaty. Peace treaties have been quite tragic for that country, for which national independence has been an empty phrase.

A highlight of our stay in Budapest was Gypsy music in a local restaurant. I have enjoyed a lot of music in my time — Caruso and the old Met; Arthur Nikisch and the Berlin Philharmonic Orchestra; Strauss waltzes in the city park in Vienna; German Military Bands; Sousa and his "Stars and Stripes Forever"; Wagner at Bayreuth; Bach as played by Stokowski and the Philadelphia Orchestra — but that old Primos and his violin played on my emotions as has not been done before or since.

In Vienna, we visited the lamp factory but I do not recall that we did anything else besides some sightseeing. When we arrived at our train to leave for Berlin, we noticed that a group of people were bidding a VIP off. He was Hugo Eckener, the man who took over from Graf Zeppelin. After we were on our way we approached him for an interview. He was mild-mannered and reserved and seemed not to mind our intrusion. Like most Germans we met, he inquired into our experience in Europe and our conditions back home.

I was surprised that we did not have a through train to Berlin but had to detrain at Passau in Bavaria and remain there overnight. Immediately upon entering our hotel we became aware of animated and confused conversation in the adjoining dining room. Our porter informed us that a new brew had just been opened and was being enjoyed by the townspeople. Such was life in Germany.

Upon our arrival in Berlin, Chris had to leave for Bitterfeld, where alumina was being recovered from clay, then go on to Frankfurt, where additional pilot-plant work was in process. While he was away I had ample time to make personal contacts and visit lamp factories and local research laboratories.

My first call was on Geheimrat Mathesius, who greeted me cordially like a dean welcoming an old graduate back to campus. While bringing him up to date on what I had been doing, I made a brief but favorable report on his son, Walther. I had to keep in mind that one of his sons had fallen in the war, and that Walther had been on the other side. He then posted me on metallurgical advances in Germany since I left, and especially on a titanium steel that he was promoting. The Geheimrat was a grand old man whose character and stability typified the features of his countrymen, which made Germany so productive in peace and so formidable in war.

My next visit was to the familiar metallographic laboratory where Doctor, now Professor, Hanemann and Fräulein Schrader greeted me. When we got down to business I learned that they were working on the problem of martensite of quenched steel. He had developed a working hypothesis which postulated that martensite is a phase, epsilon, in a metastable equilibrium. I inquired, with a smile, if that was Arnold's hardenite that he postulated. We had discussed it back in my student days. His wry smile told me he didn't subscribe to such wild ideas. Fräulein Schrader had made a beautiful collection of photomicrographs for his paper entitled "Über den Martensit." I thought it was worthy of publication in English and, with the cooperation of my assistant, Gus Pirk, I translated his paper, "On Martensite," for ASST.

There was a sequel. Some years later he tested his assumption of a metastable equilibrium by means of recording the martensite formation by high-speed photography. This was the work he did with Dr. Wiester, who told me about it when I met him at Krupp in 1945. They found that martensite forms practically instantly, or between two frames of the film. Dr. Hanemann discarded his concept and had the results published.

I was fortunate to find another of my Charlottenburg teachers in town, Professor Guertler. When I called on him in his new laboratory he was busy and gave me a sheaf of data sheets to peruse. They covered corrosion tests of a large number of nickel-molybdenum alloys in various media. They showed a decided improvement over stainless steel and have become a valuable series of alloys for use under severe corrosive conditions in the chemical industries.

Later he explained his method of finding new aluminum alloys that might be amenable to the precipitation-hardening process. He called it a "clarification cross" method. By it he found the pseudobinary series with the magnesium-zinc compound, Mg_2Zn. Upon studying the series in detail he located alloys of remarkable strengths. They were the forerunners of an important class of high-strength aluminum alloys today. To satisfy my interest in them, he referred me to Dr. Sander of Theodor Goldschmidt GmbH in Essen and suggested that I call on him.

Dr. Guertler explained how age hardening was discovered, or at least was first encountered, in aluminum alloys. The aluminum plant at Düren received a complaint that a customer's tests of a shipment disagreed in strength and ductility and hardness from the values reported by the factory. A few days had lapsed between the tests. Wilm took it from there and developed Duralumin. Earlier cases of age hardening were known but not recognized, but it was Wilm's contribution which brought the process out in the open as a new heat treatment for hardening alloys.

A third interesting development was a hardened, tarnishproof silver alloy. He assumed that it would be valuable for tableware. It probably would be, but cutlery people turned it down because there was too little silver in it for it to be labeled sterling. He had no more success with it than with his heat treated cast tungsten carbide balls, because cemented carbide soon preempted the field.

During the 1920s, advanced research in metal physics was off to a serious start. Groups in England, Germany, and the United States were using

X-ray methods in studies of the mechanism of plastic deformation and subsequent structural changes on heating in the temperature ranges of recovery and recrystallization. Both single-crystal and polycrystal materials were used in these studies. This activity prompted premature speculations on the causes of work hardening. A particularly active and productive group was in Berlin–Dahlem.

Professor Heyn had succumbed to the pressures of the war and internal strife shortly after he had become director of the Kaiser Wilhelm Institut für Metallforschung. Professor Bauer had taken his place in logical sequence. The men I wished to see were Polanyi, Sachs, Schiebold, and Schmid.

Dr. Polanyi was at the KWI for Physical Chemistry but was then studying the kinetics of reactions between gases and not single crystals. George Sachs had already established himself as an able investigator in theoretical and applied mechanics and was using the new techniques to good advantage. He came to the United States in the 1930s and became well-known as a sound and prolific contributor in physical metallurgy.

Drs. Schmid and Schiebold, as well as Dr. Sachs, were studying plastic deformation of metals, by judicious selection of test material, careful experimentation, and painstaking analysis of the results.

I noted that the Dahlem group, in contrast to some other investigators, did not try to explain the effect of plastic deformation on work hardening, although that was a favorite pastime in the 1920s. They aimed simply at describing the structural and lattice changes revealed by the experimental evidence. It can be said, however, that the earlier amorphous-metal hypothesis was rejected and that fineness of grain, although a significant factor, was not as acceptable as some type of lattice alteration. I had dinner with Schiebold and we had a long discussion of the interpretation of Laue diagrams in this connection.

The accumulated knowledge gained here and elsewhere was immensely inspiring, so much so that I wrote voluminous reports back to the lab. For the first time we had concrete and reliable data on the processes of plastic deformation and preferred orientations. I regarded the publications as special reports to me that I could use in my own work. With tungsten, the maverick, I was strictly on my own.

Looking back, I believe that the major contribution to an understanding of plastic deformation on the atomic scale, and its effects on the properties, was the demonstration of modifications of the lattice, although still of uncertain character. It is likely that attempts to clarify the nature of lattice alterations, as shown by X-ray technique, led to the initiation of dislocation theory a decade later. Any further discussion would lead into the historical development of the dislocation theory, although tracing the ideas formulated by Tammann, Czochralski, Mathewson, Jeffries, and Archer, and others during this period would be interesting.

My principal interest in Berlin took me to the lamp factories for discussions with their technical people. I spent many hours on those visits and many more on reports back home. The main attraction, I confess, was the new cemented tungsten carbide that I discussed with Karl Schroeter at the Osram Lamp Works. That turned into a project of the first magnitude. Its story is related in the next chapter.

In my student days I had become reasonably well acquainted with the old Germany and with a fair number of its people. In 1925, Germany was a different country and the people were bewildered and uncertain about the future. That story is long and was told to me frequently, probably because I was a foreigner and an American. The main points were the Communists' reign of terror at the close of World War I and their attempts to wrest control of the country as they had in Russia; currency inflation, which caused evils and disorders of which I had been unaware; control of half the steel industry and other segments of the economy by a small minority; shortages of food and lack of jobs; and the Versailles Treaty and its consequences. I was told about all of these, and also of the greatly needed and appreciated help that they had received from the United States in time of deprivation.

One day, at lunch with executives of one of the lamp factories, I broached the subject of inflation — by then hard currency had been reestablished. Thinking of some sort of hedge, I asked one of them what he would do if inflation returned. He replied calmly, "I would shoot myself. Never again."

Upon finishing my work in Berlin, I left for the Rhineland and England for a few more visits. On my way to Essen I enjoyed a brief interlude, once again, in Goslar. The town was interesting, but the highlight was another visit to the ancient palace of Barbarossa. Then on to Essen to visit Krupp and Theodor Goldschmidt GmbH.

Dr. Edward Maurer, co-director with Beno Strauss of the Krupp research laboratory, and Max Moser had published on notched-bar testing, always an interesting subject for discussion. Maurer was an able and forceful person, and an experienced and straight-thinking metallurgist as well. He gave the impression of being the practical type, yet, much like Harry McQuaid, he had a sound grasp of basic principles.

Coming to notched-bar testing, Maurer explained that Moser (at the testing laboratory) was studying the effect of test-bar width (from one-quarter to full width) with the standard 30 mm × 30 mm bar. He and Mailänder were studying the effect of temperature. We know now that in the first series, the transition from ductile to brittle fracture was due to a change in the stress field, and in the second, to a change in the strength properties that controlled the fracture mode. I had already induced Dr. Moser to present his results in an ASST paper in 1924.

At that time, Krupp was a leader in the production of wrought stainless steels. The straight-chromium, or cutlery, type of Harry Brearly was widely known for its hardness and corrosion resistance. However, I leave it to Carl Zapffe to give the history of stainless steel. Dr. Maurer told me how he developed the austenitic, or "18-8," type. As a student, under Le Chatelier, at the Sorbonne, he found that the addition of manganese to high-carbon steels made it possible to preserve the high-temperature austenitic form of steel at room temperature upon quenching. (Hadfield had accomplished the same result with his manganese steel.) Maurer utilized the same principle by adding nickel to the chromium steel and produced a new and widely used austenitic stainless steel.

It was not that simple, however, and Krupp had to learn the hard way. The new steel was found to be compatible with beer, which led to the installation of stainless steel equipment in a brewery. Then one day, everyone was horrified to learn that a storage tank had disintegrated. Thus was the phenomenon of intergranular corrosion discovered. The problem was handled by drastically reducing the carbon content and quenching the steel from a high temperature to keep the carbon in solution.

Dr. Maurer left Krupp later to assume the chair of metallurgy at the venerable school of mines in Freiburg. He remained in East Germany after World War II.

When I was in Essen, I attempted to discuss Dr. Guertler's aluminum-zinc-magnesium alloy with Dr. Sander. The Theodor Goldschmidt firm was very secretive about it; consequently I received no information.

The Kaiser Wilhelm Institut für Eisenforschung in Düsseldorf was the fundamental and experimental research institute for the iron and steel industry, and was my next stop. Let me explain that I had time to make extra visits because I was awaiting word from Chris that he had finished his business and was ready to return home. My failure to hear from him gave me an opportunity to become posted on European developments, and, besides, I figured that what was good for me was good for GE.

I was acquainted with the work of Dr. Franz Wever in which he had classified alloying elements according to their effects on the phases of iron — in particular, whether the added metal expanded or restricted the austenitic field. That provided a very useful generalization that soon became widely adopted. It was further elucidated by Köster and Tonn by using three-dimensional models. He also had studied the effects of excessively fast quenching rates on the transformations of medium-carbon steel. Quenching rates of up to 10,000 °C per second from a high temperature were not able to inhibit the transformation of austenite into martensite. Wever's work summarized and extended the classical paper of Portevin and Chevenard on the temperatures at which structural transformations occur with various quenching rates. Both investigations indicated that the failure to inhibit the martensite formation meant that the austenite-to-martensite transformation belongs in a different category than other known solid-state transformations. Incidentally, it has been examined intensively for many years by metal physicists. Step by step, work in this field has advanced the arts of steel alloying and heat treatment greatly, to the benefit of consumers and users.

Dr. Wever was another experimenter using the new X-ray technique to determine the mechanics of plastic deformation and the establishment of the resulting texture. He used a special test specimen that he deformed in a precisely controlled manner. He knew better than others, except for single-crystal investigators, just what he had done externally, and was better able to correlate the end result, internally, with the mode of deformation. I thought his work so timely that I induced him to publish it for AIME, promising that I would translate it and see it through publication.

We continued our discussion during a walk along the Rhine, but changed to a more general conversation as we returned to his apartment. Mrs. Wever prepared supper, after which we enjoyed an evening of music and literary talk.

After Düsseldorf, I stopped in Paris for a brief farewell at IGE headquarters, and then took the boat train for London. A surprise awaited me. Sir Robert Hadfield had learned somehow that I was in town, and his secretary inquired if I would drop by his office for a visit. Would I make the acquaintance of an outstanding Man of Metals, a Hall of Famer? Of course it was GE, and not me personally, that was the attraction. I assured Sir Robert's secretary that I would be delighted, and we made the appointment.

When at Minnesota, I had a graduate student from the Pittsfield plant of GE who would discourse on Hadfield's silicon steel at the drop of a transformer. According to him, Hadfield was the greatest metallurgist of all time. I had also consulted for a firm that produced Hadfield's manganese steel, and I was aware, too, of his excellent work on cast armor-piercing projectiles during World War I. I approached this meeting with some idea of the man and his works.

During an interesting introductory session, Sir Robert presented me with an beautiful and highly prized pearl-handled pocket knife. In return, he asked me for a penny. I though to myself: "This man of great wealth wants to be paid for it, but only a penny!" As I handed him his penny, he must have noticed a look of surprise on my face, because he volunteered,

"To avoid passing on bad luck, you know." However, and in spite of that penny, I very soon broke every blade in it. That is the shape it is in now, still a precious souvenir.

We then turned to the business of the day. Sir Robert had developed an austenitic stainless steel for high-temperature service. He explained that steam-turbine manufacturers had used it successfully for blading and that he would like to have GE consider it for its turbines. I told him I would take it up with our turbine engineers as soon as I returned home and would then reply. We had no doubts about the quality of his steel, but our answer had to be no. GE turbines run at 3600 rpm, requiring a blading material with much higher strength than that of the austenitic steels to avoid flutter fatigue.

My principal mission in London was to visit lamp factories. Our affiliate was the British Thompson-Huston Co., and the other major producer was the British General Electric Co. At the latter, I met a young man, Dr. C. J. Smithells, who was studying the peculiarities and perversities of tungsten. He published some of his results and then brought them all together in his book "Tungsten." GE published very little on tungsten, so for those who were interested in the subject it was fortunate that Smithells filled the gap, except for work on single-crystal tungsten.

These visits were more or less routine. By this time, all lamp factories were quite similar inasmuch as automated equipment was being used in Europe in spite of the lower labor rates. The operations were similar to those with which I had been familiar at Nela Park and our own factories. Automatic processes placed a premium on uniformity of materials, especially on the tungsten filaments and the sealed-in wire. Manufacturing costs and lamp quality benefited thereby and were reflected in the prices to the ultimate consumer. The use of lower filament temperatures in Europe, or lower efficiency of light production, surprised me, although the desire to increase lamp life may have been responsible. Bennie commented that you would have to look for their lamps with a candle.

During this London visit I had the pleasure of afternoon tea with Colonel and Mrs. Belaieff. I had met him at Ghent and had corresponded with him after he had returned to his post at the St. Michael Artillery Academy in St. Petersburg. He had inquired about what he called "reefs," or grain-boundary segregates, as shown in my 1913 paper. He had also noted that the International Congress for Testing Materials was to be held in Russia, in 1915. His letter was dated March 14, 1914, only a few months prior to the outbreak of World War I.

Both Colonel and Mrs. Belaieff spoke excellent English. That was just fine for me because the only word of Russian that I ever learned was *nyet*, although ever since Dr. Hanemann told me about their metallurgical literature I had wished that I could at least read Russian.

Both had had to flee their homeland after the Communists had taken over, and had left a large estate behind. Mrs. Belaieff assured me that they would both be shot if they were to return. They finished their lives in exile, in London and in Paris. I listened to an interesting account of life in Russia by Mrs. Belaieff, and he recounted his experiences with Chernoff, his teacher and the father of Russian metallurgy. Chernoff must have been a keen observer of the Darwin type, because he noted the recalescence of steel by eye as it cools down, and was the first to devise the quantitative relationship between the percentage of carbon in steel and the critical temperature. That accomplished, he established correct heat treatment for steel. He knew that steel must be quenched rapidly to his points *a* and *b* (200 and 450 °C) for hardening. He might recognize *A*1 and *A*3 of Osmond, and possibly A' and A'' of Portevin and Chevenard.

Colonel Belaieff was prominent for his studies of Damascus swords, meteorites, Widmanstätten structure, and the crystallization of metals in general. He wrote a concise account in "The Crystallization of Metals," summarizing his lectures at the Royal School of Mines, where he was a visiting professor.

He had also studied the structure of pearlite, the eutectoid of steel. For this he carried his microscopic examination to quite high magnifications. That brought up the current work of Lucas of the Bell Laboratories and a discussion of legitimate or useful magnifications. Abbé had shown, many years ago, that the resolution of the objective, and hence the useful magnification, is limited by its numerical aperture. Lucas exceeded that limit, apparently without realizing it. Actually, there is some merit in using empty magnifications to see details more clearly, as when a slide is projected on a screen. I had used that device to show the structure of a kalchoid at 3600-diameters magnification. I should add that Lucas continued his high-power microscopy by using ultraviolet light, with its shorter wavelength, to secure higher useful magnifications.

The National Physical Laboratory is in Teddington, a suburb of London. It had long since become a world-famous metallurgical center under Carpenter and Rosenhain of the metallurgical department. Dr. Walter Rosenhain was the superintendent in 1925. I had met him and Mrs. Rosenhain at Ghent, and him, again, when he came to the United States to give the annual Institute of Metals Division lecture of AIME. We had also had him in Schenectady for one lecture at the research laboratory, and another one uptown, at Union College. He was an outstanding Man of Metals, an imaginative and skillful experimenter with great ability and persuasiveness as a lecturer. When on his feet at society meetings he was formidable in debate as he marshaled his facts to support his position. Like some others of that time, he carried speculation beyond the legitimate limits of basic knowledge. However, he surely heightened polemics and excited audience interest. His confrontations with W. H. Hatfield were classics, and were reminiscent of those of Stead and Arnold, judging by the written accounts.

In Schenectady, Dr. Rosenhain had told me how he just barely missed discovering radioactivity. As a student, he was working with pitchblende in his native Australia when Becquerel announced that discovery. After reading about it, he checked his own materials and verified that they, too, were radioactive.

Dr. Rosenhain greeted me cordially and escorted me around his laboratory. At afternoon tea I met the staff under casual and informal circumstances. Later, we drove to his home, where I met Mrs. Rosenhain again. On the lawn, Dr. Rosenhain recounted the scene of a giant Zeppelin falling in flames over London, a most awesome sight. Indicative of there being no dull moment, when we checked on trains back to London we learned that the last one had left. I remained overnight, as was, and returned to London the next morning.

Dr. Rosenhain posted me on a forthcoming *conversazione* of the Royal Society and gave me an invitation for admission. It is an evening affair at which firms and institutions exhibit new developments in science and technology to a selected audience. I recall seeing an instantaneous photograph of a bullet in free flight with the airstream clearly delineated like Schlieren figures. Dr. Hatfield exhibited samples of Stabrite, a stainless steel that had been developed by his Brown-Firth Research Laboratory in Sheffield. Dr. Rosenhain was exhibiting his specimens of Neumann bands in steel, which he said must be twin bands. He had read a paper that afternoon on that controversial subject.

Dr. Rosenhain was busy at his booth, and since I had heard his paper I concentrated on Dr. Hatfield's exhibit. He was a relatively young man for his position, yet was already well known in steel circles. His was a demonstration of the acid resistance of his steels and not a sales pitch, although I have learned in technical selling that the former technique is most effective for the latter. I thought he did a good job of selling. To top if off, he inivited me to join him afterwards for a social hour at his club.

In my eyes, though I am not an authority on such matters, his club was an example of a venerable and refined London club. The atmosphere was dominated by doormen, headwaiters, waiters, a lounge room for drinks, wall decorations, and comfortable chairs. The only part I did not enjoy was Scotch and soda, which I cannot drink. Our conversation turned easily to metallurgy in England and the United States and to my experiences on the continent. He invited me to visit his laboratory when in Sheffield.

I presume Sheffield is the most famous steel center in the world. Its site attracted the industry centuries ago to take advantage of the waterpower of the streams that flow down converging valleys. The character of its products, being largely based on quality steels, has contributed substantially to its reputation.

Sir Robert Hadfield had announced my visit to his plant, which gave me easy entrée. The chief metallurgist met me and took me for a guided tour during which he explained what they were doing, but not how they did it. I was especially interested in their melt shop, but I was politely told that visitors were not permitted to go there. When I encounter that situation I am undecided about the reason for the policy, whether it is to hide poor practice or some new and improved practice. In this case, I felt sure it was not to hide poor practice. During the war their cast armor-piercing projectiles were as good as our best wrought product and better than those of our second and third producers.

I also took advantage of Dr. Hatfield's invitation to visit the Brown-Firth Research Laboratory. He had assured me that there was nothing like it anywhere. I could say at least that we had nothing like it in the United States, because our steel industry hardly recognized the significance of research and development as a separate function. I was surprised to witness such a large-scale program even though it was supported by two firms. A not inconsiderable amount of the effort was put into several grades of the newer stainless steels. It was about this time that Dr. Hatfield had charge of the Heterogeneity Committee of the Iron and Steel Institute and its large-scale program on steel ingots.

Dr. Hatfield apologized for being unable to invite me to dinner because of a previous engagement, but suggested that I ride with him up into the heather country and take a train back to town. On the way he explained how he determined the strength of steel needed for rear-axle shafts of automobiles. He had a shaft made of mild steel and subsequently measured the torque angle that it acquired; from this measurement he calculated how strong the shaft would have to be to withstand the torque elastically.

On the train, I was seated in a compartment with a young couple chatting continuously. I became slowly aware that, although they were speaking my language, I could not understand some of their expressions. Dr. Desch explained later that they came from a region up in the hills where many of the old Anglo-Saxon words and phrases were still used. There are other instances of the survival of ancient languages. When I was in Göttingen in 1912 my partner at lunch was a student of languages. He told me that a small sector in southeast Switzerland still used a language that was basically

Latin. Also, in Liège, I was told of a similar case in southeast Belgium — another instance of the effect of the early Roman occupation. So Belgium actually has three languages, and Switzerland four.

My visit with Dr. Cecil H. Desch was especially gratifying because it showed me why he was so highly revered as a person and so respected for his scientific attainments. I knew him by reputation in general and by his book, but more personally from my little Scotch friend of Berlin days, Sarah Leitch Elbert, who knew him when he was professor of metallurgy at the University of Glasgow.

At this time he was professor of metallurgy at the University of Sheffield, following T. O. Arnold. A highlight of my visit was an exhibit in a glass case. It contained precious steel specimens of Henry Clifton Sorby, whose pioneering work on the microstructure of steel led to the science of metallography. Dr. Desch then told me about Sorby's life and work and especially the small impact of his first publications in the 1860s on steel metallurgy. Not until after a lapse of two decades did he pursue the subject further, whereupon, and with a nudge from German sources, the new discipline was tardily and slowly adopted. One of Sorby's specimens exhibited the diffraction coloring that he called the "pearly" structure, from its similarity to mother-of-pearl. Howe gave it the well-known name of pearlite. The younger generation may be unaware that Osmond honored Sorby by using his name for the best structure of heat treated steel, which he called sorbite. It was regrettable, though inevitable, that the name became obsolete in our terminology as a consequence of the further development of his science.

As a guest in their home I enjoyed the warm hospitality of Dr. and Mrs. Desch. Our conversation covered metallurgy in part but also turned to topics far removed, showing the breadth of their interests and also his encyclopedic knowledge and remarkable memory. One time, he recalled my work on the kalchoids and my discovery of a new phase equilibrium in the copper-tin alloys. He told me that Colonel Neville was still alive and would welcome a visit from me. I commented that much as I would enjoy meeting the man who, with Heycock, had first established the copper-tin diagram, my time and my commitment to GE would preclude it. I have regretted my action since then. The work of Heycock and Neville in nonferrous metallurgy and Roozeboom in ferrous metallurgy gave the first insight into phase equilibria in metallic alloys and the correct method for depicting alloy systems. I missed my opportunity to meet one of that trio.

My visit to Sheffield was short but enjoyable, and I was soon back in London and visiting Professor H. C. H. Carpenter at the Royal School of Mines. He was a college professor and an investigator. A forebear was the ironmaster Henry Cort, who introduced iron puddling and new poling practice back in the times of the French Revolution and Napoleon. As I contemplated this visit, I recalled a predecessor in the chair of metallurgy, Sir William Chandler Roberts-Austen. He was a leader at the turn of the century in awakening interest in treating metallurgy on a scientific basis. It was he who invited Roozeboom to present his classic paper to the Iron and Steel Institute on the iron-carbon constitution diagram from the standpoint of the phase rule. He also included a passage in his presidential address which I should like to quote: "Nature has endowed steel with some of her most valuable secrets; it is our task to discover them to make better use of it."

Professor Carpenter (afterwards Sir Harold) and Constance Elam were among the first to use single crystals as test materials for studies of plastic deformation. On this visit I was favored with a discussion of their work on aluminum. Up to a point, deformation progressed very much as I had observed for zinc crystals. Although the face-centered lattice of aluminum has three equal sets of slip planes, deformation starts on only the one set that is most favorably oriented with respect to the direction of the tensile load. As deformation progresses, simple block movement is observed, although with no corresponding serrated edge. Concurrently, the lattice rotates and brings a second set of slip planes into an approximately equal orientation for slip. From that point, bilateral slip occurs and produces the chisel type of fracture. Calculations show that the flow stress is higher on the second (or latent) slip plane than on the primary slip plane. Why? This was a problem in work hardening.

The year 1925 was the centennial of Thomas Huxley's birthday. I mentioned that to Professor Carpenter. He replied that it was true, but that the commemoration ceremonies were not as enthusiastic as a man of Huxley's stature deserved. I mentioned that Tom Huxley, along with Darwin, was one of my heroes, which brought forth the issue of *Science* that treated his life and works comprehensively.

My stay in London closed with a meeting of the Iron and Steel Institute, with papers and discussions, banquet, toasts and speeches, and then I was on the *Olympic* headed for home. The practical result of those five months was not a hydroelectric plant at Priest River Rapids, nor an aluminum plant based on clay, nor improved knowledge of lamp metallurgy. A partial result was the broadening experience of many contacts with leading people in my field. Of course it was a grand experience for me, but the major result ultimately, and one that was completely unanticipated, was Carboloy and the timely establishment of our cemented-carbide industry prior to World War II.

All trips to Europe come to an end. Looking back, Chris and I felt that Philips in Holland was doing the most advanced research, both pure and applied. Chris became excited about a new X-ray tube, while Van Liempt's work on tungsten metallurgy appealed strongly to me. In Paris I had interesting visits at the Conservatoire with Professor Eon on aluminum and with M. Cournot on cobalt high speed steel. That reminded me of the occasion in 1912 when I was checking references for my dissertation at the Berlin public library. Next to me, two men were discussing a patent application on cobalt high speed steel. One man commented, "You see, there is nothing on cobalt." In Milan, we noted that Mussolini had his gendarmes at the shipping office, collecting from four cents to fifty cents for each lamp. In Berlin, I was especially interested in a new titanium steel that Geheimrat Mathesius was promoting. It was made at Hoesch Werke AG and had excellent properties when made correctly, which was a problem in itself. At Osram I ran again into the puzzling assumption of a water-vapor effect on the recrystallization of tungsten. These are just a few highlights of a memorable trip to Europe in 1925. Back home, I became acquainted with additional Men of Metals and some who were not, although they were prominent in other fields.

Thomas A. Edison was with us for a couple of days for what was to be the hanging of a plaque in his honor. Apparently he did not approve something about it, because he decided against it. My sole association with the famous inventor was through a former assistant who liked to discuss metallurgy with me. In reply to my question he told me that Edison got his full quota of sleep, only it was dispersed over intervals.

Another non-Man of Metals celebrity was Charles Proteus Steinmetz. I had occasion to see him a few times and have agreeable memories. He was attentive to our discussions but seemed to be preoccupied. One of his ideas was the measurement or estimate of the temperature inside a steam turbine when in full operation. Obviously not a simple task, he introduced fusible alloys at strategic locations and observed later whether they had or had not fused.

Steinmetz had his own method of getting his salary, if it can be called that. When he needed money he simply went to the bank, wrote a draft on GE and took his pay. He used that unique system to pay for his house. The money had to be accounted for somehow, so Ben Fisher was told to put it on his expense account. Ben was in charge of our Mazda Service. When asked to do it, he exclaimed, "A house on my expense account!"

I never heard it mentioned, but Dr. Langmuir was a graduate metallurgist from Columbia. I have a photograph of a class from the School of Mines showing him along with Campbell, Stoughton, Clevenger, a few more, and their teacher, Henry Marion Howe. Langmuir is the only metallurgist who has won the Nobel Prize. Having forsaken a metallurgical career, he asked me to help out a couple of times on metallurgical problems he had been asked to look into. One of them was leaky gray iron castings, or how to use chaplets properly. A contribution to welding was his "flame of atomic hydrogen." Its use to melt steel scrap didn't succeed, but as a welding torch it made excellent welds. I examined the first weld that was made with it and found it to be exceptionally clean and of good quality. Ben Weinman developed the torch into a useful tool that was the best available for special applications until the inert-gas-shielded arc became available.

Dr. Coolidge taught me a lesson that I have tried to remember. When he first asked me to see him on a metallurgical problem I approached the meeting with the attitude that a non-metallurgist wouldn't know enough metallurgy to ask the tough questions. He disillusioned me immediately by his exposition of his problem and perception of its fine points. Never again did I enter a discussion with a top-notch scientist with a feeling of superiority — metallurgically speaking, of course.

On the occasion of Dr. Coolidge's eightieth birthday, five divisions of the company, whose business and welfare were to a significant extent dependent on his contributions, honored him at a private dinner at Nela Park. Clyde Williams, then director of Battelle Memorial Institute, was the principal speaker. Zay Jeffries was an honored guest, as were Dr. A. H. Hull and several others. I was the guest of the Carboloy Division. After all, Dr. Coolidge had added power metallurgy to the metallurgical arts. It was typical of him that in his recognition of the honor he commented that at eighty he was old enough to reminisce but that he thought, also, he was old enough to know better. I thought it a pity that he chose not to pass on to others the fruits and experiences of his memorable career. That has now been corrected with the biography "Coolidge — Yankee Scientist," by John Anderson Miller.

While at the research laboratory I became much better posted on the scientific method than ever before. This came from the staff and lectures at our colloquia, but especially from Dr. Saul Dushman. As I have mentioned, he helped me in analyzing experimental results and in plotting them to discover the law that correlates the variables. I had noticed that plots of data usually included a curve along with the experimental points. In German publications the curve represented the law and the plotted points gave evidence that they obeyed the law. Superficially, it appeared that we followed the same practice, but our curves were simply smooth curves through the plots of the data. Saul advised plotting the data in different ways in an attempt to get a straight-line plot. That was the way I treated the data of the stress-rupture tests, and made it clear that a straight line is far superior to a curve for plotting data.

Saul was heavily involved with electron emission, gas-metal reactions at high temperatures, and other work with the gas state. I asked him why he didn't include work on metals and he replied that he picked the easy one first. However, he did think about metals, as his ASTM paper on cohesion showed.

One of our notable visitors was Professor Kotaro Honda, Japan's dean of metallurgists who was later president of the Tohoku Imperial University. He came over for the 1924 convention of ASST and also visited Schenectady. When Honda was in Boston, George Waterhouse invited a number of fellow metallurgists to meet him. The photograph of those present at the luncheon showed a cross-section of the fraternity and was reproduced in *Metal Progress* for March 1940. In addition to myself, those shown were as follows: Herbert J. French (U. S. Bureau of Standards), Robert M. Bird (G. F. Pettinos), D. J. McAdam, Jr. (U. S. Naval Experiment Station), F. C. Langenberg (Watertown Arsenal), Edgar C. Bain (Union Carbide and Carbon Corp. Research Laboratory), Paul D. Merica (International Nickel Co.), Charles E. McQuigg (Union Carbide and Carbon Corp. Research Laboratory), Albert Sauveur (Harvard University), George K. Burgess (Director, U. S. Bureau of Standards), Kotaro Honda, George B. Waterhouse (Massachusetts Institute of Technology), F. C. A. E. Lantsberry (Jessop Steel Co., Sheffield), John A. Mathews (Crucible Steel Co.), Zay Jeffries (Aluminum Co. of America), Herbert M. Boylston (Case School of Applied Science), Oscar E. Harder (University of Minnesota), Edward D. Campbell (University of Michigan), and E. D. Campbell, Jr. The following men were unable to attend: A. E. White, Bradley Stoughton, C. H. Mathewson, O. W. Ellis, Carl Benedicks, William Campbell, John Howe Hall, H. A. Schwartz, H. S. Rawdon, and Henry Fay.

When Professor Honda visited Schenectady it fell to my lot to be his host. As I took him around to visit the various departments I found that his English made communication difficult. Then I had an inspiration. I recalled that he studied in Germany. I soon found that German was no better than English as a language medium. We finally secured a man from IGE who had grown up in Japan. He functioned wonderfully well as guide and interpreter.

Moser's tests of notched bars of various widths intrigued me to the extent that I decided to do something with the principle. Instead of using the 30-mm test bar, I used our standard 10-mm keyhole-notched bars with widths of 5, 10 and 20 mm from several of our standard steels. Gerald Brophy ran the tests. The half-width bars always broke with a ductile fracture, but the double-width bars sometimes broke in brittle fashion. Thus armed, I approached Mr. Clapp, manager of the GE testing laboratories, to try the double-width test on some of his production steels. He approved and asked me to work with George Wright, his metallurgist. We soon repeated the findings of Gerry's tests — for example, that a good steel is ductile even with the double-width bar.

Shortly thereafter I came upon some excellent test material. The turbogenerator of the Ford power station in St. Paul blew up, causing tremendous damage. (I saw the plant later, when it showed aftereffects as bad as severe World War II bombings.) All parties concerned appointed a high-powered commission to determine the cause of the explosion and to issue directives for avoiding similar failures in the future. After a long investigation, the final report gave no cause or explanation of the catastrophe and hence could give no guidance for avoiding another one. It was clear on one point — that it was initiated by the failure of a retaining ring that threw the entire turbogenerator out of balance when rotating at 3600 rpm.

George and I went to work. We secured samples of the failed ring and ran the series of multiple-width test bars. We knew by then that a properly heat treated steel gave twice the impact value at double width than it did at single width, but that a poor steel gave low values. We were not surprised that the double-width bars from the failed ring developed brittle fractures with low impact values. Of all of the inspections and testing, that was the only evidence that showed why the failure occurred or indicated how to avoid similar failures. If the commission had included standard Charpy bars in its tests it would have missed the point, because the tests with the single-width bar were satisfactory.

I expected that the Turbine Department and others who were involved would heap praise on us for what we had done, and out of gratitude for showing how to increase the safety of turbogenerators by using our test. Instead, the department felt abused when George rejected an expensive forging. George was responsible for accepting those parts and he held his ground when the test said reject. There were many Sauls in those and later days who swept evidence under the rug and remained Sauls.

Part of my work in the interests of Carboloy consisted of finding substitutes for tungsten carbide. One day when I was going over early laboratory records I ran across the name of Dr. Matthew A. Hunter and his work on titanium as a prospect for lamp filaments. It proved to have too low a melting point, much lower than had been assumed. His work was very well done, yet there was no foreseeable use for that rare metal. He was then on the faculty at Rensselaer Polytechnic Institute and remained there until he retired. During World War I, he worked on replacements for alloys that had formerly been imported from Germany. That led to his position with Driver-Harris as consultant and director of research on electric-resistor alloys.

The Institute of Metals Division of AIME had inaugurated annual lectures by invited guests, as I have mentioned in connection with Dr. Rosenhain. Dr. Desch was another from England who, like Dr. Rosenhain, came to us at Schenectady. At about this time, ASST started its Campbell Memorial Lectures and invited Dr. Guertler to deliver the first one. While here he addressed several chapters and other groups on his hard silver alloy, the highly corrosion-resistant nickel-molybdenum alloys, and the alumnium alloy with zinc and magnesium. That reflected the inventor in him. As a pedagogue, he delivered a few more lectures on general topics.

A grand and venerable leader of metallography was Professor Albert Sauveur of Harvard University. He was probably the first to practice that science in this country, and with his paper at the 1893 World's Columbian Exposition, in Chicago, was the first to publicize its usefulness. Shortly after his industrial experience at the South Chicago Works of Illinois Steel he entered the teaching profession, where he brought along younger men as he contributed personally to the development of his chosen field. He also established the periodical *The Metallographist*, through which he spread the gospel. His bearing was always reserved yet pleasantly gracious, while I can testify that he and Mrs. Sauveur were ideal host and hostess in their Cambridge home. He was one of the precious few to whom we owe a debt of gratitude for fostering our science when it was in its infancy.

An entirely different personality was Dr. Robert F. Mehl. His light commenced to shine in the 1920s and continued to shine until his death. He has followed Roberts-Austen by uncovering the secrets of metals and alloys that we may better understand and use them.

Bob tried government work at the Naval Research Laboratory and industrial work with the American Rolling Mill Co., but found his correct niche at the Carnegie Institute of Technology. It was there as head of the Metals Research Laboratory that he was able to implement his own ideas on research projects. His publications on diffusion and the Widmanstätten structure were selected by Cyril Smith as a classic contribution to metallurgical literature. I cite this to emphasize the quality of work done under Bob's leadership.

It is one thing to devise and carry out an important project but another to get the message across to others. Bob had that dual capability. Whether he used the written or spoken medium, with delightful mastery of language he expounded on his subject with great clarity and retained audience interest. In private conversation he was outspoken and to the point. One time, I was assigned the subject of Cauchy relations for a symposium. After I had given my paper Bob said to me, "I didn't know you knew anything about the Cauchy relations." I confided, "I didn't until two weeks ago." In another two weeks I was back to normal, but for about a month I was an authority on that subject.

In many of the characteristics that count, Champion Herbert Mathewson was quite similar to Bob Mehl. Student of Tammann and longtime professor of metallurgy at Yale, he too delved into secrets that nature had enshrined in metals and he made good use of what he found. He also shared his findings with others. His papers were well written and reflected gentility of thought, while his talks were pleasurable expositions of his subjects. One also became aware, with the passage of time, that many of his students were not only gifted but also well trained. He attracted good material. For the record it should be noted that he was one Yale man who knew of Willard Gibbs.

Matty (I always had to avoid calling him Christy) was a first-rate scientist who came along at a favorable period when the art and science of physical metallurgy needed a boost. He summed up his philosophy one time: "I would like to urge that efforts to discover why the things that we observe do occur, not only are good for the soul, but if carried far enough, always have practical significance." He scored both ways. He investigated the plastic deformation of zinc quite thoroughly and published his findings. He also carried on to a practical significance by converting the normally brittle zinc into a ductile metal from which millions of battery cups were cold drawn. I daresay that did his soul some good too, but the mill processes were proprietary and he did not publish them.

Radiology got its real start as a nondestructive test during the 1920s. The one who promoted it was Dr. Horace H. Lester, chief scientist at Watertown Arsenal. He applied it first for castings and later for welded joints. I became acquainted with him when he was demonstrating stereoscopic radiology. By viewing the two X-ray photographs, the defects in a casting were just as clearly seen as though the metal were transparent. I thought it was an impressive demonstration of the ability to look inside a metal, but the method appears not to have established itself as an inspection method. Dr. Lester was the leader in radiographic inspection by his work for the Army Ordnance Corps and the ASTM.

Dr. Lester's broad concern was with many ordnance problems. His principal interest was one of understanding them basically, to the end that a firmer basis could be found for the solutions. A crowning achievement was the Horace Hardy Lester Army Nuclear Research Reactor. He initiated it, hence its name to honor him and his long service to our government.

Dr. John A. Mathews, vice-president of Crucible Steel, attended society meetings religiously. When he presented a paper on his favorite subject, steel, we all listened attentively because we knew he spoke out of a rich experience and with the voice of authority. It was from him that I learned that oil quenching results in more retained austenite than the faster water quenching.

Dr. Mathews was a pioneer in electric steel melting in this country. That was when he was with the Halcomb Steel Co. in Syracuse, New York. With Howard Stagg, he made the first commercial heat of steel in this country in Heroult's electric-arc furnace.

His closer friends were well aware of his warm and friendly personality, while all of us recognized that he was searching out the secrets of steel and applying them in works practice. I recall that he frequently had some young man with him at a society meeting or a banquet as evidence that he was interested in youth as well as in steel.

Late in the 1920s, a significant innovation occurred in our steel industry — U. S. Steel established a research laboratory. The story had it that Judge Gary (chairman) gave in to outside pressure. Apparently, the board agreed reluctantly. Fortunately for them and the staid old industry, they selected a man to head the enterprise who was exceptionally well-qualified for it, Dr. John J. Johnston, professor of physical chemistry at Yale University. The New York Times commented editorially on the new laboratory and noted Dr. Johnston's qualifications as director.

Dr. Johnston was a capable and experienced scientist who worked well with people. To anyone who knew U. S. Steel and the steel industry of those days, he must also have been courageous and willing to accept a challenge. U. S. Steel continued to support the new division, which is the convincing testimonial to John Johnston and the able staff that he gathered. Two of his principal assistants, Ed Bain and Jim Austin, succeeded him ultimately and became vice-presidents of the company.

One of the bright young men who came into focus in the early 1920s was Robert S. Archer. He was a student and assistant of the blind professor E. D. Campbell, at the University of Michigan. His first important connection was with the Aluminum Casting Co. in Cleveland, as principal assistant to Zay Jeffries. That was when and where the team of Jeffries and Archer started. Together they contributed importantly to the production and processing of aluminum alloys. Of even broader significance to the development of physical metallurgy, they wrote "The Science of Metals." In view of Zay's position and reputation, one was inclined to attribute the book largely to him, but Zay told me one time that Bob had contributed substantially to the thinking that went into it.

Bob was thought of as an aluminum man, but even before he really got started he made a fine contribution. The phenomenon of blue brittleness, or brittleness of some mild steels at "blue" temperature, was well known. In fact one of Adolf Martens's practical demonstrations for us was punching a hole in basic Bessemer boilerplate when at that temperature, and then driving a tapered plug through it. The expansion produced cracks. Bob diagnosed the major culprit as nitrogen in the steel. Fettweiss, in Germany, came to the same conclusion independently. Köster then followed with a more complete study of nitrogen, and Fry carried on to the development of the Krupp *Izett,* or "Always Tough," steel by the fixation of nitrogen with an aluminum addition. If Bob had remained with steel, I believe he would have gone that far personally. He did return to steel later, but that is another story.

Trum Fuller pointed out the dual role, both good and bad, of hydrogen in metallurgy. The mention of Izett reminds me that Adolf Fry demonstrated the same to be true of nitrogen. While it is notorious for making steel and weld metal brittle, it also has the ability to surface harden an alloy steel with no significant dimensional change. Thus Fry developed Nitralloy steel in the 1920s and low-temperature nitriding in ammonia gas. Ludlum Steel became the licensee in this country, and Dr. Victor Homerberg of MIT became their special consultant for the process. It is always easier to describe a process than it is to make it work, and nitriding was no exception. The shops that were doing the work ran into problems, so Victor had a busy and interesting career out of nitriding.

This does not exhaust the Men of Metals of the 1920s, either here or abroad, yet it gives a fair sample of the state of physical metallurgy and of the people involved. A lesson this experience taught was that research, to be fruitful, requires an inquiring mind and the use of established principles as the guide. Invention, which made out quite well in the past by the empirical or Edisonian method, was giving way to experimentation and the scientific method. It sank into my consciousness that research is less a game of chance than a game of skill, as someone has put it.

As an example of established principles, those of the phase rule and heterogeneous equilibria were the first to be adopted fully into metallurgy as it evolved into a science. Along with thermal analysis, it was my observation that they were largely neglected in the early decades of this century, though I must mention Professor Mathewson as an exception. This brings up one of my last functions at GE — that of a teacher.

A group asked me to instruct them in the phase rule and constitution diagrams. The subject matter is not difficult, yet, to master it, one must drill conscientiously on the rules. It is also excellent practice in logical thinking, because the rules are rigid. We held weekly meetings at which I proceeded much as I had at Minnesota. We had little fines for tardiness and other infractions that went into our pool. When I left, the group converted the pool into a going-away present of a Sheaffer pen-and-pencil set. I enjoyed using that pen for many years until the ballpoint pen came in. I still have it as a souvenir.

It is fitting that this chapter close on Dr. Whitney. He was keenly curious and alert at all times and expected, or at least hoped, that we would maintain the same attitude ourselves. That is certainly reasonable for a research staff. After all, a research laboratory is no ivory tower where one plays as he draws his pay. It is more like Wellington's playing fields of Eton, where one meets stiff competition, except that it is competition with Nature, and she hides her secrets cleverly.

Dr. Whitney could have been the one who originated the slogan "Think"; he expected us to think. He surely originated "Talk." In a talk on "Talk" he pointed out the dividends that come from talking over one's problems with associates. First you must formulate the problem in your own mind in terms that will be understood by others. Then during your talk you will get questions to be answered and points that require clarification. After a few experiences of this kind you will crystallize your concept of the problem and decide how best to tackle its solution.

Dr. Whitney was, without competition, the founder of the research laboratory in industry. He was superbly equipped to do so. His biography would be excellent reading for high-school science students, for guidance and especially for inspiration. An outstanding characteristic of Dr. Whitney, a confirmed disciple of Francis Bacon, was his devotion to experimentation. That was his way to resolve a question or to secure information. He retained his devotion to the principle all his life. As late as 1956, in a letter to me when he was eighty-eight years old, he wrote, "This is a wonderful world and I find no two of us are alike when we get under the skin. That is a fine idea of Nature, for it insures the trials of so many good experiments. We are in a period when more communication experiments must be tried and the world made more concentrated and consistent." He was mindful of the future.

Although we cannot all make his fabulous contribution, we can at least learn from his experience and profit from his example. I have often wondered if E. W. Rice, then chief engineer, acted intuitively or actually visualized the outcome when he induced the young professor at MIT to leave the academic life in Boston to blaze a new trail in industrial research. It is easy to pick a good man from a graduating class but how do you pick a Bob Feller, while still on the vine, before someone else does?

I was fortunate to have that GE tour of duty for reasons that are implicit in these experiences. I was also fortunate to be with a company with such broad interests that a new scientific or technical advance has a favorable chance of becoming a valuable property. Dr. Guy Suits, vice-president and director of research of GE, has expressed this thought nicely. He stated that a new discovery in the laboratory is the answer to a problem. The job then is to look around to find the problem. Dr. Whitney was of the same frame of mind, in that he was ever ready to follow a new lead. And he could see a dollar about as far away as anyone could.

Chapter 4

The Carboloy Story, 1925-30

My development of Carboloy and the establishment of the cemented-carbide industry in the U. S.; Carboloy patent suits; the Krupp effect; Zay Jeffries' contribution.

THIS story should be told for the historical record, if only briefly, because it covers the origin and establishment of the important cemented-carbide industry in this country. It began early in 1925, when Dr. Whitney sent Chris Dantsizen and me to Germany to investigate a process for making aluminum from clay. It progressed steadily from then, until cemented carbides were in daily use by our manufacturing industries in the late 1930s.

As we shall see, fate started to play a dominant role because Carboloy and our cemented-carbide industry came from my being in Germany at that particular time. Although attending to the chemistry and metallurgy of the clay process was our first order of business, we had free time to visit European lamp factories.

As a lamp and tungsten metallurgist I had worked with tungsten carbide a little, and I was also aware of attempts in Europe to use it as dies for drawing tungsten wire. Consequently, I inquired into this development at each lamp factory. The information was not promising, because the product was inherently weak and unreliable. At one plant, however, I was told that the Osram Lamp Works in Berlin had an entirely new approach to the use of tungsten carbide that did appear to be promising. But other European factories were not using the Osram process, and that made me skeptical, although not incredulous. I decided I must investigate it on the off chance that it really was different and new.

Our tour reached Berlin finally, and I was soon at Osram with Dr. Karl Schroeter, the inventor of the process. The consequences of that visit went far beyond my fondest dreams and ultimately induced General Electric to enter, and even to establish, our cemented-carbide industry.

The gist of the Osram process was the use of a fine powder of the monocarbide, WC, and cementing or bonding the particles into a strong body by cobalt. That was done by pressing the mixed powders cold and sintering the parts at high temperature in dry hydrogen. It is not as simple, however, as that sounds. The methods of the prior art were modifications of the Moissan process, which produced both W_2C and WC in cast form. The product was quite variable and unreliable, as I have mentioned, and was later superseded by the Osram Hartmetall.

When I returned to Schenectady, I decided to do something with cemented tungsten carbide, but it was not until later in 1925 that I was able to start. When I did, the only ones at the research laboratory who knew about it were my two assistants, Gus Pirk and Ted Schermerhorn.

The reason for this secrecy was a special one that came from an earlier experience. I had worked on the strength and behavior of tungsten and other metals at elevated temperatures and I wished to use that knowledge to investigate steels for elevated-temperature applications. That was an unexplored and unorganized field at the time.

I knew enough of thermodynamics to realize that the efficiency or cost of electric-power generation would be improved if the steam plant could operate at higher temperatures. Incidentally, a steel I had in mind was molybdenum steel, which was extensively used later for steam plants. So I wrote up a request to do the work, which went up — across — and down to the Turbine Engineers Department. The reply came back later through channels. The answer was "no." The explanation was to the effect that other companies would develop new steels and our turbine engineers would take advantage of them. It is only fair to point out that GE has become much more receptive to requests of this kind.

I was greatly disturbed by that experience, and I felt that I would get nowhere if I requested permission to work on a new tool material, a commodity that others were producing and that was also nonelectrical — so I classified my work as top secret.

Schroeter's process was straightforward on paper, and his disclosure would be adequate for one who was thoroughly skilled in the art. At that time, however, there was no such art here, and I soon found that there were many pitfalls to overcome and much to learn to produce a satisfactory commercial product. (How dry is dry hydrogen? How much sulfur could I tolerate in sugar carbon?) The many process details are well known now and scrupulously controlled.

That was a trying period, for I felt sure Dr. Whitney would drop in on me some day and ask, "Why are you working on carbides, and, by the way, what are you charging your time to?" I was working on borrowed time, and I had research assignments, too, but I finally managed to make a series of ten lathe tools that tested satisfactorily in our machine shop. "Ginger" Adams, foreman, gave me his blessing: "Doc, I think you've got it." That was balm in Gilead to me. The work then came out in the open and others were called in to help as Dr. Whitney allowed me to continue. I was soon committed to the exploitation of this new and remarkable material.

Several phases of the production and use of tools had to be mastered before we were in a position to put Carboloy tools into our shops for production tests — controls of raw materials; processing to secure a uniform and reliable product; tests of product quality prior to releasing a lot to tool manufacture; making and dressing tools; and correct usage in shop practice. I will note here only that the Rockwell A hardness scale and the transverse rupture test were developed during this period as product controls. The Rockwell A scale for testing hardness, with its lighter load of 60 kg, avoided excessive breakage of the diamond brale, as well as complaints from the front office, while the transverse rupture test was my best test for strength. Both were adopted for production control.

At the start, I tried Carboloy tools in our local shops personally. Through I was an innocent abroad I needed firsthand contact with actual machining operations. My first application was on finishing cuts on induction-motor armatures. The laminations were tough silicon steel and were separated by abrasive mica sheet. High speed steel tools could not take even one smooth cut. Warned that I had been given a difficult job, I selected my prize tool, No. 21. The first cut was clean and true to dimension and there was no visible wear on the tool. We were able to machine a fair number and at a much faster speed before the cutting edge was lost. Furthermore, Carboloy machined dry, so the drying operation was avoided as contrasted with the standard tool, which required a water lubricant or coolant. I felt really good, but bad news was yet to come.

Just a few feet away, somewhat larger rotors were being finished. The machinist told me, "Sure, I'll try your tool, but I don't think it will work on my job." This time Carboloy did no better than high speed steel. I asked him why it didn't, but he just smiled and said he just didn't believe my tool would handle his job.

I was puzzled and suspected that something was wrong but did not know what, so I decided that I needed the help of a machinist who understood such matters. That is how "Smitty" came in on a full-time basis. W. E.

Smith was a skilled modelmaker in our machine shop, and he later played a very important role on the staff of the Carboloy Co. by indoctrinating plants in the United States with the correct use of the tools. I learned from him that the tool must be set up properly in the lathe to avoid rubbing off the cutting edge and dulling the tool prematurely. I decided that the second machinist had deliberately dulled my tool. I was learning.

The next appplication likewise demonstrated the great superiority of Carboloy. It was that of machining the outside diameter of cast iron spiders for induction-motor rotors. The surface was chill hard with burned-in sand, and the spiders gave the tool a severe pounding. My tool was very successful and created considerable interest inasmuch as it machined many more spiders per grind than was customary. The foundry had been pressured to provide extra-soft iron to ease the burden of the machine shop, but soon discovered that it could supply harder castings that were less costly to make. Carboloy helped both shop and foundry.

The machinists on these tests witnessed a direct comparison on their own jobs of the new-art tool and the best of the conventional tools. They were amazed that someone from the research laboratory could come in and show them a vastly improved performance.

About this time, I happened to ride on the streetcar with the man in charge of tools for our Schenectady shops. He had heard of my tests and asked to have more tools tried. In response, Dr. Whitney set up a small production in the research laboratory under Bob Palmer and Al Cooley. Lee Weller was given the job of placing the tools, giving instruction in their dressing and use, and compiling records of performance.

The success of that operation led to the transfer of Carboloy production to our Works Laboratory under Al Danekind. That became a full-scale operation for all of our shops and demonstrated the revolutionary potential of Carboloy.

Concurrently with the decision for the research laboratory to supply tools to our shops, someone got the idea that we should charge for them. That is where the controversial figure of $1 per gram, or $454 per pound, came in. Bob Palmer, in charge of the machine shop and other facilities, Al Cooley, in charge of Carboloy production, and I were asked to set the price.

After we figured the cost of making the tools, the cost of application-engineering service, and so on, we settled on the figure just mentioned. The shops marveled at the price we asked, but they went right on buying them because they were worth it when properly applied. Carboloy not only reduced their costs but also turned out better work, and the machinists received higher pay rates.

A good example of the value of Carboloy tools was that of forming transmission-line insulators in the green state. The porcelain shop was using discarded files it received free. The substitution of Carboloy tools, at $454 per pound, increased productivity so greatly that those files were no longer used. I never did learn what was done with them after that.

At this stage of the Carboloy Story, we were able to devote more time at the research laboratory to studies that came from our experience with both process and product. I will mention one that contributed a new technique and an improved product and that is now incorporated in standard commercial practice.

A characteristic of the "cold press and sinter" product was porosity. I conceived the idea of pressing Carboloy at a high temperature principally to eliminate the pores. That was the "hot press" process on which I was granted the basic U. S. patent in the field. It is used in this country and abroad to make tools and dies, but its production is limited as compared to the Schroeter process. It also proved to be especially useful when investigating other carbides, nitrides, silicides, and borides, where I found it to be quicker and more positive in producing test material.

In the United States, the hot press process is used for large parts and for small quantities of parts principally because of the lower cost of making the graphite molds for hot pressing. I have found that it is more extensively used in Europe, where methods and equipment have been more highly developed. This was also found to be true by Gregory Comstock, who studied cemented-carbide production in Germany and Austria shortly after World War II. He reported, in 1946, that the hot press process was in a high state of perfection in those countries as an effective production method. Curiously, a few years ago my eye caught a report of the hot press method as a new invention that introduced "new techniques in powder metallurgy."

The microscope is indispensable to the metallurgist who seeks an understanding of his materials, processes, and finished products. Unhappily, there was then no known way to prepare Carboloy suitably for microscopic examination. I used boron carbide from Al Merrick of our electric-furnace department to lap tools, but I could produce no better than a finely pitted surface for microscopic examination. We used electropolishing regularly to prepare tungsten for etching, but, as I recall, it was not satisfactory for the two-phase Carboloy.

Then one day I was in our meter plant in West Lynn exploring possible uses of Carboloy. There I observed sapphire meter bearings being polished with diamond dust in oil. I arranged to have several of their regular sizes sent over to try out. They also included a finer size which they recovered by centrifuging the waste oil. The latter ran about 0.6 μm in diameter and produced an excellent polish on Carboloy. Etching then revealed the true microstructure. For the first time it was possible to see the size and shape of the WC grains and the occurrence of cobalt which produces the necessary strength. I published my method as a paper for the ASST and reproduced photomicrographs in my 1930 Institute of Metals Division lecture for AIME. Karl Schroeter, of Osram, developed a similar technique at about the same time.

These property tests and metallographic examinations have stood the test of time and are in common use by the industry.

Patent matters were critically important in subsequent events and especially because our patents justified GE's decision to establish the industry in the 1930s. However, in 1926 and 1927 the sole concern of our executives was the use of Carboloy in GE shops. I had paid very little attention to patents until Mr. E. S. Morrison of our patent department came over to me one day to discuss the Schroeter patent. Although it was assigned to GE, he informed me that all we had was rights to use it in lamp manufacture. That restriction was spelled out in those contracts with the European lamp factories that Gerard Swope, president of IGE, had arranged. But our executives wished to expand our rights to include general shop rights.

Later, someone in the patent department had run across another patent, and Mr. Morrison came over on that one. He announced that he wished to read a claim to me, to wit: "A tool for working hard metals made from a tungsten carbide with its pores saturated with a metal." Noting my surprise and chagrin on hearing a claim which covered Carboloy, he comforted me by confessing that it was an Osram patent to Heinrich Baumhauer, and likewise assigned to GE. That restriction posed some knotty problems for the future development of Carboloy.

I made a Baumhauer tool according to his patent, which, upon being tested, showed that it had the great cutting efficiency of Carboloy. A later microscopic examination showed that it was true cemented WC "with its pores saturated with a metal" — cobalt.

After that experience I was convinced, as were others, that Baumhauer's was a pioneer patent. He was the first person to increase cutting efficiency far above that of conventional tools, or sufficiently so to revolutionize machine-shop practice. In spite of that, it was not used widely because Schroeter invented an improved process for making the product.

In addition to our contractual limitation of the use of Carboloy to lamp manufacture, we learned that Krupp, in 1925, had acquired rights from Osram to market Hartmetall, as their Widia, throughout the world, including the United States. That was a dark and critical period for Carboloy and its future, but negotiations with Krupp made it legal ultimately for GE to market Carboloy in the United States. Krupp retained the right to export to this country, and also to license two American firms, Firth Sterling and Allegheny Ludlum.

Looking back, I believe it was well that I had paid but little attention to patents, because if I had known of that contractual restriction on our use of the Osram patents I might not have had the audacity to start the work. If I had not, Krupp would have dominated our cemented-carbide industry, and with Hitler running Germany, the United States would have been in bad shape in preparing for and prosecuting World War II.

The name Carboloy has proved to be a good one. People are surprised to learn its genesis. When I started work I numbered the carbide lots: Carbide 1, 2, 3, etc., and the WC and Co lots: Carbide Alloy 1, 2, 3, etc. As the pace quickened, abbreviation soon set in and the lots were called in succession Carb. Alloy, Carballoy, and finally Carboloy. The name was a natural, and GE registered it as its trade name for cemented carbides.

This brings us to the Carboloy Co. and GE's entry into, and its leading role in establishing, the cemented-carbide industry in the United States. That story should be told briefly and for the record.

The success of our tools in GE shops proved that Carboloy was revolutionizing machine-shop practice. It seemed clear to me that Carboloy tools should be made available to industry at large and that GE should shoulder the load. I realized that I was not the one to initiate a move in that direction, but in any case, I had a plan.

I had known Zay Jeffries since 1918 as an alert metallurgist who would correctly evaluate the merits of a new tool material and who was also an experienced and valued executive. He was then top executive for science and technology in our Incandescent Lamp Department at Nela Park, and also director of the Cleveland Laboratory of the Aluminum Co. of America. My aim was to get Zay interested and initiate a move to have GE make and market Carboloy. So one day I posted Zay on what we had done with Carboloy, and the conversation soon turned to an "impossible" job he had of machining pistons of LOEX, an aluminum-silicon alloy. This conversation resulted in a decision to try a Carboloy tool.

When I arrived at his office Zay asked Bob Archer, his principal assistant, to take me to the machine shop, where some pistons were ready for testing. The first one was machined with their best tool to show why they had a problem. It was a total failure because the piston finished out of dimension, and with a barrel shape and a rough surface. My Carboloy tool then machined one piston after another, true to dimension and with a smooth finish. Bob was enthusiastic, and when we returned to the office Zay looked up and commented, "I see the answer in your expressions." He was convinced when Bob showed him a finished piston.

Zay then wrote a historical letter to Dr. Whitney in which he pointed out, in effect, the desirability of marketing Carboloy tools and checking the patent situation. Dr. Whitney reacted positively by forwarding copies to Mr. Swope and to Mr. A. G. Davis, vice president of patents. There followed the period of soul-searching and negotiations with Krupp that I mentioned. Phil Mallory, president of his P. R. Mallory Co., had also become infected, and the outcome was the Carboloy Co. as a joint venture (I almost wrote "adventure") of the two parent companies. Phil became president and Zay was his right-hand man. There is much more to the Carboloy Story than space permits here, but some phases should be added for the record.

Phil was an able and energetic executive whose interest in new horizons characterized his approach to business. As for Zay, I need only say here that he knew more about economics and finance than any other metallurgist, and that he knew more about metallurgy than any other financial man. They were a good team and they got the Carboloy Co. off to a vigorous and optimistic start. We at Schenectady helped by providing publicity, advice, and service.

That fall, 1928, the annual convention and metal show of the ASST was being held in Philadelphia. I got the idea that a technical paper on Carboloy would be an excellent opening gun for spreading the news of the new tool material. It would create interest and inspire confidence because it came from a strong and respected firm, and the matter would be open to discussion. That paper was "Tungsten Carbide — A New Tool Material."

The Carboloy Co. exhibited its products at the show, with Smitty present in all his glory as one well versed in Carboloy practice.

Our experience at GE demonstrated clearly that a major effort would be required by the Carboloy Co. before a new and unusual material could be used profitably by industry. That would call for strong financial support for opening home and branch offices, establishing manufacturing and distribution facilities, and — a sine qua non — training and using many application engineers to introduce correct Carboloy practice. As a measure of the magnitude of the last task, the Carboloy Co. trained approximately 10,000 men in industry during 1936 and 1937. The gospel was also spread by the periodical *Carboloy*, to the first issue of which Al Danekind and I contributed.

Perhaps the experience of Carboloy's W. G. ("Dick") Robbins will illuminate the need of extensive shop training. Dick had been selling imported carbide tips, or inserts, in Detroit at $45 per pound. That experience induced him to join the Carboloy Co., and he told me why. Briefly put, his customers had to make and use the tools themselves with no assistance. As a consequence, so many were misused or broken that he could not generate business. He simply lacked the necessary backing and, with material at $45 per pound, he could not afford to train and use tool experts. During the introductory phase, cemented carbide could not just be sold over the counter and caveat emptor, nor could the new industry be established on that basis. I hope I have made the point strongly enough that apprentice training was necessary — it was also costly.

Carboloy's price of $454 per pound was controversial, to say the least, but it fell far below the cost of training customers. Naturally, as the consuming industries learned how to use cemented carbides, the expense of training fell off and with it the price.

For several years during the depression, the company lost $1000 a day, until it was in debt to GE by about $1.5-million. After I left GE, Zay, as president, used to keep me posted on the Carboloy operation. He once told me how he went to New York to present Carboloy's financial needs to GE President Gerard Swope, and Owen D. Young, the chairman of the board, and had come away with additional funds to keep the enterprise going. That must be a tribute to the courage of those involved and to their faith in both the product and the marketing policy.

An enterprise of this magnitude, the development of a stable and profitable industry at great expense, meant a great risk if competing firms were privileged simply to make tools and sell them at a reduced price to customers who had been trained in their use by the Carboloy Co. Protection against this risk could come only from the patent grant and use of the monopoly that our patent law provides.

In this connection we were all convinced that Baumhauer's was a true pioneer patent and that those of Schroeter and Hoyt were basic-improvement patents. In that light, if they were ever involved in litigation we felt that a court would surely interpret them broadly, so striking was the distinction between the new art and the valueless old art of utilizing tungsten carbide.

The depression of the 1930s saw a change in the cemented-carbide business. P. R. Mallory dropped out of the Carboloy Co. and one of the licensees gave up. Krupp dropped out later, but apparently for a very good reason. In the late 1930s Krupp agreed, surprisingly, to stop exporting Widia to the United States. An experience I had when with A. O. Smith could very well provide the real reason.

We were asked to supply some stainless steel pressure vessels to a U.S. plant of General Aniline, a German-owned chemical firm. They had always imported their equipment from Krupp and had great confidence in the quality of the material they received. However, Hitler had banned the export of strategic metals, which threw them on our market. If Hitler stopped the exportation of stainless steel, he would surely keep Widia at home for use in rearmament.

I have recently run across additional evidence from a German source on this point. During our invasion of France in 1944, Hitler gave an order to his western command that the supply route for importation of tungsten from Spain must be kept open *at all costs.* From all of this, it follows that Krupp had good reason to relinquish the right to export Widia to the U. S.

Coming to the merchandising practice of the Carboloy Co., ownership of a patent and its monopoly does not release a company from obligations under our antitrust laws. GE had recently won a Supreme Court decision on its methods for marketing lamps, and the Carboloy Co. used similar practice for Carboloy in accordance with advice from able attorneys. Despite rumblings from outside on this point, that practice was perfectly legal.

In a federal court case in 1940, the Carboloy Co. charged a competing firm with infringement of its patents, to which the defense alleged unlawful merchandising practice through its use of patents by the plaintiff. That defense was denied, since the court ruled that the Carboloy Co. "has done no more than was reasonably and properly necessary to enable it to protect its efforts to make a profit from the manufacture and sale of the patented material." On the other hand, in the same decision, by some weird interpretation of the evidence, the court ruled that all of our patents were invalid because "The record does not show that any of the patents in suit, or any patents in the prior art caused a revolution or a sudden great development in the cutting tool or wire drawing art." Instead, cemented tungsten carbide was held to be the "result of the normal evolution of the cutting tool art"! (That ! is mine.)

Actually, a vast and detailed experience with both old and new arts proved with great clarity that the old art was of no practical value and that Baumhauer, Schroeter, and Hoyt had initiated a true revolution. We have here a textbook case of a *revolutionary* improvement or breakthrough in an art.

Viewing Carboloy's history in retrospect, with its significant and beneficial contribution to the cutting art, it is amazing that GE should have been confronted with complaints and vicious criticism. The following incident is an outstanding case.

During the war, in 1943, a hearing was held by the U. S. Senate Subcommittee on War Mobilization dealing with our cemented-carbide industry and Carboloy in particular. In an opening statement Senator Bone said, "The safety of the nation and the welfare of its people are dependent upon the progress of science and the useful arts." He also quoted a statement of the President, "Patents are the keys to our technology, technology is our key to productivity, productivity is the key to victory."

When I read that, I thought it was a preface to praise of the Carboloy Co. for its vital contribution to our war effort. Instead, Senator Bone continued, "Apparently, no great effort was made to stimulate the use of it [Carboloy] at the very moment when Hitler was building his vast panzerized and mechanized machine to overwhelm the world. In my judgment, without cemented tungsten carbide the German military machine could not have been built in the time that it was built and in the efficient fashion that it was built."

That was a fine tribute to cemented tungsten carbide but hardly an objective and truthful appraisal of the Carboloy contribution to our war effort. As we know, it was critically important in greatly accelerating our production of war materiel and in providing carbide-core antitank ammunition. Nor did Senator Bone have any support of his allegation that the Carboloy Co. was reluctant to stimulate the use of its products. He simply ignored facts or the great effort that had been made to do so and the establishment of the new and important industry.

Very similar subject matter and very similar argumentation were involved in a Government antitrust suit against GE and the Carboloy Co. in 1947. Technical people can gain important lessons by following litigation and argument in patent suits. The Carboloy cases are no exceptions. As Mr. George I. Haight, an eminent trial lawyer, told me when I was with A. O. Smith, "The courtroom has rules all its own, so forget what goes on outside." This case is a good example.

The Government argued that Baumhauer was invalid for failure to "particularly point out and distinctly claim" the invention and also "as fully anticipated by prior art." The Government had subpoenaed me as a witness "on the question of patent validity," so I expected to be questioned on Baumhauer. I disagreed with both of its assertions and I felt that the Government attorney knew or suspected my position. In any event, when I was on the witness stand he neglected to ask me whether Baumhauer adequately pointed out and claimed his invention or whether he was "fully anticipated." I had made the tools and I knew. I wish I could also have testified on their identification as "cemented tungsten carbide" and on the difference between Baumhauer and prior art.

The Government attorney also alleged that the Baumhauer process was "impractical, noncommercial, and unused." That needs to be corrected if only in justice to the inventor. I for one, and probably the only one in the United States, made and tested the tools and, had Schroeter not invented a better process, Baumhauer would be "practical, commercial, and used." Again, I was not asked to testify on that point by either side. As for the GE attorney, I presume he could not question me during cross-examination on subject matter that the Government had not raised on direct examination.

Although the technical content of the Government's case was replete with misunderstanding and false statements, this is not the place to consider them. However, there are two points that warrant comment.

The Schroeter and Baumhauer patents originated in Germany and were written with looser and broader statements than we require to "particularly point out and distinctly claim" an invention. In Germany an application for a patent is open to public inspection and objection so the scope and coverage can be settled before the patent is issued. A simple translation into English can unwittingly result in a U. S. patent that is sensitive to attack. May this not have helped the Government in its presentation?

The second point involves "distinctly claim." Baumhauer knew what he did and said so, and he knew what he got and described it — an operative tool. He couldn't describe his product beyond that, because the necessary metallographic technique was not then known. The best he could do was to say that the "pores were saturated with a metal." Incidentally, that was a phrase that gave rise to legal argumentation and technical confusion. I have wondered how a judge is to differentiate between the essential and the nonessential, the true and false.

If Baumhauer had waited until all the details were known, he would have run the serious risk of having someone precede him in getting a patent and, besides, a German patent can be written in more general terms for early filing. A trial court should be aware of this situation and rule accordingly when and if the facts are in the record. This is especially true for a pioneer invention.

To be more specific, during the early work on Osram's Hartmetall and our Carboloy the facts had not been developed for a complete description of product and process details. This was confirmed for Germany in a review of cemented carbide since 1926 that appeared in a June, 1969, issue of *Nachrichten* of VDI, the German Engineering Institute. There it was brought out that although some basic principles were known in 1923, "a large number of details had to be cleared up before production could be started in 1926."

This experience also has a lesson, I believe, for those who advocate the "first to file" principle for our patent law. Unless adequate safeguards are provided, it will be possible for one who rushes to be first, but with inadequate knowledge of his invention, to end up with a vulnerable patent.

Leaving the technical content, the marketing practice of the Carboloy Co. was again challenged in this case. That was in spite of the fact that it had long been abandoned, although an earlier federal court had decided that it was legal when it was practiced. In 1948, a year after the Government trial closed, the Supreme Court, then sitting, modified that earlier federal court decision and the trial court now felt constrained to rule retroactively against the defendants. The judge did say that "practices which must be held to be improper under the law today were, at the time General Electric obtained rights twenty years ago, thought to be perfectly legal." He explained that although laws cannot be applied retroactively, that does not hold for court decisions. GE was caught in that unhappy trap, one of which others might well beware.

As though in rebuttal, Mr. Charles E. Wilson, then president of GE, pointed out that "there would have been no tungsten carbide industry at the time of World War II had not our own course of action made the United States independent of Germany." He also stated that "the creation and development of this industry by these men is a sturdy and inspiring example of public service born of private enterprise, and characterized by hard work, ingenuity, investment, research, risk, and courage. . . ."

I believe the historical record will outlast the legal record. We did start the new industry in this country by our work at Schenectady, possibly when my tool No. 21 took that first clean cut on an induction-motor armature. From then and there, a continuous development progressed from the research laboratory through the works laboratory and the Carboloy Co. to the present operation of the metallurgical products department in Detroit. Incidentally, that would not have happened if we had not believed that we had patent protection. Curiously enough, the Government had the temerity to insist that we "well knew" that our patents were "invalid at all times." That should teach a neophite the treatment he may encounter in a courtroom.

I have been asked if the advent of cemented carbide aroused much interest or if people were alert to what was happening. The answer is yes. A technological revolution of this magnitude, with its easily predictable and beneficial effect on our manufacturing industries and our economy, attracted great attention in the daily press and in requests for papers and talks. We also had visitors at the research laboratory seeking information. I recall one in particular, which showed how widely the news had spread.

One day, the Reverend Robert J. Ball was announced. He was the chaplain at the Vermont State Penitentiary. He had me tell about Carboloy and show him various forms of it. He was attentive and asked some good questions. On closing I said, "Now I have a question for you. How is it that you, a man of the cloth, are so interested in Carboloy?" He replied with a twinkle in his eye, "Well my people have been asking me a lot of questions about it and I just wanted to be posted. You see they want to keep up to date on the tools of their profession."

Other visitors came on business. One was Gregory Comstock, as a representative of Firth Sterling, to secure help and information. He became fascinated with powder metallurgy and remained in the field the rest of his life as teacher and industrialist. Another was Art Focke, who came on missions from our Cleveland Wire Works to test lots they were producing for the Carboloy Co. George Sieger was Phil Mallory's right-hand man, and he came frequently on special assignments.

Machining steel proved to be a surprising problem. We used a nickel steel log to evaluate the cutting efficiency of various experimental and production tools. The test tools performed favorably, but when I tried a

Carboloy tool on a steel shaft in production it failed badly. Of course we handled some jobs, but we really had a problem with steel.

Floyd Kelley of our metallurgical group had developed special skills, and he devoted some time to Carboloy. One of his contributions was that of attaching Carboloy tips to the steel shanks by copper brazing. He also demonstrated that cemented tantalum carbide is superior to cemented tungsten carbide on steel. Other workers found that titanium carbide is also excellent for steel, and various grades of cemented carbide today use both, or double carbides, for these applications.

As additional testimony to the potency of the cemented carbides, we have the revolution they caused in the machine-tool industry. In the club car one time, as I was on my way to Detroit to inspect a job at the Ford Motor Co., I overheard a man comment to his neighbor, "This Carboloy is raising hob with our machines. They are stepping up the cutting speed and want much stronger and more powerful drives."

For his Model A, Ford was machining the clutch face on the cast iron flywheel. This was done in a huge plant on Bullard Multimatics. Upon experimenting with Carboloy, Ford speeded up the operation by replacing the original three-horsepower motors with thirty-horsepower motors and then beefing up the parts that failed. This change also enabled him to use a harder, fine-grained iron, improving the quality of the clutch. Ford had planned to increase production at that time and had a duplicate clutch plant on the drawing boards that was made unnecessary by these modifications. That increase in productivity was repeated many times throughout our manufacturing industries. I have preserved a record of tool performance on the Ford flywheel job that brings out the advantage of using cemented-carbide tools, as well as the difference between carbide grades. The tool materials and the number of clutch faces finished per grind with fast machining are reproduced in the table:

Material	*Number per grind*
Stellite — conventional tooling	3
Carboloy — high-cobalt grade	45
Widia — low-cobalt grade	450
Carboloy — hot-press grade	1180

I had an interesting discussion with Dick Robbins, then president of the Carboloy Co., prior to World War II. He cited many ways in which Carboloy was accelerating production and lowering the costs of producing many military items, but he was especially keen on his carbide-core ammunition. It had exceptional penetrating power and proved later to be necessary in use against the German panzers. Not only were we able to prepare for war more quickly, but the timely availability of materiel and supplies shortened the war and reduced its cost and our casualties.

The benefits can be summed up in two words — increased productivity. Broken down, that meant corresponding savings in labor, plant, equipment, and transportation, and also in improved products. It requires little imagination to visualize the enormous impact the cemented carbides had prior to and during World War II. As Senator Bone put it so nicely, we could not have built our war machine in the time we did, nor as efficiently as we did, had we not had carbide tools and learned how to use them.

Industry is now supplied by a number of cemented-carbide plants that are doing very well by a quality product. They must, if they are to remain competitive — for the same reason that one can get good baked beans anyplace in Boston.

Forty years later, in 1965, when I visited GE's Metallurgical Products Department, it had a large CARBOLOY sign out front — courtesy of the Carboloy Division. I witnessed the high state of the art from laboratory and inspection, or quality-assurance department, through production and inventory. Things had changed since 1925, and I was a sort of ghost. Zay and Dick were no longer there, Bennie was no longer producing Carboloy, Smitty was wrestling no more with difficult jobs, and Taylor, the ingenious inventor, had retired. When I entered the office of the librarian, and she saw me, she must have guessed my feelings as she said, "I'm the only one left." The second generation was represented by E. J. Weller, the son of Lee.

As a result of this visit, I wished to have some indication of the size of the cemented-carbide industry and its over-all benefits, but reliable figures are hard to secure. The following may at least reflect its size and utility.

According to Norman Zlatin, there were more than twenty million machine tools in the 1967-1968 period. He also estimated that the yearly cost of labor to operate them was about $40-billion. How much those figures would be increased by lower productivity if only old-art tool materials were available is not known. If they were, I am sure the difference and its effect on our machine-tool and manufacturing industries would be astonishing.

E. M. Trent, in *Metals and Materials*, of the Institution of Metallurgists, has summarized estimates of the cemented-carbide industry by stating that the world production was probably about 6500 tons in 1966 and was increasing by an annual rate of 15%. Of this amount, over one-third was in the United States and was valued at about $100-million.

The reader will have noted that there were many ifs in the story of Carboloy, such as my trip to Germany in 1925 on a different mission; scheduling the trip for the winter and spring instead of later in the year; my distracting experience with improved steam-plant steels; the success of my "top secret" program; my success in interesting Zay Jeffries in Carboloy and having his strong hand to guide the Carboloy Co.; plus a few little other ifs. However, without the backing of GE, all of this would have come to naught.

The Baumhauer patent is null and void as a consequence of self-serving legal argumentation and a court decree, but in metallurgical history it rates with the epoch-making inventions. Baumhauer deserves the honor that goes with a pioneer contribution. Schroeter rates the credit that goes to an invention that is the basis of a large and important industry. Both deserve the technical equivalent of a Nobel Prize. If I may be impersonal, Hoyt, with an assist from Lady Luck, deserves a pat on the back for a valuable contribution to the art and to the industry.

Addendum

In the recent biography "Zay Jeffries," the author, W. D. Mogerman, reproduces two letters of Zay's which are pertinent to the Carboloy Story. I should like to include them here for their historical content. The first one, dated June 18, 1926, was to Dr. Whitney after my demonstration of Carboloy on his aluminum alloy pistons. It displays Zay's interest in the new tool material and its commercialization as well as his concern for the patent situation.

Dr. W. R. Whitney
Director, Research Laboratory
General Electric Company
Schenectady, N.Y.

Dear Dr. Whitney:

I am writing to extend my congratulations to the Research Laboratory on the splendid progress which Dr. Hoyt has made in connection with cutting tools. We had an unusual opportunity to give one of these tools a test here in Cleveland. We are working on high-silicon aluminum alloys for automobile pistons. When the silicon is as high as twenty percent, the alloys are practically unmachinable with any known tool on the market with the exception of diamond. We have tried all kinds of cutting tools including Stellite. Dr. Hoyt's cutting tool is at least twelve times as good as the best tool we have tried before, and probably is more on the order of one hundred times as good.

Hoyt told me something of the patent situation, and particularly the situation with respect to Osram. I have had considerable experience in the steel tool field, and it is my impression at the moment that this development will do as much for the cutting tool industry, as compared to high speed steel, as high speed steel did in comparison with ordinary tool steel. I realize that this is saying a great deal for the possibilities of the new cutting tool, but nevertheless this represents my present opinion. I would urge that the patent situation be cleared completely for the United States, because there is immediate business of considerable magnitude available.

Sincerely yours,
Zay Jeffries.

The second letter, dated August 14, 1964, was written to an executive of GE's Metallurgical Products Department in Detroit. In it, Zay comments on an account I had written on the development of cemented carbides and the antitrust suit, in 1947, of the United States vs. the General Electric Co. and the Carboloy Co.

It was timely to get Sam's story. No one else could have written it. Furthermore, it constitutes the keenest analysis of the situation that I have seen. His treatment of the antitrust case can be regarded as classic. I have known that the Government's position was not only wrong, but really underhanded, and we all know that only the General Electric Company could have made cemented carbide a United States industry at a date early enough to immeasurably help the war effort for the Free World. However, as long as we know that, and have not only no feeling of guilt but a feeling of great satisfaction that we performed a service for the Free World that no one else could perform, I suppose we should no longer worry about some technicalities in an antitrust suit. The General Electric Company, after the Judge Knox decision, offered to appeal if either Stearns, Robbins or I requested it. We all said, in effect, "Let's pay the fines and go ahead with constructive work." I offer Sam my congratulations.

Chapter 5

With A. O. Smith Corporation, 1931-39

An exciting experience with men and products; the metallurgical function; the Lincoln patent suit; research and works practice.

THE morning of January 1, 1931, I was eating breakfast on a New York Central train as it sped around the south shore of Lake Michigan. Outside, it was cold and white; inside, I was well fed and comfortable. I was on my way to Chicago and Milwaukee to join the A. O. Smith Corp. General Electric was behind me and I was about to begin an intensely fascinating episode.

The change in atmosphere was sudden and drastic — from that of a large concern that was owned by many stockholders, and was governed from afar by a few directors and top executives, to that of a relatively small and self-made company that was owned and governed by an individual. I do not imply that GE was a soulless corporation, as one might hear from atop a soapbox; rather, the difference was one between a corporate entity and a personalized entity.

The owner and president of A. O. Smith was Mr. L. R. Smith, the grandson of the founder. He was known to us all as "LR," a mark of respectful intimacy and a bond that seemed to diffuse through the organization and tie us together as a team. (We never called Mr. Swope, our president at GE, by his initials, nor did I ever have any contact with him.) In an interview, LR had described his current projects that were in the development or pilot-plant stage, but mentioned production only briefly.

By way of explanation he commented, "We do these things because we like to, and feel we are doing good, too." A brief statement of what they had done will verify his opinion and will also illustrate what he liked to do.

A. O. Smith started by making hardware for baby carriages and bicycles. At the turn of the century the company commenced production on steel frames for automobiles and soon became the principal supplier to the industry. During World War I, their production of most of the aerial bombs led directly to the invention of the first arc-welding electrode that could join steel with a truly satisfactory weld. The availability of that electrode led, in turn, to the production of couplings for oil-well casing, pressure vessels for oil refineries, and line pipe for transmitting natural gas. In the meantime a huge, automatic frame plant was built. To the visitor it ranked as the eighth, surely the tenth, wonder of the world as the 400-foot machine functioned smoothly and rhythmically. Some called it a white elephant, yet it remained in production for several decades. LR had correctly foreseen the expansion of the autombile industry and planned his capacity to keep pace. As I became acquainted with the A. O. Smith operation I had no doubt that LR was enjoying himself and also doing good. Some introductory comments will show the unusual and creative business he fostered.

I was introduced to the company with a tour of the shops. The most impressive sight was the frame plant, housed in a building by itself. Only after several visits could I encompass all that it was doing. The incoming flat stock was checked for dimensional accuracy, trimmed, cold formed edgewise to produce the hump over the rear axle, and then formed into a channel section. Additional parts were fed in at stations along the length of the machine, whereupon all were riveted together, and to the side bars, in an ingenious riveting machine. To the confusion of the skeptics, the operations could be modified to accommodate changes in frame design.

Almost equally impressive were the still shop and the pipe plant. At the still shop, I saw them arc welding huge plates into vessels for oil refineries. At the pipe plant, I saw them flash welding forty-foot lengths of line pipe at an effective rate of more than fifty feet per minute. Verily, I though, those plants should offer a fruitful field of operation for a metallurgist, with so much steel being fabricated and so much important welding being done, although I was not directly involved until later.

Although both of those firsts by A. O. Smith were tremendously valuable contributions to industry, during World War I another fruitful innovation gave birth to modern arc welding. The company was making aerial bombs by cutting blanks from strip steel, forming them to shape, and welding the seam. The problem proved to be the welded joint. Bare-wire welds split open when the bombs were dropped, and gas welds were too slow and costly to make. Only the English Quasi-Arc electrodes made it possible to go into production. Then the roof fell in when the supply ran out, a dilemma that had revolutionary consequences. Out of it came the first electrode that adequately protected the arc from deleterious impurities and produced weld metal that was as good as the parent metal. I was told about this development by Orrin Andrus, who had participated in it. Since then he has written the story for the *Welding Journal,* for the fiftieth anniversary of the founding of the American Welding Society.

An investigation developed the information that the blue asbestos for the Quasi-Arc coating was mined in South Africa and shipped to England for electrode manufacture. The rods were then shipped to Milwaukee. German U-boats were responsible for stopping bomb production. True to his makeup, LR said, "I want to make my own electrodes and from materials that I can buy here in Milwaukee." He turned the assignment over to his cousin, R. S. Smith, with Orrin as his assistant.

Orrin went across the street to a grocery store and bought a roll of wrapping paper, and thence to company stores for a supply of water glass, or sodium silicate. Finally, he secured wire rod from the rivet shop to use as core wire. All the ingredients were within a stone's throw of LR's office.

In brief, he made the "paper" rod that produced a hydrogen and carbon monoxide arc atmosphere by pyrolytic decomposition of the paper, a carbohydrate. A U. S. patent was issued to R. S. Smith.

The Smith electrode was improved by Dick Stresau, a research engineer, by substituting wood flour for paper and extruding the coating on the core. He also added slag-forming constituents and alloys to the coating when necessary. The new electrode not only revolutionized arc welding, but it also led directly to the new pressure vessels and line pipe.

In the 1920s the demand for gasoline placed tremendous pressure on the refineries to keep pace. The conventional stills for processing crude oil were then made by either riveting or by water-gas welding. Neither method provided joints that would permit refining at the higher temperatures and pressures needed to increase the gasoline output. The new electrode had already been used to weld couplings for oil-well casing, and had demonstrated that the welded joints were equal to or even better than the steel. That experience induced LR to try pressure vessels. The first vessels were pressure tested to destruction to demonstrate their strength to prospective customers. Almost immediately, the new vessels sold like the proverbial hotcakes. The substantially higher operating temperatures and pressures greatly increased the gasoline yield and reduced the amounts of less valuable products. I believe I can correctly state that this not only started the new era of welded pressure vessels, but that they, along with welded ships, bridges, penstocks, storage tanks, and so on, indirectly pay silent tribute to the Smith electrode.

A second major use of the cellulose type of electrode was for the manufacture of transmission-line pipe. The oil fields in the Southwest were wasting vast quantities of natural gas into the air, losing a valuable fuel and natural resource forever. The reason was simple economics — the cost of laying a conventional pipeline to consuming centers was prohibitive.

The pipe of that day was lap welded of low-carbon steel. The joints were so deficient in reliability that an excess of steel (heavier gage) had to be used to compensate for the low allowable joint efficiency. The arc-welded pipe was made of stronger steel, with a much higher joint efficiency, and in lengths that were half again as long. With lower steel weight to purchase and ship, and fewer field joints to make, natural gas was transmitted economically from Texas to the north. The new industry was beneficial for steel producers, the petroleum industry, the contractors who laid the lines, the gas-transmission companies, the ultimate users of a superior fuel, and, of course, for A. O. Smith.

In 1931, arc-welded pipe was still being made, but production was soon shifted completely to flash welding at the new pipe shop. The method was faster and cheaper, and the lengths were one-third longer than the arc-welded pipe.

When I arrived on the scene, the technical basis of A. O. Smith operations was in excellent condition. LR had learned that putting science to work was exciting and profitable. With that experience back of him, and instead of engaging in what many would consider the pleasures of life, he had decided on more of the same. A few words to introduce some of the staff members are in order.

The man who had over-all surveillance of research, development, and engineering was Rudolf Furrer, or Rudy to everyone. He had demonstrated outstanding ability in designing the automatic frame plant. Although still a young man, he was an experienced and smoothly efficient executive. Rudy and LR formed a strong team. Neither shied away from tackling new projects if prospects were favorable for economic solutions.

Rudy had attracted several strong men for his "cabinet." Among them were Ed Brown and Forrest Nagler, formerly department heads at Allis-Chalmers; Robert S. Archer, formerly of Alcoa; and Bill Kepler, formerly of Newport News Ship Building, now head of our engineering department. Inasmuch as we were planning a steel mill to supply most of our requirements, we had experts on various phases of the program from ore to finished product. To that end, Ed Brown had recruited some promising talent in Europe. There were Harold Etherington, an English metallurgist on fuels and furnaces, who became an authority on nuclear reactors for ACF Industries and Allis-Chalmers; Louis Reeve, an English Ph.D., who had just completed a study of the blast furnace and pyrometallurgy and who, as an authority on weldability, returned to England later as chief metallurgist at Appleby-Frodingham in Scunthorpe; Dr. Herbert H. Harris, a steel metallurgist, who became chief metallurgist of Babcock and Wilcox at their Glasgow plant; Dr. Konrad Hofmann, open-hearth superintendent at Krupp, who was later with the South Chicago plant of U. S. Steel, the German government steel plant at Wolfenbüttel, and the Brazilian steel plant at Redondo; Dr. Werner Knapp, rolling mill superintendent at Krupp, and later works manager of the Huckingen Works of Mannesmann Tube; and Hermann Moll of Neuwied, Germany, an open-hearth authority who developed the A. O. Smith gas-fired blast furnace. Ed had also recruited talent for the second echelon, among whom were Hans Jaeger of the Krupp research laboratory, who had completed his Ph.D. dissertation on stabilizing stainless steel; Hans Groebler, who became technical director of the German Society for Sheet Metal Fabrication; Dr. Hans Langebeck, a physical metallurgist who worked on steel quality; and Dr. Fritz Wienert, an ore mineralogist who became an authority on the gaseous reduction of iron ore.

Several of these men were Germans. Until they acquired sufficient English, I was called in to act as interpreter during sessions with our Englishmen and our Americans, and for general discussions with individuals. An interesting experience of that kind was a visit to the Ford steel plant in Detroit when I accompanied Drs. Hofmann and Knapp. They were well pleased with what they saw, but were surprised by the excessive cleanliness of the shops, and by the practice, at Ford, of putting inexperienced, although reliable, men in as department heads.

Bob Archer had already established himself as one of our top men in metallurgy. His field had been aluminum, but his interests included metallurgy broadly, with emphasis on research, production, costs, and economics. Later on, he became chief metallurgist at the South Chicago Mill of Republic Steel and then vice-president of Climax Molybdenum. For advanced, although sound, thinking, and penetrating logic, Bob was one of the best of his kind, and a great credit to our staff and to our profession.

Although I acquired a staff eventually, when I arrived I had only one assistant plus several laboratory technicians. A recent Wisconsin graduate greeted me with a friendly smile that reflected a genial disposition and a warm personality. Merrill Scheil has lived up to his early promise by making notable contributions to both metallurgy and his profession. I have always thought that his success was due to desire (that is, hard work), and his innate ability, a splendid combination. We became close friends, as well as collaborators on numerous metallurgical problems.

I acquired another home-grown assistant, Franklin Roach, a specialist on spectroscopy. One of his achievements that helped in our studies of welds was a spectroscopic method of determining local concentrations of elements. At that time, it was our test technique for that analysis. He left A. O. Smith and resumed his career in astrophysics and, later, became associated with the Air Force study of unidentified flying objects at Boulder, Colorado.

Of course the welding department performed a very important function in welding research and electrode production. Louis J. ("Doc") Larsen was in charge, with an able assistant in John J. ("Jack") Chyle. When Doc left to enter consulting, Jack took over until he retired. Both Jack and Merrill became presidents of their national societies. Chuck Wheatley should be added to this list. He was on the manufacturing side and, although in direct charge of tests and inspection, functioned broadly in procurement, shop practice, and sales. He became more widely known for his work on the ASME Boiler Code.

Shortly after my arrival, I transferred from the University Club to the Milwaukee Athletic Club, at the invitation of T. McLain (Tom) Jasper. Tom was a Cornishman (and never got over it), who had been graduated from Wisconsin and then remained in this country. When World War I broke out, he returned to his mother country and became colonel in a venerable artillery regiment. Upon returning to the United States, he joined the staff of the Materials Testing Laboratory of the University of Illinois, under Professor Herbert F. Moore. He found his career there, in theoretical mechanics and mechanical testing. In Milwaukee, Tom broke new ground, the design and safe practice of the fusion-welded pressure vessel. The problem boiled down to determinations of the safe operating temperature and the allowable stress.

Tom introduced a simplified test for determining the maximum stress that could be supported indefinitely at about 900 °F. It was adopted by other experimenters later, and became known as the stress-rupture test for elevated-temperature testing. His test has been confused with a similar test that I devised for stress-rupture testing at GE. Although the technique was the same, Tom was concerned solely with the safe operating, or elastic, stress, whereas my test involved and permitted plastic deformation. Tom made some other contributions, as we shall see later.

We had an exceptionally fine analytical chemist in Dr. John J. Thompson of Michigan. He, like C. E. Plummer and Dr. Dorothy Hall Brophy at the research laboratory in Schenectady, had been a student of Professor Willard of the University of Michigan. Those three were the best analysts in my experience, and were all a great credit to their teacher. Professor Willard was one of our consultants, at the behest of Bob Archer, and was always a welcome visitor, although most of us met him only at lunch.

Another development that was made before my arrival was enameled line pipe. The plant where it was made also had to be seen to be believed. The skelp was 0.45% C steel at 1/8 and 3/16 inch wall thickness. Forty-foot lengths were flash welded at the pipe shop, and delivered to the enameling shop. Alignment for welding forty-foot lengths was not quite adequate for such thin stock, but upsetting the edges cured that. After surface preparation, the pipe was sprayed with frit on the outer surface and passed through a huge furnace to fire the enamel. After it emerged from the furnace, the next step made the process economical and also fantastic. A feature of the design was the requirement of high strength to compensate for the light weight or thin wall. That strength had to be obtained by quenching the enameled pipe, a problem by itself. Water or oil quenching was too drastic for the enamel, but a pilot program had demonstrated that quenching in molten sodium nitrate was satisfactory. Like quenching gun tubes in oil, the pipe was quenched vertically in a huge tank of the salt. When not in use, the molten salt was pumped into a storage tank, where it was allowed to solidify.

Field tests demonstrated the worth of this pipe by showing that it resisted both corrosion and abrasion. The enamel surface had another advantage — it was amenable to the cathodic protection that Orrin Andrus developed.

Enameled pipe had been developed before my arrival and had moved into a full-scale plant that was being prepared for production. Metallurgical control was in the charge of a young man whose name was to become well known in the metallurgical world. Walter Jominy, like Bob Archer, had been a student and an assistant of Professor Campbell of Michigan. He had just directed a large-scale study of heating steel for forging, under the auspices of the American Gas Association.

Enameled pipe was ready for production at the time of the deep depression in 1932, when the market for capital goods was at its low point. In spite of its great potential for line pipe and its proven economic worth, sales for all line pipe fell to zero. The enamel shop could not be started up for orders of less than five hundred miles, so the program had to be abandoned and was never reactivated.

All was not lost, however, because the results and experience gained from the development program were valuable assets. Under the guidance of Wesley Martin and Waldo Higgins, A. O. Smith had valuable frit formulas that were later used in other applications.

The first attempt to use enameling for another product was too late in development to be used. With the renunciation of Prohibition, we quickly started to produce steel beer barrels. We got out quickly, too, as other companies got into production, but not before we had made a large number of them. Steel barrels raised the ire of the coopers who had always made the conventional wooden barrels. That, and the weight of the steel barrel, gave me an idea — why not make an enameled, lightweight steel barrel and encase it in wood for handling and strength. With the help of an old hand at forming steel, Merrill and I developed a three-piece barrel that was made by copper brazing the two heads to the body. The trick was to maintain a tight joint as the parts were heated for brazing. We solved that problem by controlling the steel and the various operations. But success did not come until our twentieth run. Our barrel had positive advantages, and enamel would have been a better liner than pitch, which had to be used for steel, but we were getting out of the business.

The use of enamel did not end there. We knew that enamel, or glass, was compatible with beer, and glass-lined beer-storage tanks were next on the program. Enameled tanks had been made before, but several parts had to be joined mechanically for the large sizes. Those joints were bad because they could not be sterilized. Our tank was made full-size, in one piece, and was all-welded.

Right away, the serious problem developed of heating a large tank to the enameling temperature without bad sagging over the supports. Fortunately, Doc Larsen got the idea of supporting the tank at each end and allowing it to support itself as a stiff member. That worked, and we were in business.

We also had a steel problem. The market provided excellent enameling stock for household appliances, but we could not afford the extra cost. We had to learn how to procure and use cheap steel. That was a typical situation for Chuck Wheatley, who was an old hand at steelmaking, and was well acquainted with our suppliers. We, in metallurgical research, had studied enamel defects, fisheyes, and so on, so we defined what we had to have. Working with Chuck and the enameling laboratory, we finally arrived at a good solution of a steel which, when fired, would take the enamel and practically avoid enamel defects. I learned that there are many fine points in enameling. I also learned that many tiny blisters can form on the surface of sheet steel as a result of pickling. They were avoided by cleaning the sheets by sandblasting. We had the fun in the laboratory and Chuck had the work in procurement, but we made a good team.

Our production experience and laboratory control were further utilized for the A. O. Smith glass-lined water heaters. This also led to another enamel problem: really hot water (more than 180 °F) can ruin ordinary enamel. I worked closely enough with the enamel laboratory to appreciate the large and detailed study needed to solve that problem. As is well known, this project developed into a fine business, after a special plant was built in Kankakee, Illinois.

In the meantime, A. O. Smith acquired Mr. William Heath as a vice-president and a top man in management. He became interested in the possibilities of a glass-lined silo, and it was not long before he had one in operation on his farm. They are those solid-blue silos one sees when driving through farming country, or when riding cross-country on a train.

As for me, I don't own either a brewery or a farm but I do own a large A. O. Smith Permaglas water heater. I ordered the large size, because I can hold the temperature down, or slightly above the tolerable temperature, and still provide enough hot water for the household, even when we have guests.

During the late 1920s, LR planned a new research-and-engineering building that was also to house management and sales. It was a unique structure and was widely publicized in the architectural literature on account of its design features. The outside was practically all glass, cleaned from a traveling platform suspended from the roof. The window cleaners on the platform manipulated it up and down and around the building. The offices and laboratories had removable panel walls for flexibility in providing necessary floor space. On the ground floor we had two large laboratories, the welding laboratory on one side and our furnace room on the other, with a large display room in the center. The entire building was air conditioned.

A. O. Smith was purchasing about 100,000 tons of finished steel per month which LR wished to make in a mill of his own. Thus was established a staff of ore and steel-mill experts along with the necessary laboratory and engineering support. Professor C. K. Leith, the leading economic geologist of the University of Wisconsin, was our principal ore consultant. One of the functions of my laboratory was the examination of samples of iron ore for amenability to direct reduction processes.

Work on the steel program, from ore to finished product, was an exciting experience, but the heavy hand of 1932 fell upon us. That was the year when the bottom dropped out of our market and, with that, the profits that supported our research and development programs. Many good men had to be released, but it didn't take them very long to secure new positions since competent personnel are always in short supply.

That was also the year of our Lincoln patent suit. The Lincoln Electric Co. was, and still is, a leading supplier to the welding industry. But they were marketing welding electrodes that A. O. Smith felt were infringing our Smith and Stresau patents. This was a major suit, calling for careful and detailed preparation. It fell to my lot to handle the metallurgical phases of the suit. Before we had completed our work, we had studied the metallurgy of steel welding more thoroughly than had ever before been attempted.

LR engaged George I. Haight of Chicago to take direct charge of the case. Mr. Haight was an exceptional person. After graduating from Wisconsin, he had started practice in criminal law, which gave him valuable trial experience. He switched to patent law when he became disillusioned with criminal procedure. Incidentally, the judge in the Illinois Bell Telephone case allowed him the largest fee ever granted — well over a million dollars, and that was back when a dollar was worth almost a dollar. He was also selected by the U. S. Supreme Court to handle testimony in hearings on the use of Colorado River water. His acceptance of the assignment was at great financial sacrifice, because the maximum attorney fee was held to five hundred dollars. He died before completing the hearings. As if his busy practice were not enough to occupy his time, he was also head of the Wisconsin Alumni Research Foundation, of vitamin fame, and trustee of the Chicago and Northwestern Railroad when it was in financial difficulties.

At a dinner in his honor, the president of the University of Wisconsin, Dr. E. B. Fred, conferred on Mr. Haight the title of "No. 1 Alumnus." I learned some of the reasons for his status when I worked with him — his thorough preparation and understanding of details, a great display of energy, in and out of the courtroom, and an unusual combination of knowledge and wisdom. I believe I can sum it all up by saying that he was one of our great trial lawyers.

This case required us to determine the reasons for the superiority of our cellulose coating, and to demonstrate the differences between it and bare-wire and prior-art coatings. Oddly enough, we also had to show that sulfur was bad in weld metal. Doc Larsen welded test plates with all of the electrodes and supplied specimens of each to the metallurgy laboratory for our examinations. They, and the mechanical tests, clearly demonstrated the superiority of our electrodes, both paper and extruded, as well as the equivalence of the Lincoln Fleetweld. I brought much of this together in a paper before the ASM.

I knew of no prior investigation of the kind I would have to undertake to carry out Mr. Haight's instructions, which were to make it thorough enough to stand up in court. I finally decided on a two-part program — to determine the metallurgical characteristics of each weld-metal type, and then to determine the reasons for the characteristics of each. The former included strain-aging embrittlement, hot shortness, the McQuaid-Ehn test for grain size and structural normality, density, the usual mechanical properties, and notched-bar behavior. The studies included macrostructure, microstructure, chemical analysis, foreign inclusions, sulfur prints, and gas analysis for the effects of the arc atmospheres.

We had a field day making those fascinating studies and examinations, especially since we were able to show for the first time why weld metal from the cellulose-coated electrodes was so superior to prior-art weld metal. One surprising observation was the similarity of the fused metal on the welding-rod tip to the deposited metal.

The mechanical tests confirmed the already well-known superior strength and ductility of the new weld metal. I also had notched-bar tests run; they showed good notch toughness of our weld metal versus none for bare-wire weld metal and low values for the prior-art weld metal for coated electrodes. Our job was to ascertain why. These matters belong to the history of weld-metal development and need not be discussed here. A few points, however, may be worth mentioning, especially because our findings were new at the time.

The chemical analyses showed that weld metal contains as much sulfur as mild steel does. When we made sulfur prints of a section through the weld and parent metal, however, the weld metal did not reveal its presence, although the stock did. We tried every conceivable method of sulfur printing, but with the same futile, negative results. We were never able to determine why we failed to develop a print, but I speculated that the sulfides might occur as a submicroscopic dispersion that might somewhat increase the strength.

The question of porosity was puzzling. We knew that the arc atmosphere of our rods was about 50% hydrogen, and that the molten metal must dissolve a substantial amount at the high arc temperature. We also knew that some of our welds contained wormhole porosity, but why were the welds not much more porous? I examined the etched cross-section for clues and I believe that I found one. About 90 to 95% of a layer of the deposit was columnar, and had solidified from the bottom up. Only a thin skin at the top of the layer was equiaxed. The weld metal was deposited at a temperature far above its freezing point, which kept a molten pool over the solid-liquid interface and permitted the gas to escape as solidification progressed.

The question of metal transfer across the arc came up, whether by droplets or otherwise. I proposed passing a probe quickly through the arc to collect a sample of droplets that might be there. Our physics department constructed a probe for Doc Larsen; he published some of the results he obtained.

Our conventional chemical analyses showed some differences between the different weld metals but could not clarify the considerable differences in properties and characteristics. The usual analysis of bare-wire weld metal might pass for Armco ingot iron, for example.

A strong clue to the reason for the brittle character of the bare-wire weld metal was its known high nitrogen content, as reported earlier by Bill Ruder. We knew that something was missing from our chemical analyses, so we initiated gas analyses for oxygen and nitrogen. This was done by Louis Reeve using the carbon-crucible, vacuum-fusion method of the National Bureau of Standards, which gave the total amounts of those gases.

I had Louis run some of our weld-metal samples for which we had results from the McQuaid-Ehn and other tests. When I tried to correlate his results with the properties and characteristics of the weld metals I got nowhere, however hard I tried. For example, we knew that oxygen makes steel abnormal. Our McQuaid-Ehn tests proved that bare-wire weld metal is abnormal, yet our weld metal was normal in spite of its high *total* oxygen content. It became clear that I could not use total oxygens and nitrogens in my testimony at the trial. Only if the gas content were very low, for our weld metal, could I testify with assurance that the arc protected the deposit from air. But even our excellent weld metal contained sufficient oxygen and nitrogen to lower the quality, unless rendered innocuous. The manganese and silicon contents compared favorably with the analysis of a silicon-killed steel, but oxygen was well above that in a good commercial steel of equivalent quality. If I were to use gas analyses, and that was to be a feature of my testimony, I needed evidence to show that our high oxygen content was not harmful.

To make a long story short, I asked Louis to try to differentiate between iron oxide and other oxides that might occur in the weld metal. I recalled that Wüst had used a tin-antimony alloy to lower the melting point of a steel sample for determining carbon by combustion. I suggested that Louis try that technique to determine oxygen as iron oxide at a low temperature, at which the more stable oxides would be unaffected. The temperature could then be raised to determine the amounts of oxygen that occurred in the other oxides.

Thereupon, Louis developed his fractional vacuum method of gas analysis, but he used pure tin to fuse the sample. He started with samples of the various oxides, using, for example, an oxidized Armco ingot iron for iron oxide, to determine the temperatures at which the oxides were reduced. When he ran our weld metal, he found that practically all of its oxygen occurred as silica and none as iron oxide. He also found large amounts of iron oxide in bare-wire weld metal and in prior-art weld metal. Louis published his work in an AIME paper.

At last it appeared that we were getting someplace — our weld metal had its oxygen fixed in an innocuous form, and was deoxidized, in contrast to prior-art weld metal. Ours also had a generous amount of metallic silicon, which we calculated by subtracting the silicon in the silica by vacuum fusion from the total silicon by chemical analysis.

Our metallographic examination was puzzling in that we could not detect a postiive difference between bare-wire weld metal and ours. Most of the inclusions were very small and appeared dark, although our weld metal was cleaner. That was when we started to use polarized light to examine foreign inclusions. With crossed Nicols, the iron oxide inclusions remained dark (opaque), but the silica inclusions were illuminated (transparent), and produced a "starry night" effect against the black background. Each inclusion had an optical cross with concentric rings. That is a better description of the doughnut-like inclusions that had attracted Howe's attention. Merrill and I published a paper on our use of the new technique, showing by examples that it makes the identification of foreign inclusions easier and more positive. This technique for distinguishing opaque and transparent types has since been incorporated in the table for the identification of foreign inclusions.

These analyses and microscopic examinations brought oxygen into harmony with our various tests of quality and properties. Two puzzles remained, however: the origin of those silica inclusions, and the origin of the metallic silicon in our weld metal. I might be asked questions on them on cross-examination.

The electrode core wire and the steel plate, or parent metal, contained too little silicon to provide the amount in the weld deposit. That left the sodium silicate of the coating as the only possible source. At the high temperature of the arc, the hydrogen and charred carbon from the cellulose could reduce the silica and produce metallic silicon in the weld metal. Explaining the silica dispersion was not as simple, though I felt I should make the attempt.

I had to speculate that both silicon and oxygen were dissolved in the fused core wire, and that the oxygen precipitated as silica when the weld metal solidified. When we examined the fused end on the electrode, we found the same silica inclusions.

We had an even more puzzling case at this time. Walter Jominy was working on a high-speed forging process, and he brought in for examination a forging that had a bad surface area. When we sectioned it, we found those same transparent silica inclusions well below the surface. Between them and the surface, we found opaque iron oxide inclusions. Walter told me that the surface of the bar stock from which the forging had come had been in contact with a silica patch of the refractory of the heating furnace. Merrill and I published a photograph of this area, taken with reflected polarized light. The steel contained no silica inclusions, which raised the question of how they got from the silica patch, through solid metal, to the location beneath the surface, as our photomicrograph shows. A somewhat similar process is the growth in size of the thoria particles in a tungsten filament as the lamp continues to burn.

I dare not say that our speculation is correct, because I recall Kettering's comment that logic is a process by which we can reach wrong conclusions. We did, however, receive support from an unexpected source. Merrill came into my office one day with a publication in his hand, and announced, "These fellows have got something here." It was a paper by Sims and Lilliquist, of the American Steel Foundries, on inclusions in cast steel.

Their experiments had shown that inclusions, or their constituents, were dissolved in molten steel and did not form as such until solidification occurred. Clarence Sims followed up this line of investigation, especially on the effect of aluminum additions on the form and location of manganese sulfide inclusions. It would be interesting if someone would quench molten weld metal and steel, as Professor Duwez reported in his Campbell Memorial Lecture, to check this behavior of foreign inclusions.

I made an additional observation that may be of some interest. I was curious about the amount of iron oxide inclusions that I saw in bare-wire welds, as compared to the percentage of oxygen in the metal as iron oxide. When I probed a polished section with a sharp needle, under a binocular microscope, I discovered that all of the inclusions were hollow. Those were not oxidized pellets, as they had been described, but were porous pellets instead.

Before I was ready for the trial, I had one more matter that needed attention and that was to be a crucial point in our case. At the time of the original R. S. Smith patent application, an interference arose with a Cravens application on plaster of paris, or calcium sulfate, as an electrode coating, bad as that may seem. To get the coating on the core wire, he made a slurry and dipped a cotton string in it, and then wound the saturated string on the core. Welds made with the Cravens rod were completely worthless, because their oxygen and sulfur contents made them hot short as well as weak and brittle. We did secure some excellent photomicrographs of sulfide eutectics in our work on Cravens.

This created a sad prospect for the Smith patent: that of a completely valueless patent, seriously limiting the scope of a basic patent that covered a very valuable and revolutionary product. When the Smith patent was being prosecuted in the Patent Office, back in World War I days, the disastrous effects of the calcium sulfate were not realized. Consequently, nobody knew that the Cravens patent was inoperative and strictly a paper patent. R. S. Smith's patent was a legitimate paper patent.

As a result of Cravens, Smith agreed, in his innocence, to limit his patent to paper, yet it did have the effect of anticipating the use of wood flour by Stresau and invalidating the patent.

The trial was immensely interesting. We all worked hard, and Mr. Haight hardest of all: He put on a magnificent case. Our evidence of the novelty and superiority of both Smith and Stresau cellulose coatings over prior-art coatings was overwhelming, and valid in all instances. It was neither refuted nor substantially disputed. Furthermore, Fleetweld was proven to be the equivalent of Stresau.

While we were at the trial, which was held in Cleveland, we worked from seven o'clock each morning until two o'clock the next morning on the days' records, and on testimony we were to give. The only time for relaxation was at leisurely dinners at a dining room in the railroad station on Public Square. Mr. Haight was a wonderful raconteur on those occasions. His topics ranged from his law practice to big-game hunting in the Canadian Rockies to his experiences in London law courts to the Wisconsin Alumni Research Foundation. The food was excellent, adding to our enjoyment of those sessions.

To me, the expert testimony for the defense was unbelievable, it was so naive and full of error. It might have been a tactic on the part of the Lincoln attorney to get something in the record to confuse a nontechnical court that had no expert to rule on what was true and what was false. An example may illustrate what I mean. The Smith patent covered a reducing atmosphere of hydrogen and carbon monoxide. A prior-art patent claimed the use of a carbonate in the coating, and was used by the defense to anticipate Smith. When the expert testified on this point, he claimed that the carbon dioxide of the carbonate was a reducing gas! It takes a good attorney to counteract quickly testimony of such irresponsible character. After recovering his composure, presumably, Mr. Haight questioned him in cross-examination and asked in what way carbon dioxide was a reducing gas. The alert reply was that it was reducing because it reduced the carbon content of steel — an assertion that an oxidizing agent is its reverse, a reducing agent.

The U. S. District Court in Cleveland and the U. S. Court of Appeals in Cincinnati ruled in favor of Lincoln Electric. The Smith patent was held to its restricted scope, and the Stresau patent was held to be invalid over Smith. Inasmuch as prior art was either restricted or invalid for a patent grant, the Lincoln Fleetweld was not held to be in infringement.

That was the fate of two pioneer and basic U.S. patents on the welding art that initiated the modern era of metal joining. For the first time in history, two pieces of steel could be joined by arc welding with a deposit that made them, in effect, one piece. In principle, Stresau became the popular AWS E6010 electrode. The utility of arc welding is too well-known, and its use too widespread, to require further exposition of its merits. At home, our shock at the loss of what we thought was an airtight case soon wore off. The Great Depression was upon us, and we who remained on the payroll had important developments in progress to bring in new business.

In conclusion, it is only fair to say that Lincoln merchandised the new electrode in splendid fashion. Their production, distribution, and sales activities were in good hands, and they taught the new art to all concerned by means of schools and publicity. Looking back, I would have to say that Lincoln, at that time, was in a better position than we were to develop and expand the industry. From that standpoint, the decision had a favorable and salutary effect on the welding industry.

We derived an important bonus from our work on the Lincoln suit. We defined our procedure for characterizing weld metal and steel: combining chemical analysis, notched-bar tests, metallographic examination of foreign inclusions, fractional vacuum fusion, plus other tests such as McQuaid-Ehn, Fry's test, electrolytic extraction of inclusions, nitride anneal, etc. This technique enabled us to diagnose the quality of steel and to determine whether it had been properly made.

At about the time of the Lincoln suit, we began to feel the effects of competition in pressure vessels. We could not squeeze much out of our shop costs, in spite of exhaustive studies, which meant that any important saving would have to come from purchased steel. Steel was also the largest item by far in the cost of pressure-vessel manufacture. Rudy asked me to look into steel procurement with that in mind.

The specification that we used for nearly all our heavy plate called for a discard of more than 50% of the ingot. The argument of the supplier was that steel from the heart of the ingot would be top quality. That sounded like a good idea, because pressure vessels were used in highly critical service. However, it also meant that we were paying for the production of over two tons of steel (less the scrap value) for each ton received in Milwaukee. We had a second specification that was used for cheaper plate for less critical applications. I studied those two specifications in detail.

I found that the acceptance tests for the expensive plates, from the heart of the ingot, were much less restrictive, on the basis, I presume, that the 50% discard from the top, bottom, and sides would leave only good steel. The less expensive plates utilized much more of the ingot, but the specification required more searching acceptance tests. Finally, I concluded that we would reduce our steel costs substantially by procuring only the less expensive steel of more certain quality. That was a good introduction to the way to use technical information in steel procurement. I found that Chuck Wheatley agreed with my analysis and conclusion. From then on, we used the cheaper and better steel.

We had much more experience with skelp procurement for pipe and casing, mainly because we found that not all the steel that met our purchase specification was of good welding quality. Once in a while, our pipe shop superintendent would call me to come out to his shop, that he was closing it down.

As I have mentioned, our pipe was all made by flash welding in forty-foot lengths in forty-five seconds. After straightening and flash trimming, the pipe was hydraulically expanded a few per cent when it was inside closed dies to round up the pipe to the correct dimension, and to increase the strength by cold working. An equally valuable objective was the test of the longitudinal weld seam.

We realized that we had to locate and scrap all pipe that would be likely to leak in service, or else we would run the risk of field failures. I was called in when leaks, or other defects, were discovered in excessive amounts.

This situation called for an investigation of production flash welds. I found that leakers, or rejects, were caused by either flat spots in the central weld line, or by segregations in the outbent fiber of the flash. The former were machine defects, and were our responsibility, while the latter were steel defects, and were the responsibility of our supplier. On the occasions when I was called in, the steel defects ran as high as 10 to 15%, which was more than our pipe shop could live with.

Though all our skelp was purchased to the same specification, it was steel from one of our two suppliers that was giving us serious trouble. Steel from our other supplier was satisfactory, and was responsible for less than about 1% rejections. Our only hope of securing a reliable supply of acceptable steel was to induce the first supplier to avoid those harmful segregations. To that end, I sent several samples to their chief metallurgist for his examination, along with my diagnosis and discussion. I was really trying to be helpful. Once they knew the source of our trouble, and since we were probably their best customer, I felt they would correct their steelmaking practice.

I arranged to have their chief metallurgist come to our plant to discuss causes and, hopefully, cures. Of course, he was accompanied by their contact man for A. O. Smith. I presented our evidence of the phosphorus banding and the oxidized slag stringers that initiated the defects. Instead of discussing the evidence, he dismissed it with the comment that what I saw as outbent fiber was really Fry lines. That was so absurd that I could hardly believe he was serious. The man had surely never tried to produce Fry lines in their open-hearth steel, and I doubt if he ever saw Fry lines, because they are so different from outbent fiber. I had produced Fry lines in weld metal, as I had described in an ASM paper.

That turned out to be a bad day for his company. Instead of entering into a hopeless argument with him, I made up my mind on our only profitable course of action. I went to Vice-President Rae Bell, who was in direct charge of our huge volume of steel purchases, and advised buying no more skelp from that source. I explained that if they didn't know what it was in their steel that caused our excessive rejects, we couldn't expect them to remedy the difficulty. Rae's agreement with my recommendation resulted in a loss of business for one mill and a gain for the other. Oddly enough, I ran into a strikingly similar situation with steel from those same two mills at a later time, when I was consulting.

I believe that this experience teaches some lessons in steel procurement. Both suppliers were major steel companies, which makes one wonder why one would supply good steel consistently and the other would not. A lesson here is "Know your supplier." If a supplier uses practice that you know produces steel of good quality for your application, that is a favorable situation, because you will not have to argue him into using it — and paying for it. In other words, if we had continued to purchase steel from that first mill, we would have had to include quality clauses in our purchase specifications, and the mill would have added an extra to its price.

What about costs? I am no cost accountant, but materials, labor, fuel, and overhead must be about the same for good and bad steel. We certainly paid the same, while the use of poor steel incurs extra costs in fabrication, and is prejudicial to good customer relations. Intelligent procurement weeds out suppliers of poor material and keeps the purchaser off the "preferred list" of unknowledgeable consumers who catch the cats and dogs.

Before I finish with these lessons, and to be fair, I know from experience that other grades of steel from that same mill have been produced to high quality standards for the applications. One shipment that I examined, recently, was about the best structural steel that has come to my attention.

During those times, Chuck Wheatley and I had numerous contacts with our suppliers. We in the laboratory determined the metallurgical requirements that must be met, as well as points on which we could safely make concessions that would be beneficial to the steel company. Chuck knew how to use this information in arranging purchases to secure what we had to have, yet without paying for something we did not need. Of course, we took into account the requirements of our customers as well as of our fabrication.

This problem of procurement loomed large for Super-Yield casing, an extension of the High-Yield casing program. The High-Yield casing was limited to the larger sizes that could be handled in the flash-welding machines. That restriction limited our market to the upper section of an oil-well string, which uses the larger sizes. Rudy conceived the idea of hot reducing the larger casing to smaller sizes so we could sell complete strings. He had a simplified version of a Pilger mill made on an old lathe bed for a trial run. Although it was a homemade machine, it reduced the casing as we hoped it would. When I went in to give my report, Rudy saw on my face that the answer was positive. He issued orders at once for a new hot reducing mill for production. We, in metallurgy, followed the early runs to take temperatures and observe the performance.

I got into the act by proposing that we heat treat the casing by quenching it after it came out of the mill. The idea came from my earlier experience with quenching splicebars to make use of the finishing temperature. I discussed it with Rudy, and he sent me over to tell LR about it. LR became enthusiastic, because he was aware of the possibility of improving steel by heat treating it. I then read an expression on his face which seemed to say, "What are we waiting for? I'll get this started at once." I explained that although the idea sounded very promising, a commercial process would require the steel to be uniform from heat to heat, and from month to month. The grade of steel that we were buying was not one to be heat treated, and although our laboratory tests were very encouraging, I could not recommend heat treating in production until I had tested a sufficient number of consecutive heats for uniformity. He accepted that, which was unusual, because he always wanted to start at once, and asked me to do what was needed.

Chuck Wheatley arranged to have a specimen sent to the metallurgical laboratory from each heat as rolled. I followed a sufficient number of tests and then concentrated on other matters, but I neglected to stop the testing. When I caught up with it, finally, we had tested over 300 consecutive heats. They were beautifully uniform and we called a halt.

I then told LR and Rudy that I was satisfied (and pleased) with our skelp supply, and that I felt we could rely on it for production. I should add that we were buying a coarse-grain, semikilled steel, which was not a heat treating grade.

The idea of higher-strength casings for the new and deeper wells was not new. Other producers were heat treating casing by normalizing, but they had to use a more expensive steel as compared to our skelp.

Our management wished to take an additional step to make doubly sure that the steel manufacturing processes would supply a uniform product. This was, in effect, to make us the pipe division of our supplier, developing a common interest in the project. To implement this program, we jointly followed five consecutive heats through their steel mill, and through fabrication and testing in our pipe shop. We had our observers in their open-hearth and rolling-mill shops, and they had theirs in our pipe shop. The object was to record all pertinent data. We had never before been able to keep track of individual heats of steel through our pipe shop, because too many observers would have been needed.

As soon as the results of the open-hearth and rolling-mill practice were available, I secured permission to discuss them with John Chipman at MIT. John was well established as a top authority on open-hearth practice, so I felt he could give us reliable advice on the results. John studied the heat records and our supplementary slag and metal analyses and temperature measurements. This reminded me of our similar studies during World War I, but John had the great advantage of the studies of the physical chemistry of steelmaking that had been conducted in the intervening years.

John decided that all five were good heats, but that one of them was not quite as good as the others. That was before they had been processed into casing, but as soon as the pipe-shop records were available we found that both of John's conclusions were correct. That was a fine performance by John, as well as a tribute to the value of scientific studies of open-hearth practice.

There was a sequel to this episode. We were buying skelp, a low-priced product, but we were heat treating it. When our supplier's chief metallurgist learned what we were doing, he exploded. He told me that we should be paying extra for heat treating quality. I could not agree, because I believed that it was solely our business what we did with steel we bought and paid for. I was truly shocked at his reaction, because it was his company's good practice that had given them all our skelp business and, besides, he would put us out of the casing business. He had evidently forgotten that we were to be their pipe mill. Our purchases were exceptionally attractive since both open hearth and rolling mill could set up for extended runs instead of having to change the practice more frequently for smaller orders.

Merrill and I were involved in considerable experimentation in flash welding, though only to handle problems of procurement and production to meet customer requirements. I need not discuss details, in view of the advances in the art that have been made since then, although we did run into a problem with the central weld band that gave us considerable concern.

Bill Lane, my friend of Carboloy days, when he was the Los Angeles representative, had formed a partnership to produce the Lane-Wells perforating gun for opening up the oil-well casing at specific oil strata. It was a useful device, but when we fired a bullet to perforate our casing at the flash weld, the entire length split open. To make a long story short, we were fortunate in being able to correct that condition.

It was no doubt inevitable that I should run into brittle fracture and notched-bar testing again. Several incidents were responsible. A length of casing fell off a truck one wintry day in Louisiana and broke in two. The fracture was a typical brittle fracture and had initiated at a crack in the heat-affected zone of a repair weld. One of our men commented that he would bet that he could take a sledge and break up the casing into little pieces — which, of course, he could not have done.

Another fracture occurred when a length of casing was being straightened. After flash welding, the pipe comes out bowed because of the shrinkage of the weld zone. A simple two-ram gagging press bends the pipe in the opposite direction until it is straight for flash trimming. This time, the fracture started at one of the serrations of the extruded flash. Our metallographic examination revealed a martensitic structure, or bainite, although "bainite" had not yet entered our vocabulary. We realized that we needed a small modification of the steel analysis because we could not tolerate that structure in the flash — we might be making more pup joints than pipe.

Another situation arose that caused us to depart from our standard flash-welding practice, although it had no bearing on casing performance. We received word in Milwaukee that one of our customers had been examining our casing in the laboratory and had observed the coarse-grain structure of the heat-affected zone. The customer felt that it was a dangerous condition for oil-well casing. Word of that kind spreads rapidly, and our sales department was apprehensive lest it damage our position. The situation appeared to be especially critical because an important competitor was advertising seamless pipe and casing "with walls without welds." (Note: That company adopted welding subsequently!)

Mr. Heath called me in to discuss this situation, which showed that he, and others, were taking it seriously. He asked me to look into it and to find out what could be done. Ultimately, I conceived the idea of forging the heat-affected zone when it was at a temperature at which the structure would be refined.

I could not be sure that a forging process would be successful, because the initial bump had to be big enough to complete the weld and to extrude the flash, and yet leave enough for forging at the lower temperature to refine the grain and not coarsen it. By the use of a timer on our experimental flash welder, we were able to split the bump into two cycles, one to weld, and one to refine the grain. Our tests showed that we were able to produce a fine-grain heat-affected zone, but we found no significant improvement thereby.

I reported to Mr. Heath that while we could eliminate sales resistance, I could not guarantee any corresponding improvement in casing performance. He and the sales department decided to modify the flash-welding process according to our experimental results. It became known as the "double bump" process. The cost of the change was minimal and, presumably, we kept peace with our customers. As long as the casing was strong enough to support the weight of a string in the hole, the principal stress was compression from the collapse pressure. In either case, that coarse grain did no harm.

This experience gave me great respect for semikilled steel — when made right, I must add. When not made right, the potential quality suffers a distinct loss. I made good use of this experience later on, when we were developing Otiscoloy at Battelle.

These and other incidents convinced me that our people should be posted on notch brittleness of ductile steel. I recalled the Heyn test as a simple means of demonstrating the difference between notch-tough and notch-brittle steel. We accumulated specimens of both types, about 2 in. long by ½ in. wide and 3/16 in. thick. I first demonstrated this to Bill Kepler, who, as our chief engineer, was the one I particularly wished to persuade of the reality of the effect. I gave him a notch-tough specimen and had him secure it in a vise. With a hammer, he gave it a sharp blow to open up the notch and bend the specimen ninety degrees. He straightened it out, and repeated the performance until the bar broke. He had felt the resistance of the steel to the blows. He then repeated the test with the notch-brittle steel. With his first blow, the bar broke completely in two and the top half flew across the room. He had experienced the lack of resistance, and his involuntary expression proved he had received the message. There followed a thoughtful pause, whereupon he commented that I must have given him a piece of brittle steel. I then had him secure one of the halves in the vise and told him to try again. This time the bar simply bent over and proved its ductility. Bill had sold himself on notch brittleness of ductile steel, which made it easy for me to communicate on the subject. Though he may have known it before, he most certainly knew then, that steel was not just steel. He also assigned Fritz Zerbe to work with me on metallurgical problems.

That was an excellent test for proving the point, but when I approached Tom Jasper to try it, I was met with, "I wouldn't waste my time." To persuade me that notched-bar testing was absurd, he showed me his plot of Charpy values versus tensile per cent elongations for a variety of steels. It was a shotgun plot, and he wanted to know what use one could make of such an incongruous relationship. To him, his plot meant that the Charpy test was useless and deceptive; to me, it meant that ductile steel can be either notch tough or notch brittle.

I was curious about his reason for not recognizing what was so clearly to be seen. I knew that Timoshenko's book was his authority on theoretical mechanics, so I looked up what was given on notches. I was not greatly surprised to read that if notches were present, it was sufficient to use ductile steel, although with no treatment of notch brittleness. I believe the author has improved on that treatment in later editions.

Another man accepted the fact of notch-brittle steel but asked me, with assurance, where were all those notches in engineering structures! Subsequent experience with brittle ships convinced him that there were, indeed, notches in engineering structures.

These and other influences made me decide to publish something on Charpy testing and on the distinction between it and impact testing, or between notch behavior and impact behavior. People were calling the Charpy test the impact test — to guard against shock in service rather than to avoid brittle failures, whether static or dynamic. An odd situation developed from this concept. One experimenter tested Charpy bars statically, and referred to his test as a static impact test! Another investigator published results for unnotched test bars in high-velocity impact tests of ductile steels. He was using a test machine in which the inertia of the parts dominated his measurements. For one test, he reported practically zero energy absorption at the maximum velocity, although its deformation and fracture were the same as when tested statically. I need not go into the absurdity of this result, although if his test bar had broken as a brittle fracture, he would have made an interesting and valuable contribution.

Almost twenty years had elapsed since my first two papers on toughness and "Principles of Metallography" had been published. In these, I had given reliable data on these points. I thought the time was ripe to discuss the situation again.

The popular *Metals and Alloys* was then under the inspired editorial guidance of Dr. Horace W. Gillett. I felt it was an excellent medium for spreading the message. In a four-part article, I attempted to tell something about the notch effect and its influence on steel behavior. I also explained that impact behavior came in a different category, and had to be studied with plain bars, with no notches. I inveighed against calling the Charpy test the impact test, because it seemed important to avoid the hidden dangers lurking in that false terminology. (Its insidious effects were to come to the surface during diagnoses of our brittle ship failures during World War II.)

I had no way of assessing the influence of those articles on metallurgists and engineers, although it must have been negligible. At least, nobody wrote to the editor that my treatment was wrong. I was, however, asked to speak on the subject, although at the same time I had become very sensitive to the danger of hurting the cause by overemphasis. If people don't believe you or are unwilling to accept easily demonstrable facts, and act accordingly, I doubt that anything is gained by trying to ram it down their throats.

I will mention just one tactic that I used at one of those talks to emphasize my point that the Charpy test was not an impact test. I had already explained that Charpy had simply used Russell's pendulum type of testing machine to determine the energy absorption in breaking his test bar. I then said that I wanted this inscribed on my tombstone under the heading "The Charpy Test":

> The guy below now lies at rest
> All free from stress and strain,
> While up above
> He done his best
> To call it a notched-bar test.

The ASTM held symposia on impact testing in those days, and some of us in the field were invited to present papers. I thought they were an excellent forum for spreading the gospel, and I accepted gladly an opportunity to try to clear the atmosphere on the meaning of the test and the use of the results.

I believe I didn't help much, at least not in the right places, if we can judge by the confusion of the day. In particular, I proposed the principle of what I called the match point as the stress condition that matches the strength properties for the brittle fracture. In other words, at the match point, or at any lower temperature, the steel is subject to brittle fracture. That holds for a test bar in the laboratory, or for a service structure, although the two match points will differ in temperature unless the laboratory specimen closely reproduces the structure in behavior.

For these reasons, I proposed the match point as the proper transition temperature (TT) of the steel and test bar. It not only produces brittle fracture similar to what occurs in service, but is also the only definitive temperature on the TT curve. Sometimes, higher temperatures are selected as the TT, but the steel deforms more or less before fracturing and changes the notch geometry or stress field, while the work-hardening effect also changes the steel properties. The results at higher temperatures can no longer relate to the virgin metal, or to the initial notch. Furthermore, the brittle fracture of service relates to the initial steel properties, essentially unaltered by plastic deformation, and to the initial geometry of the notch that triggered the break.

I have noticed that some investigators, especially in England, regard the stress and notch conditions that are responsible for brittle fracture as the central point of interest and, presumably, the best condition for a definition of the TT.

To sum up and simplify my early treatment, the match point for a given notch geometry or stress field is determined principally by the yield point, and its relation to the cohesive strength. The yield point, in turn, varies with temperature and strain rate and is principally responsible for their effects on the match point, although quantitatively their effects vary from steel to steel. Of these, temperature is the most conspicuous factor, although the high strain rate of a running fracture must be very important too. More recently, Don Clark of Cal Tech has supported this concept in an ASM paper.

I have dealt with these principles in early publications; but more recently, the sophisticated treatment of brittle fracture by fracture mechanics has developed into a rational and quantitative approach to this field and will be mentioned again.

I have mentioned that A. O. Smith pressure vessels were used in great numbers by oil refineries. A problem that developed with some of the crudes from sulfur corrosion led to the use of pressure vessels lined with stainless steel. Although a lime treatment would have reduced or possibly eliminated the attack, a liner of ferritic stainless steel proved to be the economical solution.

As I now recall, Orrin Andrus and Doc Larsen were chiefly responsible for promulgating this project, and for developing the lining method. The selection of the liner was guided largely by our experience with a modified AISI type 430 stainless steel for Du Pont. That standard type permitted a spread of 4% chromium, from 14 to 18%, which was much too wide for practical purposes. Du Pont's specification restricted this range to 16–17% chromium, and that was the steel we used at the start for our liners. There was, however, a difference in the use to which the lined vessels were put — Du Pont's were used for nitric acid storage, and ours for refining crude oil. We were to learn the significance of that difference.

We built a huge spot-welding machine with which the stainless steel sheets were attached to the plates on about one-inch centers. We first adopted that liner for some vessels for a refinery in the Canary Islands. The vessels were in good shape when they left our plant, but after they had been in service at 900 °F, we received word that cracks had developed in the liners.

I had some specimens cut from a failed liner and returned to the laboratory. Our headaches had begun. In addition to cracks in the liner, carbon had deposited in back of the liner, doing additional damage, and some of the spot welds had failed. We knew we had to solve those problems or we would lose out on lined refinery vessels. We found the cracks easily enough, but when we tested the liner, we learned that it had become brittle, and significantly harder than it was when installed. That was before very much was known about the sigma phase in iron-chromium alloys, although we did know that an aluminum addition reduced quench hardening. The returned liner showed that service at 900 °F raised the hardness, and presumably induced brittleness.

That examination started us on a continuing investigation of the ferritic stainless steels for service at elevated temperatures. This led us to a lower chromium content, or to type 405, with an aluminum addition. Merrill has studied this steel in great detail, and has published on it in *Metal Progress*.

The failure of the spot welds was easy to clear up — they had not welded. The shop introduced a test of spot welds that worked very well. A pressure of 25 psig was introduced at one end of the plate, between two spots. When the pressure at the opposite end also reached 25 psig, the spot welds were inspected. Unwelded spots were easily located, because they rose up off the baseplate, and were repaired. This experience brought us into closer contact with our vessels in service, as a couple of incidents will illustrate.

There was no need to inspect the vessels sent to the Canary Islands, but I did inspect cases of corrosion of our vessels in the United States. Having worked on lining materials in our plant and tested them in the laboratory, I was in a good position to judge the field performance. Inspection reports from the refinery were simple and direct — the liner was failing, or corroding. However, the reports were inadequate for diagnosing the cause or for deciding who was at fault: the refinery or we.

I recall a good example, one that taught me to check important cases personally. A report came in from a refinery that the liner in one of our vessels was corroding; that is, that it was no good. Shortly afterward, I was inside that vessel with light and a magnifying glass. Instead of a general attack, which I had expected to find, I discovered that the corrosion was localized between a tray and the liner. That space had been improperly caulked with asbestos, inducing crevice corrosion. The refinery was well satisfied to be posted on maintenance to preserve its equipment.

A man from the laboratory must approach a field inspection with an open mind, and diagnose the cause of the trouble exactly as he sees it. The business and reputation of his company are in his hands, not in those of the president, sales manager, or other executive. With prior laboratory tests, knowledge of fabrication practice, and general experience in examining failures, he is best prepared for diagnosis, and advice on repairs or remedy. Another case history illustrates a different diagnosis.

For some reason, which we felt must be inadequate, Standard Oil of New Jersey was not buying our equipment. Russ Irish (All-American end from Wisconsin), our New York manager, negotiated a big meeting with Standard Oil people, at their Mahwah refinery. I was there, with Doc Larsen on welding and Chuck Wheatley on fabrication. The others in our team included Russ, and Henry Needham, in charge of pressure-vessel sales. Russ had worked hard to set up the meeting, and was heavily involved in its success. He must have felt satisfied, because he staged a celebration afterwards, including an excellent dinner at the New York Athletic Club. Furthermore, we must have been successful, because Russ received a good order.

One of the vessels ordered by Standard Oil was a nitric acid storage tank, which we made of the austenitic, or 18-8, type of stainless steel. Our laboratory tests had shown it to be superior to the straight-chromium, or ferritic, type, for nitric acid. We procured the steel, tested it, fabricated it, X-rayed it, and shipped it. Sometime later, their plant inspection reported corrosive attack. Russ was greatly disturbed and called for help.

A few days later, I was inside the vessel to find out what was meant by corrosive attack, and I recalled the bit about having an open mind. Ours was technical selling, and any suspicion of subterfuge, or lack of knowledge on our part, would affect customer relations badly. I saw corrosive attack all right, but I had never seen that kind before, nor have I seen any similar type since. I examined it closely enough to discern corrosion of irregular areas, which suggested markings of some kind, at the bottom weld joint. The finger pointed directly at us. Russ awaited me as I emerged, and asked for my opinion. In reassuring terms, I told him I was quite sure of my diagnosis, that there was nothing wrong except for a few local areas of attack, and that I was confident that we could clear it up to Standard Oil's satisfaction.

Back in Milwaukee, I discussed my observations with Chuck Wheatley. He found that after radiographic inspection, his man had neglected to remove the lead markers before the vessel had been quench-annealed from 1950 °F. The locations where the lead had fused had induced the acid attack. We repaired the vessel and returned it to service.

In spite of disappointments and failures, stainless steel recorded progress in the 1920s and came into its own in the 1930s, yet it still had problems to surmount. I believe it was then that stress-corrosion cracking became critically important in steel selection and processing — it surely was with me, and remained so with Merrill after my departure.

One day, Walter Jominy came in and told me that a stainless-lined digester of ours, at the Forest Products Laboratory in Madison, Wisconsin, had developed cracks in the liner. It was the first time I had heard of that vessel, because it had gone through sales, engineering, and production as a routine job. Walter wanted us to examine it to learn what had gone wrong. Samples of the failed liner soon revealed that the problem was stress-corrosion cracking. This was not a general stress corrosion, but the development of cracks that were caused by tensile stresses in an active medium. Actually, there was no general corrosion, although I have observed that effect on stainless steel when under stress. The steel was the low-molybdenum austenitic type 316, which was adequately resistant to the liquor of the experimental digester, but was susceptible to cracking when under stress. We also learned that the liquor liner had bronze valves.

There is no point here in discussing a problem that is now covered by voluminous literature. I believe the important point is the procedure that we evolved for handling the problem for our equipment. We used the Huey nitric acid test for general corrosion resistance, and the Hatfield-Strauss acidulated copper sulfate test for intergranular attack. (The latter should now be called the Hatfield test, Merrill has informed us.) Those tests are made on unstressed specimens, but, obviously, we would have to test our materials when under stress. For that we devised a simple horseshoe type of specimen that was stressed by wiring the two ends together when under tension.

Finally, we developed a systematic test program that included the variables of the steel as received from the mill, one or more welding conditions, heat treatments, and other pertinent fabrication processes. A collection of our horseshoe specimens would be exposed in the customer's plant at one or more critical locations. Steel and fabrication would be determined by the results.

At the time when hydrogenation was being introduced, we were able to use this technique quite effectively. An important customer became interested in new equipment, and we in their problem of steel selection. During a session at their plant, we discussed the details with their chemists and engineers. It developed that they had received from an independent testing laboratory corrosion reports on which they were relying for the selection of the steel to use. By that time, I had worked closely with Henry Needham on steel selection and our testing procedure. For example, International Nickel put out a handy collection of plain disc specimens for plant tests. They appeared to have merit, but I had told Henry that we could not rely on tests of that kind because they failed to include stress. After Henry and I had reviewed the testing-laboratory reports, he brushed them aside and said something to the effect that we should forget them and get down to business. That surprised our hosts, and they wanted to know what he meant. He explained that their pressure vessel could not be made and operated in a stress-free condition. He then asked me to tell about our experience with stress-corrosion cracking.

We had a different problem with General Aniline. They were a German firm, with plants in this country, and had always procured their stainless steel equipment from Krupp. At a session at their plant, we were told that because Hitler had banned the export of strategic materials they had to shop in this country. I could glean from their side comments in German that they were uneasy. They had relied completely on Krupp, and were not sure that we could handle their problem. Henry, too, became uneasy, because the requirements were new to him and he could not be sure that we could do as well as Krupp. I managed to get a lead to the crux of the problem, and was also given a sample of the Krupp steel that had served satisfactorily.

I decided that this was a contamination problem, and not a regular corrosion problem. Back at the laboratory, I abraded a section of the sample and tested it for corrosion, whereupon it turned out to be plain 18-8 stainless steel, and not some super quality. All that we needed to do was to passivate our stainless liner. As was the custom of the laboratory, we prescribed the treatment and then followed it in production.

The columbium-stabilized stainless steel, type 347, was a very popular steel because, unlike ordinary type 304, the welded joints were not susceptible to intergranular attack in the heat-affected zones alongside the weld deposits. It was a valued item of the stainless steel line, because we could put the vessel in service without first heat treating it. The titanium-stabilized steel, type 321, was equally satisfactory when made right, but that was the rub. Too much of it in those days was "dirty," and lacking in corrosion resistance. We were well acquainted with the care that had to be exercised in making a satisfactory addition of titanium in our Alloy No. 10, so were not surprised.

Some of our applications were better served by the high-molybdenum type 317 than by either the straight 18-8 or the type 316, and it became a favorite. Its drawback was its tendency to become sensitized. About that time, Russell Franks of Union Carbide published a paper on columbium-stabilized type 317, reporting excellent stabilization. We thought we had our answer, but when we tested commercial samples we found no improvement. When I tried to secure better material, our supplier said that they would meet our specification for composition but would guarantee nothing on stabilization. We knew that the microstructure was decidedly different from that of 18-8, but we knew little about delta ferrite and the sigma phase, surely not enough to correct the situation and secure a stable structure. We were, of course, dependent on the suppliers.

A good example of laboratory support for a new application of type 317 stainless steel came with the use of highly corrosive phosphoric acid as a catalyst in an oil-refining process. This experience started when I was on vacation. One morning Bill Kepler telephoned to say that they were working on some catalytic chambers for Standard Oil of California for use at a refinery on the Bahrain Islands in the Persian Gulf, and that they had set up a date for me to meet Kenneth V. King, their chief materials engineer. Sales and engineering had been working on it, and now they had got around to me and Kenneth. I asked Bill what I was to do, and he replied, "Sell him our liner." When he told me that I was due in Kenneth's office in San Francisco the next morning, it was quite a shock to me, a train man. He told me to come to his office that afternoon to go into the details, then leave on the *Hiawatha* for Chicago and catch the sleeper plane for San Francisco.

I wasn't tuned in to such fast traveling, but it worked out just as Bill said it would. Not only that, but I went via Los Angeles. That early-morning view over the Grand Canyon was as glorious a sight as any on Cinerama, and we also circled Hoover Dam, and got a view of Los Angeles and California from the air. By the time I arrived in San Francisco, I had practically forgotten pressure vessels, liners, and phosphoric acid.

I was met by one of our West Coast salesmen, who took me to Kenneth's office. He brought me quickly down to earth with his first remark, "So you are the people who stick the liner on here and there." I came back with,

"Yes sir, and we're proud of it." That got us off to a good discussion, which pleased me, because I recognized that he was an experienced and well-qualified engineer. I knew that our competition was trying to persuade him that only a completely bonded liner would be satisfactory, so I told him all about our liners, how we attached them, how we selected our materials, how we tested them, and how we processed them in fabrication. I emphasized that we knew that our liner was satisfactory before it was attached, and that we did not alter its condition afterwards. Kenneth knew all the questions, and I knew — well, let's put it this way, I knew enough of the answers to satisfy him. I believed that Kenneth would choose the organization that he felt had superior understanding and control of the production of those vessels. Fortunately, we had had considerable experience in producing vessels for the oil and chemical industries.

We faced a major challenge on account of the severity of the chemical environment. We started at once on the selection of steels for resistance to phosphoric acid as measured by the corrosion rate. The tests were fairly voluminous, and exhausted our supply of acid before they were completed. That point, unfortunately, came at the New Year holiday, and, in our haste to secure the test results, we had to renew our supply locally. A quick check of the Yellow Pages located a chemical house from which we could secure more acid. Not long after the tests had been started again, the laboratory phoned me and asked me to get right over, that something had gone wrong, New Year or no New Year. The new acid was attacking the steel much faster than the first acid had, which was evident from the speedy blackening of the acid solution and the excessive weight loss. This acid was made by one firm, while our first lot had been made by another. We had to stop testing until we could secure additional acid from the original supplier. Both were labeled "chemically pure," so we had proceeded with the new supply quite unsuspecting that "chemically pure" didn't mean just that. I strongly suspected that some impurity was responsible for the rapid attack. This experience taught us to test specimens in the customer's plant whenever possible.

I should explain the need for our great haste. We had engaged one of the very few freighters that could transport our vessels from a Gulf port to the refinery. It was scheduled to be in port on a specific day, and every day's delay would cost us $4000 demurrage.

I neglected to say that we received the order, so I must have sold our liner to Kenneth. There is also an aftermath. I had often wondered, since then, how our vessels had performed in service, but did not learn until a few years ago. A California man who had just returned from the Bahrain refinery assured me that they had performed satisfactorily. All of us had worked really hard on this job. We had not only to produce satisfactory vessels, but we had to work against a firm delivery date. There is something about working under stress with sizable stakes that is exhilarating, but we seemed to thrive in spite of, or possibly because of, the pressure. I was glad to learn that we had also worked well.

A notable achievement in pressure vessels was our multilayer design, masterminded by Tom Jasper. It greatly reduced the wall thickness or weight, as compared to the solid-wall construction, and did not fragment if overloaded to destruction; therefore, it was a safer design for containing high pressures. The individual layers were made of much stronger steel, and the method of fabrication produced the autofrettage effect of putting the inner layers in compression. These vessels were useful as accumulators for hydraulic presses, and reduced the safety hazard because of their close association with the press.

I made a little contribution myself: I used an austenitic stainless steel for the inner liner for use with corrosive charges at elevated temperatures. A higher expansion rate of the liner when the vessel was heated gave some concern, but a simple calculation indicated that the thin liner would simply go into compression and would be prevented from buckling by the internal pressure.

Those were pioneering days in the introduction of stainless steel for large solid and lined vessels for the oil and chemical industries. Progress was made possible because of improved melting and processing by suppliers, judicious selection and processing of materials on the part of fabricators, and the great improvement in welding technology. At A. O. Smith we learned that sound laboratory control was a necessary prerequisite for successful and continuing participation in the business. We were solidly supported by management in this function.

We kept in close contact with the suppliers of our materials, and with our customers. Two men stand out in my memory for their helpful discussions of our requirements and also as go-betweens with the suppliers. They were Herbert French and Dr. Vsevolod Krivokok. They were both well posted on the state of the art and never tried to oversell the product. We discussed our problems freely with them because they obeyed the rule against misuse of confidential information.

Herbert had a good record as a metallurgist and inspector during World War I, and as a specialist on steel and heat treatment on Dr. Burgess's staff at the National Bureau of Standards. As an executive of the International Nickel Co., he was not selling to us in the usual sense, but he was well posted on stainless steel and its properties and characteristics in design, fabrication, and end use.

He usually came with something new to discuss, or advice to give, on the austenitic types of stainless steel that contain nickel. We were especially interested in his comments on obscure points that crop up during the early development of a new field. I am sure he helped both his company and the consuming industries.

The other welcome visitor was Dr. Krivobok of Allegheny Ludlum; he was also on Bob Mehl's staff at Carnegie Tech. He was able to post us on steel manufacture, or on what we could reasonably expect from our suppliers. Instead of restricting ourselves to the usual customer relationship, we tried to develop a spirit of cooperation with outside technical personnel, and to let our purchasing and their sales people handle the commercial details. That worked very well.

On some of my visits to Pittsburgh, I enjoyed special evenings with Krivy in his bachelor quarters. One event was a Russian dinner. Although he had been in the United States since World War I, when he was an inspector for Russian procurement, he still had a lot of Russian in him. He had secured an authentic Russian cookbook and had learned how to prepare his native dishes, although I suspected that they were somewhat Americanized. In any event, a dinner at his home was a treat. After dinner, we would enjoy an evening in his excellent music library. It was there that I first became acquainted with the music of Rimsky-Korsakov.

One time, Krivy recalled some of his younger days in Russia. His father was a general in the Czar's army and had wanted him to enter a military career, but it backfired. At the military academy he could not accommodate himself to the strict military discipline, and he was sent to the United States on a military mission. After the Bolsheviks took over, he remained here, and became a metallurgist under the tutelege of Professor Sauveur at Harvard. His father was a loyal Russian who fought the Communists but was left in peace in southern Russia after the takeover.

Occasionally, I used to meet another man with whom I enjoyed swapping ideas on what could be said about metals, *à la* Bunsen. He was Joe Emmons, the chief metallurgist at Cleveland Twist Drill. He was knowledgeable on metallurgy, including his drills and the materials that they machined. And he had an inquiring mind.

It was about this time that Joe announced a molybdenum-bearing high speed steel. Since then, it has become a very popular steel, or series of steels, in which one per cent of molybdenum substitutes for two per cent of tungsten. It had its ups and downs for a number of years, until metallurgists learned how to handle it.

Joe had acquired an assistant by then. When I was giving the Priestley Lectures at Penn State, I became aware of a bright young student who asked intelligent questions and showed an active interest in the subject matter. He was Ralph G. Kennedy. After Penn State, he went to the National Bureau of Standards and then to Cleveland Twist Drill. He lived up to his early promise, and ultimately became chief metallurgist.

In spite of the exodus of many of our technical personnel in 1932, we continued a research program on a greatly reduced scale. One project was the development of a mud turbine to drive oil-well drills. The mud was pumped down to the bottom of the hole, where it passed through the turbine to which the drill was attached. It was then carried back to the surface with the chips from the drilling operation. The materials phase created metallurgical headaches.

Another project was the electric melting and refining of copper by passing it as an electrode into a conducting flux. The copper was melted upon entering the flux, and was refined as it passed through. I regarded it as an adaptation of a much earlier Russian process for welding steel castings. I thought that the latter was a legitimate Russian first, although not on their recent lists. It is similar to the Electroslag and the Hopkins processes of this type.

We also developed a cast iron that could be hot worked. When we tried to secure a patent on it, we discovered that Timken Roller Bearing and another party involved us in a three-way interference. A settlement resulted in Timken becoming the ones to commercialize the product as their "Graphmo."

Our major research project of those days was "Smith Alloy No. 10." Bob Archer had instigated work, in our experimental foundry, on heat-resistant alloys for use as resistor elements for electric heating. The initial work included a variety of alloys, although with emphasis on chromium and aluminum as the most promising alloying additions to an iron base. One series was high aluminum and low chromium, and another was the reverse.

The high-aluminum alloy, with more than 20% aluminum, was a good alloy but was very brittle at room temperature. I recalled that I had helped Swanger of the National Bureau of Standards to convert the brittle rhodium into a ductile form by utilizing experience from tungsten technology. I had Hans Jaeger try that technique on the high-aluminum alloy and it, too, became ductile in the fibrous condition. We gave up further work on it, however, in favor of the superior high-chromium alloy that became our Smith Alloy No. 10.

Alloy No. 10 was basically a low-carbon, iron-base alloy with about 38% chromium and 8% aluminum. That is about the maximum useful alloy content, and it is at, or close to, the maximum melting point. The German and Swedish alloys, with lower alloy content, were inferior to Alloy No. 10 in maximum operating temperature and in heat release for rapid heating, but had the advantage of ductility in the wire form. Alloy No. 10 could probably be drawn into ductile wire, but we lacked the proper equipment for the purpose.

The melting point of a wire resistor is an important property, because it controls the heat release. One time, we equipped a furnace in our heat treating shop with Alloy No. 10, making it possible to bring the furnace up to temperature in about half the time required by the nickel-chromium heating elements. Louis Reeve determined the melting point to be 2850 °F, which is above the iron melting point of 2795 °F. A check specimen of essentially pure iron fused before the Alloy No. 10 samples did. Each alloy individually lowers the melting point of iron, from which it must follow that the ternary alloy is at, or close to, a maximum on the liquidus curve. Unless an intermetallic compound forms, this is a very unusual occurrence.

Melting this alloy was as important as its composition, and required us to depart from conventional induction-furnace practice. Melting without a slag was not to be considered, for well-known reasons. The development of satisfactory slag practice raised two problems: our slag must be correct metallurgically for the alloy, and it must not wash down the side of the crucible as the charge circulates. I checked constitution diagrams of slag components, and stopped when I found that equal parts of lime and alumina fused at the alloy melting and refining temperature. They produced an excellent white slag, and enabled us to deoxidize the melt through the slag. It was our answer. Thereupon, Chuck Greenidge and I developed the melting practice that was patented in our names. It was not simple, especially the need to learn how to make the necessary titanium addition. A patent was also granted on the alloy, and was issued to Bob Archer and me.

We had ingots converted into wire rod at Columbia Tool Steel Co. and Hoskins Manufacturing Co. Both companies were very helpful, and earned our appreciation. The wire rods were swaged in our laboratory to ⅛-in. diameter on a small swaging machine.

We next tested the wire for performance. The ASTM life test for the nickel-chromium resistor alloy was entirely inadequate, because the time to failure was far too long for test purposes. Furthermore, it did not duplicate the constant-temperature condition in an electric furnace. We decided to test by three-minute on and off cycles that heated the wire to 2600 °F and cooled it down to room temperature. An acceptable wire had to run for 100 hours, with but very little sagging. The tests were automatic and the entire life cycles were recorded.

Merrill and I presented a paper on Smith Alloy No. 10 to ASM. We didn't expect it to replace the nickel-chromium alloy, but, as the best alloy for resistors, we visualized its principal use for electric furnaces. A. O. Smith turned it over to the Hevi-Duty Electric Co. to exploit. I learned later that Battelle had installed their furnaces in the heat treating laboratory. This reminds me that we discussed electrically heated soaking pits with Walther Mathesius at South Chicago. The practice had considerable promise, but only a full-scale operation could prove whether it was economical. Walther felt that the advantage would be its close control of temperature and pit atmosphere. He also had an excess of electric power that could be put to good use.

In due course, I learned that Vere Brown, chief metallurgist at Allegheny Ludlum, had developed a lime-alumina slag for melting stainless steel. He called it a calcium-aluminate slag. One day, Vere took me into his holy of holies and demonstrated the slag's virtues. He was doing on a large scale what we were doing on a small scale. I have read recently that M. Perrin of Ugines in France also uses this slag for fine steel melting.

During my visits to Hoskins, I met a grand Man of Metals, Albert L. Marsh, president of the company. Many years earlier, he had pioneered the development of nickel-base alloys for thermocouples. The platinum-platinum rhodium thermocouple of Le Chatelier could be used at higher temperatures, but the nickel-base alloys were more economical for use at lower temperatures. The well-known Chromel-Alumel thermocouples resulted from his early work. Marsh also invented the popular nickel-chromium electrical-resistor element that is widely used for household appliances.

While all of this was going on, our salaries were dwindling drastically during the depression years. By way of compensation, we were allowed to consult on the side. I recall one case in particular, because it had a moral.

One morning, we read in the paper that a house in suburban Wauwatosa had been blown up by a gas explosion. The Milwaukee Gas Co. was sued for damages, on the basis that its gas line had leaked gas through the basement wall and into the house. The gas company asked me to investigate the situation. The house was a very sad sight, but I found no evidence that the outside main was responsible. In brief, the evidence convinced me that gas had leaked inside the basement, and had exploded when the wife turned on the oil burner that morning. That was the conclusion of my report. From my investigation and numerous tests, I found that the gas pipe inside the house had developed the leak, which meant that the gas company was not responsible.

I had expected to testify at the trial, but the case was settled out of court. The reason? The husband was sick in bed at the time of the explosion, with a bad heart condition, and did not survive the shock. His widow was left with a badly wrecked house, and in poor financial condition. This prompted the gas company to settle the case in her favor for a substantial sum.

We had another explosion case at that time. A bootlegger had a still that blew up on him in his basement. He served a term in prison, but as soon as he was released he sued the gas company for damages!

I was not called in on that case, but I was the expert on another house explosion. The plantiff claimed that a nipple had failed, allowing gas to escape. It was an unimaginative claim, which would be raised only by one who was unschooled in metallurgy. I examined the nipple, and then testified in court that the fracture was a fresh fracture and that the nipple was sound. No flaw was visible, so I concluded that it must have been willfully broken. My testimony was not refuted, although the attorney for the plaintiff put up a gallant fight. These cases made me wonder why an attorney would attempt to build a case so blatantly, on testimony that was so palpably wrong?

During my work with the gas company, Mr. Kemen, chief engineer, became interested in my use of the Heyn test on a number of lengths of pipe from the basement. This led him to test pipe that he had in stock, whereupon he found that the variation in notch toughness was too great for comfort. From those test results, he established a standard for acceptance, and segregated his notch-brittle pipe for noncritical use. He also used the test on pipe that was offered for sale to his company. That imposed no great penalty on the supplier, because he had customers on his "preferred list" who were not so choosy.

One day, I believe it was in 1938, we had a visitor at A. O. Smith. He was my old friend of Charlottenburg days, Guillaume — as I knew him then, but now William — Kroll. He had developed a method for producing titanium metal in his Luxembourg laboratory, and was in the United States to secure financial backing to produce it commercially. After we had discussed student days and our tour through the Ruhr and the Rhineland in 1913, he explained his mission. His story and his magnificent contribution are now well known, but at that time neither we nor anyone else felt justified in going ahead with him and his exotic metal. I knew him as a capable and sincere experimenter, and I was familiar with his publications, but I could do nothing for him. Ultimately, his persistency and his belief in his product paid off, although it took the strong support of our government and of others to bring titanium to the marketplace.

About this time, we developed an interest in gold mining. Some may recall that the price of gold had been increased during the depression years, promoting interest in mines that could not be operated profitably at the old price. LR had an option on one in the mountains of southern Colorado high above Monte Vista. He sent a group of us out to see if we could contribute to the success of the operation. From my student days at the school of mines, I had experienced enough of the glamour of the profession, as well as the hard labor of the work, to clothe this mission in an aura of excitement.

Our mine was 12,500 feet high in rugged mountain country, and I had to admire the tenacity of the early prospector who had located it. In addition to having a field day in the mine, and in the mill, I had an experience on the way home that revealed something of the complexity of the steelmaking art.

Harold ("Dede") Detrick, one of our engineers and later a vice-president, and I returned home together. His former boss, when he worked at the Gary plant of U. S. Steel, was then with the Colorado Fuel and Iron Co. in Pueblo. It was on our way, and convenient for a visit. With considerable gusto, he recalled the first shipment of steel to A. O. Smith for the automatic frame plant. It was an important trial order, and they were quite conscious of the severe edge bend that the wide flat stock had to take. That prompted Gary to go all out to make a clean and high-quality steel.

The steel was finally shipped to Milwaukee, and Gary waited anxiously for word of the results. Then someone discovered that the rolling mill had mistakenly applied a heat of ordinary steel to the order. Gary instructed A. O. Smith, by telegraph, to return all of that steel and said they would ship a new lot. Much to their surprise, Gary received a telegram from A. O. Smith stating that the steel had (already) formed perfectly, and that subsequent shipments should duplicate it. Things happen like that sometimes.

During the 1930s, our concepts of what goes on when steel is quenched were radically changed and improved. The early ideas of Arnold about "hardenite," and of Osmond, Carpenter, and others about beta iron, as the structure that gives quenched steel its high hardness, had expired as more rational ideas had developed. Even at the time when they were strongly, even emotionally, debated, they were strange assumptions that lacked conviction — except of course, to their adherents. Sauveur's assumptions of the structural transformations that occur during quenching were also untenable, as I was able to demonstrate while still at Minnesota. A similar fallacy occurred in the statement of the Nomenclature Committee of the International Society for Testing Materials — that sorbite was produced by quenching steel from just above Ar_1 This may give an idea of the confusion during the early days of this century.

The structures of quenched steel and of quenched-and-tempered steel were at least named, even though they were not properly identified. They were sorbite and troostite of Osmond, and osmondite of Heyn. Those terms were useful only in that they indicated something about the treatment of the steel. Martensite, of Osmond, was also included, and it remains in our nomenclature today. The field was ripe for a more detailed and sophisticated study of the structural transformations of steel when heat treated. Ed Bain emerged as the principal architect of the new metallography of the structures of steel.

Looking back, I believe we can discern critical observations that could have led the way to our present concepts. Offhand, I would select the experiment of Taylor and White with high speed steel as a forerunner.

They found that it can be quenched to about 1100 °F and held there for an extended time, and still harden the same amount on quenching as if it had been directly quenched to room temperature. About two decades later, Portevin and Garvin, in France, and Wever, in Germany, contributed their investigations of the quenching process. Chapin, in the United States, reproduced the Taylor and White experiment, except that he used conventional hardenable steels. The temperature of the intermediate quench was considerably lower, although the results were similar.

In 1930, Davenport and Bain published their classic paper on TTT curves (time-temperature-transformation). They were also called "S curves," from the general resemblance of their plots to a capital S. Upon quenching a steel to various temperatures below the critical temperature, holding at temperature for various time intervals, and quenching to room temperature, the isothermal-transformation rate and the times of initiation and completion of the transformation were determined. The plotted data were TTT curves. Metallographic examination revealed the structures of the partly and completely transformed steels. Sorbite, osmondite, and troostite went into discard as names of structures, and were later replaced by upper and lower bainite and tempered martensite.

Fully hardened steel had a different structure — martensite — but it was incorrectly incorporated in the early TTT diagrams. Martensite is now correctly shown to start forming at a moderately elevated temperature and to continue to form from austenite as the temperature falls. This temperature dependence of martensite formation is missed when the steel is directly quenched to room temperature. This clarified the much earlier finding of Osmond, that some steels, after quenching to room temperature, still contained some untransformed austenite, and must be cooled to a still lower temperature to complete the transformation to martensite.

In the late 1930s, Ed Bain gave a memorable talk to our local ASM chapter in which he presented the modern views of the metallography of steel in a clear and enjoyable fashion. The occasion was a high point among technical talks, because Ed fitted so many of the pieces of the jigsaw puzzle into their proper places. He spoke with consummate skill, which made me feel that I was beginning to understand steel transformations.

The groundwork that Ed laid down stimulated correlative studies, especially of the mechanisms by which the structural changes occur. Foremost in this area were Bob Mehl and collaborators at the metals research laboratory at Carnegie Tech.

Bob took up studies of diffusion, Widmanstätten structure, the formation of pearlite, and structural transformations. By means of X-ray crystal-structure technique, the relationships between the parent austenite lattice and the precipitate or new lattices were determined. Bob's collaborators were men who have made their own marks in metallurgy. Among them were G. Derge, C. S. Barrett, O. T. Marzke, and D. W. Smith. They also demonstrated the quantitative effect of the austenitic grain size on pearlite formation. This work cleared up the greater hardenability of coarse-grain austenite with smaller intergranular area where the transformation starts. This shows what an able teacher can do with talented people. The field has been a fruitful one for metal physicists, not only for steel but for various nonferrous alloys as well. The details of the phase change from austenite to martensite have been much more difficult to decipher. All this knowledge has been used to develop austempering, martempering, maraging, and ausforming, as actual or potential treatments to improve the mechanical properties of steel.

In the meantime, in the 1930's, other investigators were busy on determinations of an exceedingly important property or characteristic of heat treated steels — hardenability. Marc Grossman, Al Boegehold, Walter Jominy, and Bob Archer were all heavily involved in developing the concept, and in devising methods for determining it.

A major impetus must have come from the report of Emil Janitzky, chief metallurgist at the South Chicago plant of U. S. Steel, and Miss Mary Baeyertz. They demonstrated that steels that were quenched and tempered to the same hardness had very nearly the same tensile properties. This relationship was reported for carbon and alloy steels of 100,000- to 290,000-psi tensile strength. Of what use, then, was the addition of alloying elements such as nickel and chromium, if all steels developed the same properties? That was a good question, because it was also known that alloy steels were superior to straight-carbon steels for many applications. This was not true for wire, cutlery blades, and light sections in general, but was definitely true for large parts that were heat treated. Out of this situation came the concept of hardenability, as another valuable contribution of metallurgical research to the rational selection and treatment of steel.

The TTT curves of Davenport and Bain showed that the addition of an alloying element to a carbon steel retarded the transformation of austenite to pearlite and bainite. That defined the significant advantage of the alloy steel — it would harden to martensite on a slower quenching rate, or in larger sections and to a greater depth, as compared to carbon steel.

Once this concept of hardenability was recognized, the next step was to make it useful, or to measure it quantitatively. I first learned of this from Marc Grossman, able director of research of the Illinois Steel Co. When I was in his office in Chicago one day, discussing some of the characteristics of notched-bar testing, we discovered that we were both going to Pittsburgh by day train. In the club car, we had a long discussion of hardenability, uninterrupted by the affairs of our jobs. Marc's method consisted of quenching a round bar of steel in water, sectioning it at a location that was unaffected by end effects, and making a hardness traverse across the section. A plot of hardness versus the distance from the surface then characterized the hardenability of the steel. To compare steels, Marc determined the largest bar that quenched to 50% martensite at the center. He called it the "critical diameter." Imasmuch as different quenching conditions produce different quenching rates, his "ideal critical diameter" was the one achieved with a perfect quench.

Considerable elaboration on these ideas and the testing of many steels were needed to develop a workable system for steels of different compositions and different quenching velocities. As the work of Bob Mehl had shown, the grain size of the steel as it is ready for quenching also influences the hardenability, and it, too, had to be included in the system. The end result was a table of numerical factors for each alloying element, carbon content, and grain size, by means of which the ideal critical diameter can be calculated. Finally, the actual hardening effect was determined for different quenching methods or quenching rates.

At the General Motors research laboratory, Al Boegehold was at work, in a similar way, on automotive steels. Whereas more expensive alloy steels had been used for many important parts, he substituted simpler and cheaper steels. However, for practical evaluation of different steels, a means for determining the hardenability, or depth of hardening, was needed. That came out, in the paper of Jominy and Boegehold, as the Jominy End Quench Test. Some modifications and other hardenability tests were developed, and were useful, but the test that bears Walter's name has become most widely used.

In the meantime, Bob Archer showed that the maximum as-quenched hardness depends primarily on the carbon content of steels. The hardness increment added by the alloying elements in solid solution is masked by the dominating effect of carbon. Accordingly, the tensile strength depends primarily on the carbon content and less on the alloy content, assuming adequate quenching to at least 90% martensite. It didn't take long to determine that, as a general rule, the as-quenched hardness should be about 5 HRC higher than the as-tempered hardness. This determines the carbon content, while the desired depth of hardening is determined mainly by the alloy content and grain size for a specific quenching condition.

As a simple, summarizing statement on hardenability, I should like to quote from the booklet on alloy steels by the Bethlehem Steel Co.: "Surface hardness obtainable after quenching is largely a function of the carbon content of the steel. Depth hardness, on the other hand, is the result of alloying elements and grain size, in addition to the carbon present in the steel."

I have just described one of our most significant and fruitful metallurgical generalizations. With this basic information, we can now engineer steel selection and heat treatment on a scientific (that is, a quantitative) basis. Such a rational procedure eliminates most of the empiricism of former times by providing a simplified and direct approach.

This development, stated here very briefly and simply, affected the heat treating grades of steel in the automotive and machinery industries, and in ordnance. This was reflected in changes in the SAE-AISI standard steels during World War II. One new classification was the National Emergency steels that were developed to simplify compositions and to conserve strategic alloys on the rational basis of hardenability. Another new classification was the H (for hardenability) steels, which made it possible to procure steels according to hardenability, when desired, instead of chemical limits, as was done before. Jominy values were determined for a large number of steels, and the results were reproduced for each steel as hardenability bands. The upper and lower limits of the band gave the expected spread for each steel type. The concept of buying hardenability instead of composition was sound, but the actual depth of hardening varied over too wide a range for many purchases. We must assume that a wide spread is realistic if the supplier is to avoid too many lost or diverted heats.

Another standard series of steels came from the development of Grainal by Jerome Strauss, vice-president of the Vanadium Corp. Its addition increased the hardenability of a steel, although its action was a mystery. The active ingredient was finally found to be boron, out of which came the B steels. They form a new SAE-AISI series of low-alloy steels that are treated with a very small boron addition, as a relatively cheap way to increase the hardenability.

I have mentioned Dr. Cain's work, at the National Bureau of Standards, on the determination of oxygen in steel. An improvement on Ledebur's hydrogen reduction of the oxide was described in the early 1930s by H. C. Vacher and Louis Jordon, also at the National Bureau of Standards. They fused the sample in vacuum in a graphite crucible for carbon reduction. It created considerable interest, because studies of the physical chemistry of steelmaking required reliable data on oxygen. To answer the question of reproducibility of the results, J. G. Thompson initiated a cooperative program in which several laboratories were issued standard samples for analysis and report.

We were not invited to join this program, which I held to be wrong because ours was the only laboratory at which the fractional method was being used. I wished to compare our results with good values for total oxygen, but I also wished to check for fractional determinations of oxygen and nitrogen against the various steel types of the standard samples.

After we had completed the necessary preliminaries, we obtained one each of the standard samples. Each laboratory analyzed its samples and sent the results to Thompson and had, of course, no knoweldge of the results of the bureau, or of the other laboratories.

All the results were published by the bureau, although Merrill and I published a separate and more extended report in which we compared our results with the combined best values of the others. These papers were published by AIME.

It is a matter of record that our results agreed well with the best values of the total oxygen and were the only ones that reported the combinations in which oxygen and nitrogen occurred. We had also examined the samples under polarized light, confirming the fractional results insofar as that was feasible.

The work that Louis, Merrill, and I did with fractional vacuum fusion demonstrated its ability to characterize steel and weld metal, with emphasis on "characterize." We used that term in those days to distinguish between different types, such as rimmed steel with high oxygen, non-aging steel with the nitrogen fixed by aluminum, etc.

Our mention of nitrides of silicon and aluminum, from the skepticism with which it was met in those days, must have been dismissed as being too far out of line to warrant serious consideration. Before being so bold, however, we had checked data on nitrides, so ours was not just speculation. Furthermore, within twenty years, silicon nitride in steel was being investigated by modern technique, and must now be accepted as a true ingredient of steel.

The situation with aluminum nitride is similar. We found that the nitrogen of aluminum-killed or fine-grained steels did not appear until the fourth or highest temperature fraction, so we held it to be associated with aluminum. I referred to it as a "nitrogen fixer," as well as a deoxidizer. (I believe that is a better term than "denitrogenizer," as aluminum has been called recently.) My evaluation of that role of aluminum was in accord with metallurgical experience of the time, but it was not until later that its dual role was generally recognized. We simply gave reasonable interpretations of our data. Considerable information is given on the nitrides of both silicon and aluminum in the *Journal of the Iron and Steel Institute* of March, 1967, and on aluminum nitride in the *Journal of Metals* of May, 1959.

Louis Reeve's was a substantial achievement, yet I have seen results on a wider range of steel analysis for which the fractional method returned inconsistent results. It is recognized that some elements interfere in vacuum fusion, which might be responsible for erratic results. For whatever reason, the method is obviously limited in application. However, that is not for the reason given by one critic of our paper. He said forthrightly, "It can't be done," when he was looking at results showing that it *had* been done. There is not much that can be done in that situation. For those steels and weld metals for which it is reliable, its results, combined with those of chemical analysis and various metallurgical tests, give a practical and very useful characterization of a steel or weld metal.

In May, 1939, the time had come for me to leave A. O. Smith. I had had an interesting, exciting, educational, and broadening experience — which, I am convinced, was shared by my entire staff. We were all receptive to challenge and willing to work hard to meet it — the spirit of LR and Rudy, whose examples inspired the entire organization. As the metallurgist with laboratory facilities, I was able to coordinate my function with those of procurement, engineering, production, and sales. Only by close liaison can the laboratory support and guide those divisions, and thereby acquire a responsible role in company affairs. We had accomplished that position while I was there, and I said adieu with real regrets.

I performed a simple ceremony when I left my office for the last time. My close association with Merrill for nearly ten years apprised me of his attitudes and capabilities, and of his grasp of the metallurgical function. That made him, in my opinion, my logical successor as head of the department. I took it upon myself to put him there. So, while my car awaited me outside for my departure, I called him, and met him at one entrance to my office. I had previously bade a fond farewell to the staff, and now I told Merrill: "As I leave by that door, you come in by this one and take over at my desk." It was that simple, and Rudy approved. As one door closed on me, another door opened for Merrill, giving him the opportunity to grow and develop into a Man of Metals.

Chapter 6

Battelle Memorial Institute, 1939-59

Technical advisor until 1953; establishment of the welding department; metallurgical and welding research; work on brittle ships and many other projects.

ONE bright sunny day in May of 1939, I packed my personal effects in a new Pontiac and headed for Columbus, Ohio, and Battelle Memorial Institute, where I was to be a technical advisor on the staff. I sincerely regretted leaving the exciting and congenial atmosphere at A. O. Smith, and parting company with my competent staff, yet I knew Battelle well enough to feel that it would be an ideal location for me to top off my career. That was selfish, in one way, but I also felt assured that the breadth of my early experience would make me useful in my new capacity. There was a strong pull toward Battelle.

Battelle had been in operation for ten years, and was well known as a private and independent institute that engaged primarily in research for industry. I need not relate here its founding by Gordon Battelle, with the assistance of his mother, because that has been told repeatedly. Nor do I need to trace its growth to a multimillion-dollar international enterprise. I will be content to say simply that Gordon Battelle's stated objective of providing the facilities for creative research has been amply fulfilled, as has his faith in the role of research for industry been justified. Battelle has also had its quota of Men of Metals, as we shall see.

"Gil." How much that name means to those who were fortunate enough to know him! I first met Dr. Horace W. Gillett during World War I, when I was on an assignment from the U. S. Bureau of Mines to go over some work he was doing at the Cornell Station in Ithaca, New York. He met me at the railroad station in his Dodge — moreover, a dilapidated Dodge. As we started up the hill to the campus, it made a horrendous noise, so rasping that I thought it must have come back from a tour of duty in the mud of Flanders. He was testing some bearings and completely ignored the racket, but I had to wonder what the end point of his test might be. I soon learned a lesson — don't judge a man by the car he drives.

Somewhat later, Gil switched his allegiance and became chief of the Metallurgical Division of the U. S. Bureau of Standards in Washington, D. C. He had brought together a metallurgical advisory board that met with him periodically. One time, when I was at lunch with Zay Jeffries and Gil, he told us he was leaving to become director of Battelle Memorial Institute. The trustees of Gordon Battelle's will were strictly laymen in research and science, yet by some remarkable intuition (or with good advice) they had made a wise selection. Just as Dr. Whitney had formed the General Electric Research Laboratory as a reflection of himself, Dr. Gillett placed his own indelible imprint on Battelle. The sponsor's interests were paramount; the work must be soundly done and the results must be reliable, and Battelle must operate on a truly ethical plane. Also, the sponsor has the rights to any patents resulting from his projects. It is safe to say that other institutes of this type operate in much the same way that Gil developed for Battelle.

There was a serious and constructive side to Gil, and also a lighter side. His labors at, and for, Battelle were prodigious, because of his association with numerous projects, and his supervision of all reports and technical papers that were issued. He was generous with advice and assistance to staff members, especially the younger ones, with frank and down-to-earth criticism. His phenomenal ability to read rapidly and digest thoroughly, along with his encyclopedic knowledge, made him uniquely fitted for these functions.

He could be sharp, too, in comments on mere verbiage or irrelevant passages, and especially on tone. He once told a Battelle author who was refuting certain adverse comments of a critic, that he couldn't just say the man was stupid and a liar and expect it to be convincing. He should let the facts and his arguments state his case.

Gil showed his human side in a lighter vein. One time, he was in his car at the exit of the parking lot, waiting to enter traffic on King Avenue. I was talking to him while he waited. Soon, he saw a clear space and drove out into his lane. At that moment, a worker came speeding down the same lane and could not avoid a collision. Gil had the right of way and I told him that the other car was coming much too fast. Gil ignored the implication of my comment and handed the man a twenty-dollar bill, asking if that would cover the repair of the body. With a surprised look the man, obviously a poor workingman, accepted the offer and drove on.

Admittedly, Gil would never be one of the country's best-dressed men. He was not prepossessing in appearance, and usually wore a loose, baggy sweater. They tell of the time he entered an office in which a new stenographer was at work. Upon seeing him, she asked him to empty her wastebasket, and he did.

Speaking of mistaken identity, Dorothy Sachs told of an incident that left her flabbergasted. She had charge of our cafeteria, and one day, shortly after she took over, a man with English setters came to the door of the kitchen and asked for any bones she might have. After a quick look, she said she would get some for him. As she was wrapping them up, she thought to herself that the poor man needed something more than bones. So she included some sandwiches and fruit in the package. Dorothy had heard of the famous Dr. Gillett and wished to know who he was. A couple of days later she asked to have him pointed out. She almost lost her composure when she realized that Gil was the man who had come to the door for bones. We never did learn what Gil thought of the sandwiches.

Gil's influence spread far beyond the confines of Battelle. He was acclaimed both at home and abroad, and for many years was editor-in-chief of the excellent *Metals and Alloys*. Through that medium he projected his personality to his many readers via editorials and comments. In the issue in which Gil changed from plural to singular, he sang his swan song. In a eulogy of his life and contributions, a generous sample of pithy quotes was reproduced. We can all read them with pleasure and profit.

I could go on and on, for Gil's life was full of large achievements and small incidents. His acquisition by Battelle for the brick-and-mortar period was a most happy one. I have often wondered by what technique it was that the trustees were so unerringly guided to the right man.

Of course, Battelle needed a staff. One of Gil's first additions was a man who played the dominant role in the future development of Battelle, and who took the lead in inducing industry to accept science. Dr. Bertram D. Thomas, then president, paid a worthy tribute to his predecessor in his 1963 address to the Newcomen Society, "The Legacy of Science — The Story of Battelle Memorial Institute." I should like to quote from it:

> Among the original staff members was one who had identified himself with the primary objectives of Battelle more completely and more fervently than any other. He probably saw more clearly than anyone the challenge offered by a new approach to the problem of getting the results of science into practical use. He saw that the

> principal obstacle standing in the way of revolutionary change in the world of technology was the attitude of industry toward science; an attitude compounded of the dead weight of the depression (in the early 1930s), and a lack of understanding, and sometimes of indifference and ignorance of what science could do. He set for himself the task of overcoming this obstacle. Like all men of genius, seized with an idea, he probably would have worked it out regardless of circumstance. But Battelle Institute was destined to become a uniquely effective instrument in his hands to carry out a mission that was almost a crusade.
>
> The man was Clyde Williams, who more than any other, built the organization that we identify today as Battelle Memorial Institute.

Then Bert continued a little later:

> It is difficult to categorize Williams's principal contribution to the building of Battelle and to separate it from what was in fact the building of an entirely new concept of the not-for-profit research institute.

Clyde joined Battelle in 1930, as assistant director. That gave him the opportunity to view its operations broadly, and to exert his influence on its development. In 1934, Gil decided to step down to chief technical advisor and recommended to the trustees that they appoint Clyde as director. As the institute grew, and later became international with several geographically distributed laboratories, Clyde became president. He retired in 1957, to establish his own consulting business, and Bert Thomas, his longtime assistant in management, became president.

Clyde was a businessman with a vision. He knew that research must show a profit if sponsors were to be satisfied, but also that Battelle had to have business to grow. Clyde's genius lay in combining those tenets to the satisfaction of sponsors and of the institute, yet by following Gil's guidelines in ethical conduct. Needless to say, Bert has followed loyally in the footsteps of his predecessors.

I have not yet mentioned one ingredient of success that all institutions must have in their staffs, if they are to grow and remain competitive — the will to work. In this connection, Battelle had its old pros whose dedication to their jobs inspired others. By old pro, I have in mind something more than competence and experience. There is also the connotation of ability to take the bumps and hardships that seem to infest the lives of most of us, and yet to keep going coolly and constructively. One man who personified this attribute in his work was Dr. Oscar E. Harder.

Oscar resigned his position as my successor at the University of Minnesota to become assistant director of Battelle, and to continue his career as a well-known and outstanding metallurgist. One usually thinks of a professional man in terms of his achievements, but Oscar is also remembered by friends and associates as a human being of good will, with a fine sense of humor and a friendly mien. He always had a good story or anecdote to emphasize a point, either in his office, when discussing a research project, or at lunch with friends. One of his distinctions probably made him unique — he had a Ph.D., yet did not graduate from high school.

I could list a number of his impressive achievements, such as a low-alloy steel, leaded steel for free machining, valve steels for the Ethyl Corp., and heat-resistant castings for the Alloy Casting Institute, but the one which called for tenacity in pursuing a goal, in spite of great difficulties and frequent disappointments, was his work on Elgiloy for Elgin Watch. They wanted a new and "perfect" alloy for watch springs, and Oscar gave it to them, making a major advance in watch technology. It was a very difficult job well done. I gave it a nickname, Oscarloy.

At Battelle we were alert to the possibilities of developing sponsorships, and we spent time in preparing a program to submit to a company, or group of companies. Of course, a proposal could be turned down, but I thought that a "no" placed considerable responsibility on the one who said it. Unless he had a valid reason, he would be ignoring an opportunity to benefit his company or industry. My first major attempt received "no" for an answer.

At A. O. Smith, I had had considerable experience with stainless steel. Although the industry was doing as well by its customers and users as could be expected in view of the state of the art, there was still much to be done to improve the situation for all concerned. I felt that a group research effort by the producers of stainless steel would increase and broaden its acceptance. Vsevolod Krivobok was well aware of the desirability of the program and helped in developing it, but the producers were not ready to cooperate. It must have been the top executives who vetoed the idea, because all of us who worked daily with the product were familiar with the problems that should be attacked by the research method.

I had become well aware of the superior properties of the high-molybdenum stainless steel, type 317, and I was especially interested in the stabilized version that had been demonstrated first by Russell Franks of Union Carbide. The few attempts to produce it commercially had failed to reproduce the laboratory results, and industry was denied the use of a superior material. Actually, the low-molybdenum type 316, with less corrosion resistance, became the popular molybdenum grade. I am satisfied that we could not have met the requirements of those Bahrain vessels with the type 316. Similarly, the titanium type 321 lost out to type 347, in spite of the lower cost of 321. The reason for the inferior corrosion resistance of type 321 was the difficulty in handling titanium in steelmaking. Our work on Smith Alloy No. 10 had taught us how to alloy a substantial amount of titanium without detracting from the quality. That would have led to another good investigation, especially because Chuck Greenidge, our man at A. O. Smith who had made Alloy No. 10, had joined Battelle, and would have been available to develop a reliable procedure for making type 321 stainless steel.

During World War II, Krivobok was chief metallurgist for Lockheed in Burbank, and after the war he joined Inco as their specialist on stainless steel. He was always a lively member of a group, or of society sessions, with humorous yet pithy comments. He took a leading role in investigations of welding type 347 stainless steel in thick sections to overcome weld cracking. He was a popular member of our fraternity.

One time, I told Krivy about my experience with a Communist commission when I engaged one member in conversation in German. The man seemed quite pleased to talk directly to me, without going through an interpreter, but he was cut short by the leader. Krivy agreed that these commissions were under strict discipline, and illustrated it with an experience of his own.

A former student of his at Carnegie Tech was a Russian who returned here after the war as head of a commission. They were good friends, and Krivy invited him for a social evening. After a couple of hours of reminiscing, his friend said, "Now I should like to have you meet the members of my mission." At that, he backed up to the door and, without glancing back, opened it and said, "Here they are." They had been silently awaiting that moment, ready to enter.

Later, I was chairman of the High Alloys Committee of the Welding Research Council that made possible some research on stainless steel. I could not work the all-important melting practice into the research, because our budget would not stretch that far. After serving a number of years, I decided it was time to turn the job over to someone else. My final meeting was at Niagara Falls, where I introduced as a special feature an invited talk by Russell Franks. I thought he might tell us how to stabilize type 317, but instead he surprised us with the benefits of extra low carbon in austenitic grades. Out of this came the new grades 304L and 316L, with carbon limited to 0.03% max, which avoid heat-affected-zone sensitivity or intergranular attack.

I must also mention one cooperative program on corrosion, and another on stress-corrosion cracking, with Merrill Scheil in charge of the latter. W. O. ("Bill") Binder, of Union Carbide's research laboratory at Niagara Falls, took over as chairman.

I had much better luck with another proposal. Came the war, and, with it, a great requirement for armor plate. It happens that I knew something about armor plate, as well as about the relatively meager funds that had been available to Army Ordnance for research and development. So I looked into the situation in Washington, and especially at Watertown Arsenal. The research laboratory came under Col. H. (for "Heine") Zornig, with whom I became well acquainted in later years, and then under Col. S. B. Ritchie. Both recognized the possibilities of a simplified steel composition for armor. Clyde always encouraged the staff to explore new possibilities, and I had no difficulty in securing his approval of what I had in mind.

Upon reviewing armor-plate analyses at Watertown Arsenal, I noted a great variety. I was told that each supplier had his pet analysis, but there was no rhyme or reason for the variability when they were viewed as a whole. Although Battelle had several ideas for improving armor plate, my thought was the use of a carbon-manganese steel made to fine-grain practice, with Grainal (the forerunner of boron) and sufficient molybdenum added to secure adequate depth of hardening as the section thickness increased. The advantages of this steel were that it could be made in the open-hearth furnace, and hot rolled on a production mill. The opportunity to save strategic alloys, as well as the possibility of setting up armor-plate manufacture on a true production basis, appealed to Army Ordnance. Armor plate was a specialty item at the time.

That concept came to me quite easily, because I had worked with carbon-manganese and manganese-molybdenum steels at A. O. Smith. I knew that they were good steels, when made right. We made the experimental steels at Battelle and, much to my delight, the first test samples that I took to Watertown Arsenal proved my point. Lt. Matthews conducted the ballistic tests on the local range, and expressed surprise that the simple steel performed so well. It now appeared that exotic analyses were not needed to meet Army Ordnance requirements.

This work went on for a little over a year, during which I tested various analyses and heat treatments. As a consequence, the Ordnance Corps advised suppliers that they could use the simpler analyses, provided, of course, that shipments were able to meet ballistic tests at Aberdeen Proving Grounds. I was able to suggest a couple of major steel companies for sources of supply. As I recall, one of them became a principal supplier of armor plate.

One day, representatives of the Otis Steel Co. of Cleveland called to see Dr. Gillett about developing a high-strength low-alloy (HSLA) steel for them. Gil asked me to join in the discussion. Most of the major steel companies had HSLA steels on the market, and Battelle had developed Hi-Steel for Inland. They all had better corrosion resistance and greater strength (50,000-psi yield strength versus 32,000 for ordinary sheet steel) than conventional mild steel has, and sold for 50/32 times as much. The customer saved weight and fabrication costs, which made a good bargain for everyone concerned. These steels used alloying additions of nickel, chromium, copper, etc., and were made to fine-grain practice. Most of them used a high percentage of phosphorus, to gain the dual advantage of increased resistance to atmospheric corrosion and increased strength.

Gil and I were both keen to tackle this project, and it was not long before he thought of trying a good composition that made major use of phosphorus and the less expensive alloys, and contained no strategic elements. The preliminary tests were so encouraging that we were justified in enlarging the scope of the work.

About this time, I took over the program and worked closely with the Otis management and their metallurgical and sales people. I also introduced an idea of my own. I knew from my experience at A. O. Smith that semikilled steel, when made right, is an excellent commercial, or tonnage, steel. It has good properties and a better surface than fine-grain steels, and it was cheaper — a great advantage for the sheet and strip market. Fine-grain steel uses aluminum deoxidation, and the lower ingot yield makes it cost more to convert from ingot to finished form. If we could use semikilled practice, though it went counter to what other companies were doing, Otis would have a desirable cost advantage.

Our tests at Battelle showed that the semikilled practice, usually identified by a resulting silicon content of 0.05 to 0.10% gave very satisfactory results, so one day I asked the Otis metallurgist, "Do you people have good melting practice?" "Why, of course," was the expected answer. Then I asked, "Well, how are you on making semikilled steel?" He told me that they make a lot of steel to that practice, and that he would have their open-hearth superintendent discuss it with me. On my next trip to Cleveland, I was able to compare their practice with that of Inland Steel and found it to be very good. We then followed with tests of regular production heats of the new steel and found that they were equal in quality to our laboratory induction-furnace heats. Otis then went into production of their HSLA steel, which they named Otiscoloy. After Jones and Laughlin acquired Otis, the name was changed to a J&L name, one of their Jalloy steels.

Our tests showed that Otiscoloy was not quite as good as one more expensive steel, and not quite as corrosion resistant as another, which was also costlier. We felt, however, that it would be satisfactory for about 90%

of the applications. To satisfy ourselves on that point, we decided to make a market survey.

I had to find someone to do the field work, which prompted Howard Russell, who will appear later, to suggest Hugo Johnson. Hugo had been on odd jobs at Battelle while attending Ohio State University (one of Battelle's excellent provisions) but was rather submerged in the staff. Howard said Hugo deserved an opportunity to grow, and I especially liked his comments on Hugo's patience, and his ability to interview people without making them mad. I asked Howard to send him in. Shortly, Hugo appeared at the door, big and handsome, and we became acquainted at once. He took the job with enthusiasm, and turned in a fine performance for our sponsor. That is where Hugo got his start. He kept on going and became secretary, and ultimately president, of the American Iron Ore Association in Cleveland. He has a fine home and family in a suburb and it is always a pleasure to meet him. Howard was right.

During the war a question was raised — how can we increase the efficiency of steel production and use? At a big meeting of transportation-industry people in the Pentagon, I was asked to present the capability of the HSLA steels. This meeting was presided over by Mr. Budd, president of the Chicago, Burlington and Quincy Railroad, who explained the purpose of the meeting. When my turn came, I explained the various strong points of Otiscoloy, especially its favorable composition and melting practice, and I made the particular point that only thirty-two tons of HSLA steel would have to be made and transported in place of fifty tons of regular steel. After I finished, Earl Smith, a top man in this field and chief metallurgist of Republic Steel, got up and approved all I had said. However, an old railroad man was in charge of the meeting, and he closed it by saying that he thought it better to allow railroads to continue their regular practice. So the axe fell, to the confusion of us workers.

Then the axe fell again. This time it was wielded by a metallurgist, Professor George Waterhouse, head of the metallurgical department at MIT, whom we met in Chapter 2 as chief metallurgist at Lackawanna Steel. He was the steel expert for the War Production Board and must have approved Otiscoloy, or, I suppose, I would not have made my presentation. Later, he took another look at our analysis and changed his decision, because he would not approve a high-phosphorus steel. There is no point at this late date in discussing the matter, but to many of us his rejection had no factual basis and seemed to say that either we didn't know our business or we were espousing a false cause. As an aftermath, some years ago a leading producer announced that its high-phosphorus steel had been used for 185,000 freight cars.

For those who are not familiar with the HSLA steels, let me say that the various elements contained in steel are either harmful or beneficial, depending on their amounts, and on the requirements of the particular steel. Even carbon, the principal alloying element and the one of greatest usefulness as an addition to iron, is harmful when too much is present — just as harmful as a so-called impurity element would be if in excess. The same is true of other impurities, and that includes phosphorus. When used correctly, phosphorus is a valuable alloying element; it has a very real advantage over carbon in HSLA steels, because it increases the corrosion resistance as well as strength. This reminds me of Dr. Saklatwalla, who worked hard in introducing Cor-Ten to overcome prejudice (ignorance) against phosphorus. I also recall the specimens of high-phosphorus steel that T. S. Unger showed me in 1915.

I have already told of the inauguration of modern arc welding at A. O. Smith. When I arrived at Battelle, I found no welding department. Clyde took but little persuading to agree that one should be formed, and we sent out a help-wanted call. Battelle was working on a project for the Welding Research Council, which I was able to use to get the department started.

In reply to our advertisement, we secured C. B. ("Connie") Voldrich to be our chief welding engineer. He had been a friend of mine at A. O. Smith, and had had several years of good experience as a welding engineer with the Newport News Ship Building Co. More important — he was a keen young man, with a yen for wondering how and why things worked, or didn't work.

Subsequently, we acquired Bob Williams, who had been teaching welding at the University of Illinois; his father was in the metallurgical department at MIT, and later was to become head of it. Bob is a Bostonian, but loves to tell stories reflecting the spirit of proper Bostonians. One I recall is about two old-maid aunts who were entertaining a couple of young nieces from the Middle West. They were on a vacation trip that included Boston, and they asked their aunts where they were going for their vacation. One of them replied, "Oh! We're already here."

Connie also acquired Perry Rieppel, who came from the aircraft industry and has been making us look good ever since. Perry was strong on both theory and practice, and with excellent personal characteristics. He was a fine addition to the department. His later record, with increased responsibilities and broader opportunities, shows that we made a wise choice. These three have gone on to other duties on the Battelle staff, and younger men have taken on the direction of the work on metal joining.

The work for the Welding Research Council was on the effects of carbon and manganese on welding plate steel. The results were published by the

American Welding Society, and they demonstrated that suppliers should not rely too heavily on carbon for strength nor skimp too much on manganese, or too heavy a burden would be placed on welding.

This project introduced us to the question of what weldability is, and to the problem of how to determine it. The diversity of thinking on this important characteristic of steel is amazing, as is the idea of each one who works on it, here and abroad, that his own conception is the correct one. Ernie Thum, late editor of *Metal Progress,* asked me to write a critical article on it, but one would have to write a book to do it justice. This is not the place, either. Instead, I will simply give a short account of one or two projects we had at Battelle.

One project identified cracking of the heat-affected zone (HAZ) adjacent to the weld as the critically important criterion of the weldability of the steel, as distinguished from characteristics of the deposited weld metal. We had reports from the West Coast that cracks were being found in engine mounts for airplanes, but our project got its initial impetus from two incidents that happened closer to home.

In one instance, a quarter of a million dollars worth of engine mounts had been rejected when cracks were discovered at the welds. The way in which they were found is interesting, and pertinent to the question of weldability. The chief metallurgist at the fabricator's plant usually walked along the shipping platform on his way to his office. One morning, he noticed cracks at the welded joints and rejected the entire lot. The day before, they had all passed inspection! This was a clear case of delayed fracture, and must be taken into account in any concept of the mechanism of HAZ cracking — or "underbead cracking," as we called it.

We secured our own concept of this situation by accident. We were working on a welding problem of engine-mount joints for J. B. Johnson of Wright Field. J. B. was top man in his field of aircraft metallurgy, and he had an enlightened attitude toward research. After Connie had prepared test samples, we took them over to Dayton for J. B. to examine. They were as beautiful as salesmen's samples. The next morning, Connie phoned me to come over to his office. The specimens were spread out on a table, and all showed cracks at the welds. "What happened?" I asked. "They cracked," he replied, truthfully. Fifteen hours earlier, they had been sound, at least to the eye. This was another case of delayed underbead cracking, and we could not ignore the evidence before our eyes.

That settled it, and I decided it was time to do something about correcting this situation. The importance of producing sound aircraft was clear, but I felt that I should secure the advice of an outside authority. I quickly settled on Earl Smith. Earl was not easy to contact, and finally I had to go to

Chicago; I enjoyed an excellent dinner with him at the Palmer House while I explained my ideas. Earl approved the objectives and the need of a project to determine the cause and cure of delayed cracking.

This failure of engine mounts had become so serious that I was able to secure a government-sponsored project to study it. We were to study the aircraft-quality 4130 steel welded with E6010 electrodes. That electrode is a modification of the original cellulose-covered rod, the pertinent feature of which was the high hydrogen content of its arc. Art Williamson of Superior Tube was project chairman, and Bob Archer was on the committee. Bob had told me previously that he suspected hydrogen as the cause of underbead cracking. That was at the time of profuse speculation about, and little knowledge of, the true role of hydrogen.

We had two lots of this 4130 steel and, by good fortune, one cracked badly and the other did not. That gave us good material for our program. Both had passed military inspection, both were closely of the same analysis, both had the same hardenability by the Jominy end-quench test, and, superficially, both had the conventional structure. Alike as two peas in a pod, why did one crack and the other not? In any event, this refuted the statement of a welding metallurgist that weldability is the opposite of hardenability — although, superficially, there are reasons for thinking that it might be. He had simply ignored the difference between the time-temperature cycles of welding and the prolonged heating for hardening.

Clarence Sims, Maurice Banta, and I worked on this project and published our findings in a paper presented in 1945 before the AIME. We identified two accompanying steps, or necessary conditions, of the mechanism of underbead cracking: the slow, room-temperature transformation into hard martensite of retained, or residual, austenite of the HAZ; and the precipitation of atomic hydrogen from the austenite, and its conversion into molecular hydrogen as martensite is formed. The latter form of hydrogen is insoluble, and builds up excessively high local pressures (Sims and Zapffe). Under the influence of the shrinkage stress, a crack is initiated and then grows to produce the underbead crack. We showed that if either retained austenite or hydrogen is eliminated, the mechanism was inoperative and cracks failed to form.

This project included the development of an underbead-cracking test that we described in our paper. It gives a good estimate of the relative sensitivity of the steel. In one form or another, it has been used at Battelle and by others. But this leaves unanswered the question of why one lot cracked when the other did not, although they appeared to be equivalents and were welded in the same way. There is a good lesson here that I had already learned at Nela Park, in 1920. One day, a lamp man there was expounding his theory to account for scrapping 3% of the lamps made at one of the plants. An old-timer waited until he was through, and then said, "Your theory sounds good except for one thing. How do you account for the other 97% being good? According to your theory they would have been scrapped too." Those who would account for underbead cracking by hardness, or hardenability, or martensite, or shinkage stresses alone, appear not to have been aware of this point. To conclude, we found that differences in microstructure were responsible for the difference. Then we followed up by determining the melting deoxidation and processing details for making weldable 4130 aircraft steel. This practice became standard for handling 4130 steel for engine mounts. The critical steps were cold work, followed by a subcritical anneal to produce alloy carbides rather than iron carbide. In our paper we also showed a quantitative relation between cracking and the extent of the transformation of austenite to martensite at elevated temperatures. Since then, surface hardening by rapid induction heating has been found to follow the same rules that we laid down for welding. A crucial point is the fast heating and cooling of the steel.

Some people thought we were overemphasizing retained austenite and hydrogen. They pointed out that in the East, engine mounts were made of the same 4130 steel by gas welding, yet without cracking. Also, that steels of even greater hardenability were welded by the atomic-hydrogen torch, which should be much more dangerous. We were able to refute this by pointing out that both methods were slower than arc welding, and that the HAZ cooled slowly enough to permit austenite to transform at a temperature at which the release of hydrogen did no harm. Actually, the retarded cooling of the HAZ duplicated Hultgren's method of avoiding flakes in steel rails and forgings.

Our work was followed up by Perry Rieppel and Manley Mallett. Their excellent experimental program showed clearly that hydrogen and retained austenite were truly the culprits; their paper in the Research Supplement of the *Welding Journal* should be read for details.

The way in which we became aware of prior structure as a dominant feature of our investigation may be of interest, because that aspect of weldability is ignored. Ordinary critical-point curves gave no essential difference between our two steels. Then we realized that heating, holding, and cooling occur at much faster rates during welding. When we repeated critical-point curves at an accelerated rate, we found that the crack-sensitive steel exhibited a large difference (hysteresis) between heating and cooling. The crack-insensitive steel, on the other hand, showed little hysteresis.

That was our clue. Obviously, the "bad" steel had more carbon and alloy in solution at the peak temperature on heating, which depressed the critical point on cooling. This difference is missed by the Jominy test and by ordinary heat treatment, because there is sufficient time at the high temperature for all of the carbon and alloy to dissolve prior to quenching. Following that clue, we then developed the standard practice for 4130 steel, which, incidentally, we learned had been used by the Selby Seamless Tube Co.

R. G. Baker, in the Research Supplement for July 1968, comments on hydrogen cracking, particularly that "different results could be obtained by different laboratories and fabricators, although their procedures might nominally be the same." That was precisely our experience and dilemma, but he, and apparently others, failed to take prior structure into account, as we learned we had to do. Actually, it was the secret of our success in establishing sound practice for engine mounts.

As an addendum, let me say we did not answer all the questions. Our job was to get the solution to a pressing war problem. Also, we now know that these hardenable steels are welded with low-hydrogen electrodes without cracking. Credit is also due to Clarence Sims and Carl Zapffe, who were adding substantially to our knowledge of the behavior of hydrogen in steel.

I have said nothing about "toe cracks," which form at the junction of the edge of the outer pass and the parent metal. That is another problem, and I know of no research in depth that illuminates that process as we did the one of underbead cracking.

I did have an experience that set me back a bit. It shows what one may be up against at times. Herbert French was a good friend of mine, and I was telling him, with some pride, how we had solved the knotty problem of underbead cracking. When I got as far as mentioning the slow transformation of austenite at room temperature, he stopped me with: "Your theory is wrong. Retained austenite does not transform that way. That happens only on a drop in temperature." That finished my discourse, but it so happened that Professor M. Cohen and collaborators, at MIT, had just published their ASM paper on the isothermal room-temperature transformation of retained austenite. It was a timely confirmation, and I had to wonder what Herbert thought of it. I also reflected on the reluctance we have in accepting something new, although it is good that it is that way in many, many instances.

A sad example of the influence of this human trait came from experience with the failure of many welded ships during World War II. For several decades we had had cases of catastrophic failure by brittle fracture, plus disclosures in the literature of this treacherous behavior of steel. Yet engineers and metallurgists blithely ignored the danger sign, and the lesson had to be learned the hard way. ASTM issued specifications for ship, bridge, and other structural steels, but they contained no provision for guarding against catastrophic brittle fracture. Engineers simply specified a steel by number and, in their ignorance, felt secure with tensile ductility and their factor of safety or design stress. Another feature of the lesson was the low stress that produces the brittle fracture, or catastrophic failure. It can be well below the design stress, as was discovered by tests of our welded merchant ships, and was later confirmed by laboratory tests.

On a cold winter morning in January, 1943, a tremendous explosion-like boom reverberated in Portland, Oregon. At first, the citizens feared it was a Japanese bombing attack. The real case was ultimately located — the welded tanker *Schenectady* had broken in two while moored at its pier. That was an effective eye-opener to the danger of brittle fracture of a steel that does not measure up to its application. May I reiterate that this, as well as considerable additional experience, shows that procurement is a first-order function. The present situation is considerably improved; and we at Battelle helped to bring it about.

The *Schenectady* was built by Kaiser, which brought Henry Kaiser immediately into action. The following account was told to me by Professor S. H. Graff of Oregon State University. Henry Kaiser, searching for a metallurgist to investigate the failure, telephoned my namesake, the editor of the *Portland Oregonian.* Mr. Hoyt, in turn, telephoned the president of Oregon State and was referred to Professor Graff. In a rather roundabout way, they found a good man.

A short time later, a commission arrived from the East with Dave Arnott, chief surveyor of the American Bureau of Shipping, in charge. They were startled to find Professor Graff there cutting out samples, against strict orders. They wanted to know who he was, and what he though he was doing. Oddly enough, the commission did not include a metallurgist. After explanations, Graff joined forces with it. One would have thought that with more than one hundred feet of steel ripped apart there must be some metallurgical aspect of the failure that should be considered.

Not long thereafter, I had to make a trip to the aircraft companies in California, and asked to be routed via the Pacific Northwest. Something told me that big things were ahead. Professor Graff welcomed me like a buddy and gave me the results of his investigation, including notched-bar tests for notch brittleness. The fracture of the *Schenectady* started at a notch, where the deckhouse joined the shear strake at right angles. From there the fracture proceeded transversely around the ship and made the "report that was heard around the world," as one man put it. The steel was notch brittle, at the air temperature at the time. I thought of what Sherlock Holmes would say, "'Tis very simple, Watson. There is your diagnosis."

But it was not that simple! Apparently nobody went along with that explanation. At least a couple of years or more went by before notch brittleness of the steel was taken seriously. For one thing, an old gentleman got up in least two meetings and placed the blame on locked-up stresses, "as in a banjo." He was talking the engineer's language, but had not learned the lesson of accounting for the lack of failure of many other welded vessels under equally severe service conditions and with the same locked-up stresses. I believe that a more valid reason for ignoring Graff's results, which were not published, is to be found in the disbelief in notch-brittle behavior of steel on the part of the engineering profession. That point comes up again. Furthermore, the board of investigation, which was established by the Secretary of the Navy, had this to say about those locked-up stresses, after the dust had settled. Studies had shown that they were as high as 47,000 psi, but did not contribute materially to the failures. This is from the Final Report on Design and Methods of Construction of Welded Steel Merchant Vessels, in 1946.

Another reason showed up a little later, when I was visiting the shipyard of the Bethlehem Steel Co., in Quincy, Massachusetts. Paul Ffield was showing me around, and one time he called in their marine architect, or designer. The conversation turned to steel and "notched-bar impact tests." The designer, in discussing the *Schenectady* failure, said: "Those impact [*sic*] tests of Graff had nothing to do with it, because the ship was moored at its pier at the time. There was no impact involved in the failure." I was dumbfounded, and could only think again of the evil of calling a a notched-bar test an impact test. I was also reminded of the time in 1940 when I inquired at the Navy Department what they were doing with notched-bar tests. I was told that the Navy had surveyed their problems and had found no application for the test, and so the matter had been dropped. Dropped, that is, until the *Schenectady* fiasco occurred, and the tests of its ship plate by our National Bureau of Standards forced all concerned to accept the reality of notch-brittle behavior of ductile steel. This is not to single out the Navy, but simply to cite one example of the nearly universal rejection of notch brittleness. I ran across numerous instances of that attitude.

As soon as I returned home from my visit with Graff, I tried to get a project at Battelle to show what was wrong with the bad steel, and what was right with the good steel, of which there were ample supplies. I had used this technique at A. O. Smith to compare steels and weld metals. With our metallurgical talent and facilities, I felt we could make a valuable contribution to the brittle-ship problem. Not long thereafter, I received the blow that nobody believed in a program on notch toughness of ship plate, as Clyde Williams learned when he discussed the program with potential sponsors and others.

The official investigation first followed the lead of locked-up stresses. It consisted of a large-scale, two-year test program on ships at sea, which developed two important points. One was the finding that the maximum nominal or average stress in the ship structure (the strength deck) was about 8,000 to 10,000 psi, or about one-sixth to one-fifth of the ultimate tensile strength. That really sank in. The other was that steel was the principal factor in the brittle-ship problem. At the same time, faulty welding and inadequate supervision and inspection were corrected, and design was improved, to minimize notches and effects at locations of stress concentration.

There is no good reason for me to detail the subsequent events, although from them was developed a sound engineering basis for the fabrication of merchant ships by welding. It is of interest that the dominant role of steel was coming into sharper focus. I had been invited to deliver the Adams Memorial Lecture before the American Welding Society in 1945, which enabled me to discuss the reasons why notch-tough steel should be used. The old A7 of ASTM, which was "just steel," was in the first grips of senility, but it was still too early to attempt an obituary. The steel type that would be "good enough" was yet to be determined, although I could have easily recommended the composition, melting and fabrication practice, and Charpy notched-bar values, for a really good steel. It was not that simple, however, because new specifications would have to be developed to meet the economic as well as the quality requirements, and cognizant organizations would have to be convinced of their need. However, I must have made my point on the significance of steel, because Dave Arnott asked me after the lecture if steel was not responsible for a lot of the trouble. I replied to the effect that those ship fractures ran through many feet of ship plate, and that we should use a steel with sufficient notch toughness to resist a running crack, or brittle fracture. One had but to look at a photograph of a ship that had broken in two as if it were made of cast iron, to realize that steel was the major component of the brittle-ship problem, and that it called for metallurgical treatment. Defects might initiate a crack, but the running fracture was clearly a steel problem!

Giving this lecture enabled me to meet Comfort again, after a lapse of twenty seven years. Although primarily an electrical engineer, he was an advocate of welding, and established the American Welding Society shortly after World War I. This lecture was a tribute to his devotion to the science and practice of welding, and his many contributions to the society. He and Mrs. Adams were familiar figures at AWS meetings, and always occupied seats of honor at Adams Memorial Lectures.

Our National Bureau of Standards ultimately tested a large number of samples from various ships, which represented plates in which the fractures started, through which they propagated, and in which they stopped. Very properly, they adopted the Charpy notched-bar (impact) test to compare the notch toughness of the steel of the three categories. The notch toughness was lowest in the first group, higher in the next, and still higher in those plates that were tough enough to stop the running fracture.

These test results were uncommonly valuable. For years, they gave us the only rational, or at least useful, correlation with the service behavior of structures. I would caution, however, that this correlation was valid only on a statistical basis. Not until more than a decade later did we have a rational system for use in this field.

The War Metallurgy Committee of the National Academy of Science – National Research Council (Clyde Williams, chairman) took an active part in organizing work on the brittle-ship problem. The preliminary study led to the establishment of a committee on ship steel under the auspices of the Academy – Research Council. As originally constituted, this committee failed to accomplish very much except dissension. There were too many members of the "steel is steel" persuasion on board. One knowledgeable member simply resigned, remarking that a machinist of his acquaintance knew more about the effect of notches than the whole group of dissenters did. An Englishman also resigned, saying that an English blacksmith knew more about what a notch could do to steel than the committee did. I, being more accustomed to frustration, stayed on until the committee was dissolved. When reconstituted later, the committee on ship steel was able to function effectively.

Bob Mehl was chairman and Finn Jonassen was the secretary. Finn was originally at the University of California at Berkeley and was an engineer on the ship tests at sea that I have already mentioned. Finn had the virtue of an open mind, and was always ready to listen to what was said. He was an excellent person, and his early passing saddened all who came to know him and to appreciate his worth. Our stated responsibility was setting up research programs on the brittle-fracture problem, and budgeting the available funds. Thus, we were to support studies of the causes of brittle fracture. As one man put it, "The steel is ductile, isn't it? Then how come it breaks brittle like cast iron?" We were to try to answer that.

In the meantime, I was developing an idea that I felt would yield results of immediate practical value. I held that the steel producers would benefit if we could show how to make better steel. Tests from ships, and of production heats of ship plate, showed that the temperature at which the danger of brittle fracture set in varied from below the freezing point to above 100 °F, a demonstration of lack of control of this vitally important steel characteristic. I felt that we should sponsor a project to show the suppliers what manufacturing variables had to be controlled.

The evening before a committee meeting in Pittsburgh, I asked Finn Jonassen and Max Gensamer to come to my hotel room for a preview of my program. As I recall their reaction, it was generally favorable and encouraging. The next day, I proposed a program to make induction-furnace heats in the laboratory, with which we could study composition, deoxidation practice, and rolling practice. We had several representatives of steel companies on our committee, and they wanted to know how the variables could be controlled in a meaningful way by means of little laboratory heats. I countered that melters of big open-hearth heats weren't doing so well themselves. To do what I knew we could do cheaply, in the laboratory, would be very expensive to duplicate with commercial heats. I added that we, at Battelle, had developed laboratory practice that duplicated full-size commercial heats even for semikilled practice. The proposal was accepted, and the Navy agreed to finance it at Battelle.

We had an excellent project committee, whose counsel and guidance were of great benefit. I believe the members were happy to have the opportunity to study the practice in detail, and under close control. One member was Max W. Lightner, who was in charge of applied research at U. S. Steel and later became administrative vice-president of research and technology, succeeding Jim Austin. Another was Tom Washburn, of Inland Steel, an expert on steel, as we learned at A. O. Smith. The David Taylor Model Basin took an active interest in all this work, and was the agency that translated Professor Neuber's book on notch mechanics from German into English. Their Capt. Wendell Rupp was on our committee. He was an active participant in committee deliberations, and expressed himself well and forcefully. He was also good company otherise.

Fran Boulger had direct charge of the work at Battelle. With Clarence Lorig and R. H. Frazier, a series of reports was published, giving a large amount of information on the effects of the manufacturing variables on the notch toughness of ship plate. The study was timely, in that it gave industry and specification-writing authorities the information that they needed.

In the meantime, the American Bureau of Shipping had been busy. Upon being convinced of the inadequacy of the current specification for ship plate, the bureau came out with three grades, two of which were new for ship plate. They were ultimately included in A131 of ASTM. Grade A is just steel, with the requirement for chemical analysis limited to maximum phosphorus and sulfur. Grade B is presumably semikilled steel, although no limitation was provided on silicon to indicate that more positively. The carbon is restricted to a maximum allowable amount for easy welding and good notch toughness. The manganese is given as a range, as it should be. Grade C is a fully killed steel, made with fine-grain practice. Plates of Grade C may be required to be normalized, and must be if the potential quality is to be realized. Incidentally, I have been unable to get a definition of fine-grain practice, although everyone knows what it means. Notched-bar tests, to indicate whether the steel has been made properly, were not required. Grade C is used for strength parts of the ship for provision against brittle fracture.

A few words on brittle fracture may be in order for those who have not followed it. Brittle fractures were noted very early when Bessemer and open-hearth steels were being introduced. Bessemer steel was more sensitive than open-hearth steel, and, as I found when browsing through the *Journal of the Iron and Steel Institute* (British) for 1879, Bessemer angle bars broke from "some trifling blow," especially "during the late severe weather in Chatham." Of more recent date were the failures of the zoo bridge in Berlin and the brittle bridge over the Albert Canal, in Belgium. There is one bridge, however, that has remained whole and serviceable for a long time, although made of Bessemer steel — the famous Brooklyn Bridge. At Midvale Steel, in 1916, I was told that when a freight car was being loaded at the mill, one girder fell and shattered into several pieces. The rest were shipped to the site, and I presume are still serving faithfully. Of course, it is a suspension bridge, and it is riveted.

The Three Rivers Bridge, in Quebec, was not so fortunate, although made of open-hearth steel. One day in the spring of 1950, an engineer from the public works department of the Province of Quebec was in my office at Battelle. He told me that some of the strength girders had cracked and were to be repaired. The main support girders were made of two lengths and were welded at the centers. He wished to discuss brittle fracture. On a vacation trip to Canada that summer, I looked into that interesting situation.

At the plant where the welding was done, I learned that two different steels had been applied, one of which was high in carbon and would be apt to give trouble in welding. It did. Also, I examined their mechanical tests and the results of their metallurgical examinations of the steel and the welds.

On my way to Quebec City, I stopped at Three Rivers and saw the repairs in progress. They were riveting heavy doubler plates, or butt straps, at the span centers. The joints looked strong enough to withstand even a severe earthquake. The accent is on *looked.*

In Quebec City, I met with the chief engineer of the province and his staff to discuss steel and brittle fracture. He finally got around to asking, "Was that good steel in our bridge?" From the information I had gained, I had to reply with doubletalk — yes and no. I had to say that it was good steel because it met his specification. The supplier had fulfilled his obligation. On the other hand, and as a metallurgist, I had to say that it was very poor steel. It would serve satisfactorily in some applications, but it was a poor buy for the bridge.

The very next winter, a disastrous failure happened to that multimillion-dollar bridge. All spans collapsed at their centers, and several people were plunged into the icy water below. It is only fair to say that the Province of Quebec had built a number of welded bridges that had withstood severe winters and normal loads without failure. It is to be hoped that this one failure didn't dampen their ardor for welding. Professor M. E. Shank of MIT has written an account of many non-ship failures, showing that they are really an old, although largely neglected, experience.

We had an unusual project on an unusual type of deterioration of a welded joint. One day, Gil told me he had just heard that welds in the power piping at the Springdale Power Plant of the West Penn Power Co. had developed long stringers of graphite alongside the weld. I was chosen to look into it, and I continued to look for several years.

The pipe was seamless tubing of carbon-molybdenum steel for high strength at 925 °F. I went over to Pittsburgh in a skeptical frame of mind, but they made a believer of me as I went over the evidence. The situation had to be corrected, so the Edison Electric Institute and the Association of Edison Illuminating Companies sponsored a project at Battelle. Our results were published in a series of reports or papers before the ASME.

The final answer was relatively simple and practical. Our tests showed that 0.5% chromium produced a marked stabilizing effect on the carbide of the steel, although 1% was adopted for future deliveries. There was one uncertainty, however, because we found an incubation period before graphite could be detected. After that, graphitization progressed normally. Power plants around the country sampled their lines with "boat" specimens to be ready to repair them when necessary. Many of them found graphite, with one notable exception. The joints at one plant were still sound after repeated samplings, but, with that incubation period in mind, the tests were continued. Ultimately, graphite was found, as in the laboratory tests.

Before the project was closed at Battelle, we had at least a partial understanding of this unusual and mercurial behavior. Sometimes it happened and other times it didn't. The steel was aluminum deoxidized to secure a good surface on the inside of the pipe, and we had begun to suspect that aluminum was somehow involved. Then W. A. Reich called my attention to nitrogen in this connection. He was with General Electric and had found that an induction-furnace heat with low nitrogen graphitized, whereas a similar steel melted in the electric-arc furnace and with high nitrogen did not. Apparently, nitrogen was an inhibitor, or retarder, but was robbed of that function when aluminum tied it up as aluminum nitride. Our tests of that hypothesis at Battelle confirmed Mr. Reich's experience.

From time to time during the war, Battelle was asked to look into production situations when something was going wrong. I recall one of these situations that was very critical. Four of our submarines in the far-western Pacific had been unable to use their diesel engines. They did not dare radio for help, and would have been sitting ducks if discovered. I was told that morale in our submarine service had hit a low.

I visited the fabricator in Hamilton, Ohio, to check their manufacturing practice and to learn how the diesels failed. The latter was simple — teeth were breaking off the timing gears. As I had heard so many times, everyone involved in the production of the gears had complied with the specifications. That included the steel supplier, the forge shop, the heat treater, and the machine shop. After a study of the situation, I concluded that procurement was at fault. Before I left, I revised their purchase specifications at certain critical steps. Good procurement was the answer, because no breakages were reported after the improved specification was used. Presumably, the morale of the submarine crews went back to where it should be.

The moral of this incident is that we should keep in mind that suppliers need not exceed the requirements of the purchase specification, because their function does not relate to subsequent operations or to the end use. The purchaser should provide a realistic specification. That was but one example of inadequate procurement practice.

We had an extremely interesting project from the Ordnance Corps of the Army, resulting from the query of an ordnance officer whose name I don't have. He wanted to know the cost of artillery fire and its effectiveness against a target, to get a value of the over-all efficiency. Effectiveness divided by cost gives the efficiency, the yardstick to use in planning for production and use.

The cost was not just the dollar value of weapons and ammunition. We also took into consideration the use of skilled labor, always in short supply, materials, especially strategic materials, plant and equipment, transportation and logistics, management personnel, and so on. The ramifications were many.

The effectiveness boiled down to the fire that would be needed to accomplish missions over a range of specific conditions of weapon, ammunition, and target, along with the versatility of the weapon.

I had the over-all direction of the project, but I spent most of my time on the cost phase. Our principal mathematician, Clarence Bell, had charge of effectiveness. His phase required extensive calculations, which induced Battelle to install a computing laboratory, without which the project would not have been practical. Predicting the effectiveness of artillery fire was accomplished with the methods of probability mathematics.

A side issue was the validity of our line of field artillery. Could one weapon be eliminated, to simplify procurement and maintenance, without vitiating combat effectiveness? Or should one be added to fill a gap, and would its use justify the additional cost? Those were good questions.

This project developed into far too long and detailed a study for me to do justice to it here. I will simply review it briefly, starting with the cost phase.

Studying costs meant that I interviewed experienced ordnance officers on procurement, in the chief's office and at the arsenals. Visits to the manufacturing plants produced a large volume of information and data on all phases of production. My contacts took a lively interest in the project as a way to give Army Ordnance a firmer basis for procurement, the reason that prompted it.

My other source of information was various army posts, where I interviewed artillery officers who had had combat experience, and infantry officers who had been under enemy fire. As a Civil War buff, I confess I was fascinated by these accounts, as well as by reading histories of numerous field actions in the principal theatres of World War II.

Fort Sill, in Oklahoma, is our major artillery post, where tactics are worked out and officers are trained. The commanding general at that time had been the chief artillery officer of Gen. Omar Bradley's Twelfth Army Group when we were held up at the Ardennes Forest. As an example of the helpful information I received, he told me of the TOT (Time on Target) that he organized. That is the tactic of using all available weapons, large and small, to hit the target simultaneously. Our army was making no progress until then, but, as he explained, "All hell broke loose in one big crash and we were able to advance." A TOT requires intimate knowledge of what each weapon can do to secure its maximum effectiveness. He then turned me over to other officers who briefed me on our weapons and their various targets and missions.

I was especially impressed by the emphasis they placed on logistics, which stemmed from their desire to "get there fustest with the mostest." Many other points were discussed, such as the emplacement time of heavy ordnance, the 180-mm gun versus the 250-mm mortar, fixed versus self-propelled weapons, and time fuses versus proximity fuses, although they knew that the latter were much more effective. They were less concerned about the problems of producing the various ordnance items back home, as I had to be. Information of this kind was given to Clarence.

In the meantime, Clarence was at work on field artillery, ammunition, typical targets, etc., to determine the number of rounds for the *X*% probability with which a given mission would be accomplished. With information on a weapon and its ammunition, and on a specific target, we could estimate the effect of fuse setting — on, for example, the effectiveness of fire. If that same mission were studied by firing tests on the proving grounds, the calculated and actual results would probably not coincide. The calculations, however, would indicate how best to use the weapon and ammunition. The same procedure would also compare the effectiveness of two or more weapons.

Calculations of cost, a complex value, would then give the efficiency with which the mission would be accomplished for each parameter group. Unhappily, we were not permitted to complete the cost phase of the program. It was abandoned when the surveillance of our contract was transferred from the chief's office to Aberdeen Proving Grounds and the ballistics research laboratory. At our first meeting there, we were told that the ballistics research laboratory was not concerned with production, but only with effectiveness. Complex and difficult as the cost phase was, some of us felt that the tremendous expense of preparing for and conducting modern warfare should have been studied in depth. During the war, Operations Research had been developed as a formal method of rationalizing complex systems for mathematical treatment. Our cost analysis was well suited to the new treatment.

Calculations of effectiveness of artillery fire also developed into a long and complex program that kept several mathematicians and computers busy. I believe it was both faster and cheaper than it would have been to shoot out the various missions with variations of weapons, ammunition, fuses and fuse settings, range, terrain, atmospheric conditions, logistics, etc., against each of a fair number of targets. The actual computations of the effectiveness of artillery fire gave the best technique, or combination of parameters, for accomplishing a specific mission.

In 1951, Picatinny Arsenal established a scientific advisory council to draw in outside authorities for counsel. I was invited to join, and I enjoyed an interesting association until I left for Frankfurt in 1959.

Picatinny Arsenal is pleasantly located in the foothills of the Appalachian Mountains in Dover, New Jersey. It is the arsenal of our army for making and developing ammunition and the hardware with which it is used. Col. J. P. Harris, then the commanding officer, established the advisory council for the research and development group under Don Beeman. We had distinguished people in the council, among whom were Professor Paul Gross, later president of the AAAS; Gen. C. W. Clark, later chief of research and development of the Ordnance Corps; Dr. M. Jacob, a leading authority on rocketry; Col. H. H. Zornig, a top authority on ordnance; Dr. D. C. Drucker, a top authority in mechanics and metal physics; and Professor L. F. Audrieth, a former ordnance officer and later scientific attaché to our embassy at Bonn. We were called at irregular intervals to discuss items on the agenda, under Don's leadership.

The council meetings gave me opportunities for consultations with arsenal personnel, particularly at the metallurgical and testing laboratories. In this way I was indoctrinated on hardware at Picatinny. I also went to various plants to check and advise on contracts where help was needed. Those plants were well equipped and staffed for mechanical operations of one kind or another, but not for metallurgical control. At one plant, the product was variable in quality and the scrap losses heavy. The mechanically minded personnel had done fine work on the mechanics of the processes, but faulty and nonuniform starting material had been an unnoticed variable. As soon as I called that to their attention, I received cooperation in revising their purchasing specification. From what I have since then read, I assume that production problems were licked.

Another contract interested me greatly on account of the extreme precision that was required in manufacture. The device was one that was sensitive to barometric pressure. The chief engineer at the plant demonstrated its extreme sensitivity by raising it over his head, which caused a lamp to light up. Upon lowering it to his waist, the light went off. That was Sensitivity with a capital S. The problem was that they suffered a 50% scrap loss. That is all I was permitted to tell, because everything else was highly classified.

I had an interesting case that involved impacts, or high strain rates, not to be confused with the notch effect. A component was used that had to withstand an explosion for a tiny fraction of a second, or a few microseconds. However, it was tested for acceptance by a hydrostatic test to a specified pressure. The design was based on the static test. I recalled what Dr. Desch had stated at a colloquium, back in Schenectady days, about high strain rates, which prompted me to suggest that a simpler steel of lower strength might be substituted for the alloy steel that was being used. Someone worked in an explosive test of the component, which had been made weak enough to fail in the static test. It passed the explosive test, to the surprise of all, and induced considerable interest in following up the concept.

We then tried to have Don Clark of Caltech make some high-strain-rate tests for the Picatinny Arsenal. Don was one of the very few active workers in this field at the time. He begged off, but agreed to train a man on the technique. Picatinny then started a program on its own, but I left shortly thereafter for Frankfurt and lost track of it.

Various investigators have attempted to find a fixed relationship between the effects of strain rate and temperature on the behavior of steel. The idea is intriguing. Instead of testing a specimen at room temperature with an explosive charge, one could use instead a notched-bar test at the equivalent low temperature and secure the advantage of a simple and cheaper test.

One would need to have only the correlation to select the testing temperature to use, assuming that the strain rate is known. I recall that Krupp was substituting a notched-bar test at a specified low temperature for the regular explosive-burst test of a cylindrical specimen from a gun-barrel forging. The presumption was, however false, that the two tests produced the same behavior of the steel.

The first time I encountered a major program establishing this relationship was on one of my visits to Watertown Arsenal. Dr. John Holloman explained his experimental program along with his fascinating hypothesis, but I couldn't buy it. Apparently he ignored the difference in the stress fields of a notched bar and uniaxial tension and the significance of the cohesive-strength – yield-strength relationships of Ludwik. The correlation is not between strain rate and temperature per se, but between strain rate and the temperature at which brittle fracture of the notched bar sets in as the testing temperature is lowered. This was demonstrated experimentally by A. V. deForrest of MIT, and is inherent in the match-point principle that I have advocated as the proper expression of the transition temperature. The clue to the invalidity of John's hypothesis is the circumstance that strain rate and temperature, as variables, change the yield strength by different rates for different steels.

Upon leaving Watertown Arsenal, John joined General Electric as director of the Metallurgical and Ceramic Section of the Research Laboratory, and then became manager of the General Engineering Laboratory. He had made his mark by illuminating the field of steel metallurgy, but he became less and less a Man of Metals when he was appointed Under Secretary of Commerce for Science and Technology, and then president of the University of Oklahoma. After he left Oklahoma, he became a special assistant to the president of MIT. John combines an unusual amount of talent and desire, and what has been our loss is the gain of government and education.

Battelle became involved in Viking–American archaeology one time. Mr. Arlington H. Mallery felt that he had unearthed evidence that the Vikings had come as far west as Ohio and wished to have Battelle representatives go with him to a couple of sites near Chillicothe. Bob Adams and I were selected. The more interesting site was on the side of a range of hills. Mr. Mallery said it was that of an old iron refinery of Viking origin. It was built on the upslope to utilize the natural draft to fan the fires in the same way that the ancients utilized the updraft at Solomon's copper smelter at Ezion-Geber, on the Gulf of Aqaba. I was quoted in the papers as saying "There is some interesting material here," which I think was a pretty good statement. I had learned it from my German professor, Dr. Hanemann. When presented with something new that he didn't understand he would say, "Very interesting, very interesting."

Mr. Mallery had a great deal to say about the famous Kensington Rune Stone. Taken at face value, it tells about a band of Vikings in Minnesota. The pros and cons of its authenticity have been discussed: Some have asserted that it is a recent plant, while others assume that it is many centuries old. That difference of opinion might be resolved if the condition of the surface weathering could be used to estimate the length of burial. The condition of the grooves of the characters would be interesting.

I would like to mention the Little Inch pipeline, and not just because Battelle made a fine contribution when it was converted from oil, in World War II, to natural gas. This line runs from Texas and the Southwest to New Jersey. Although it could operate safely enough on oil, its safety would be quite different when transmitting gas at high pressure. The transmission company had in mind welding a six-inch-wide reinforcement strip over the longitudinal welds for the full length of the line. Naturally, that would be a multimillion-dollar project. Worse yet, however, the so-called reinforcement would not have increased the strength or safety of the pipe.

Before going ahead with that proposal, the company approached Battelle to learn if a better method could be devised. The project was in the direct charge of Lloyd R. Jackson, who contributed handsomely to the solution, saving about one-third of the previously estimated cost and greatly increasing the safety of the line. Lloyd's procedure was simple in concept. He had them test the line, section by section, at pressures that were somewhat higher than the operating pressures. When a failure occurred, a repair made the line sound. Lloyd, like his boss Howard Russell, seemed not to seek publicity, but he was recognized as an able and effective research scientist, both by the Battelle staff and by sponsors of projects to which he was assigned.

One time, while in the shop where the pipe was welded, I noted that the finished lengths were hydrostatically pressure tested when completely exposed to the air. That showed that the test pressure stressed the pipe somewhat below the yield strength of the steel. That low-pressure acceptance test is doubtless satisfactory for some grades or applications of pipe, but switching from oil to gas would surely introduce a risk of failure. I knew from my work at A. O. Smith that welded pipe must be tested at pressures that stretch it through the yield point for adequate assurance of strength, or soundness of the weld.

The difference between liquid and gas, or between the amounts of stored energy, has been demonstrated by pressure tests of pipelines. When air is used to avoid residual moisture from the use of water, a fracture can run for miles, or much longer than it would run if water had been used.

The First World Metallurgical Congress of the ASM — what an experience! Bill Eisenman was at his best in organizing and managing this large and impressive gathering of metallurgists from the free countries. In cooperation with the Economic Cooperation Administration and the United States government, the ASM entertained several hundred foreign metallurgists on tours, and at the annual convention and metal show in Detroit, in 1951. Bill had organized it, with Zay Jeffries as director general. This was the largest project of its kind for the ECA, although in previous years approximately 5000 foreigners from the free countries had come to study our methods and then to take home reports on how to improve productivity and living standards in their countries.

The conferees were divided into groups under a manager and with two tour escorts. Each group had its itinerary, but all came together in Washington, Cleveland, and Detroit. I was tour escort with the research group with Professor Tom Hughel of Purdue University. Noel Phillips was our amiable and efficient manager. We visited twenty industrial and university research laboratories and other institutions. As Marjorie Hyslop wrote in an account of the congress, "It was a totally new, amazing, and inspiring experience." Amen.

Our hosts responded splendidly in carrying out the objectives of the visits, and they answered questions in free and frank fashion. Some of our members were particularly active in acquiring information, Georges Blanc most notably. He was director of research of the foundry division of the French Organization for Assistance to Industry. His numerous extracurricular activities consisted of visits to foundries that were not on our itinerary.

Battelle, for example, is very well-organized for visits, and had escorts for small groups and qualified personnel to discuss the work of their departments. For lunch we were taken to the beautiful Scioto Country Club. Georges did the honors afterward, and some may have wondered why a non-English-speaking man was selected; he answered that easily, responding in English to Battelle's hospitality with well-chosen and, at times, humorous comments.

Speaking of humor in a foreign language, which is a real test of command of the language, there is a reverse twist — humor in English addressed to a foreign group. While in Detroit, I happened to be dining with a group of conferees when the point came up. I announced that I would like to try an experiment. I recounted a bit of humor that had been told by a bellhop at the lodge in Bryce Canyon, Utah. He told about his duties, one of which was to bring in supplies for the chef from the Westinghouse walk-in refrigerator. One day he went to the Westinghouse, as they called it, for a side of beef. When he opened the door, and much to his surprise, he saw a rabbit sitting in a corner and, very obviously, taking it easy. The boy asked, "Hey, what are you doing here?" The rabbit replied, "This is the Westinghouse, isn't it?" The boy was puzzled, but said, "Of course this is the Westinghouse, so what?" "Well," replied the rabbit, "I'm westing." The guffaws that greeted the silly ending were spontaneous.

We had no language difficulties on the tour, although Professor Mishima, of permanent-magnet fame, had to help his friend Kokubo, and I helped at times in German.

One of our visits was to the Institute for the Study of Metals at the University of Chicago. Charles Barrett was in England for a year, and his place was taken on an exchange agreement by Professor A. V. Reynor. We had several Englishmen in our group — Dr. James Pearce, director of the British Cast Iron Research Association; John Cross, of the Royal Naval Air Scientific Service; Cyril Phillips, of the Mechanical Engineering Research Organization; William Reese, of the National Physical Laboratory; and Dr. Hubert Sutton, director of research and development for aviation materials. Occasionally, in movies, the director will have an English group engage in repartee, developing a delightful bit of polite humor. My group of Englishmen engaged Reynor in a running repartee of wit, whimsy, and banter that was every bit as clever as anything I had heard on the screen. Oh, that I could have recorded it! It would have displayed English humor in its true light as it would have dispelled the mistaken notion that the English lack a sense of humor. They just don't think that a custard pie in the face is screechingly funny.

Trust Bill Eisenman to cook up something. He arranged to have all tour groups join at Cleveland for a big show at his farm, Sunnimoor. The day was devoted to Americana, early Americana to be more precise. We had a hayrack ride to the grotto alongside a little stream, covered wagons and real Indians, a corn roast and refreshments. Bill and his wife, Millie, were everywhere, showing their pleasure as host and hostess to the metallurgical elite of the free countries.

Detroit was not an anticlimax; it was an outstanding convention of metallurgical science and technology. I would like to mention just two of the many interesting papers we heard that week. Dr. Reese reported on a project at the National Physical Laboratory on the effects of impurities of iron. With highly pure iron, the Charpy V-notched-bar test shows that the change from ductile to brittle fracture occurs over a range of only one degree Celsius. As impurities are added to the iron, the transition shifts to a higher temperature and the range may or may not widen. How is that behavior to be explained?

The other paper was presented by Connie Voldrich and Perry Rieppel of Battelle to the American Welding Society. Determining the weldability of steel was in a state of great confusion, as were criteria by which it could be measured. Battelle had a project to attempt to make sense out of the situation. The popularized bead-on-plate test of Dr. E. H. Schulz of Germany, and its modifications in this country, were shown to be unconventional tests of the notch behavior of the steel rather than of its weldability. I have long been puzzled at the large expenditure of time and money on a test whose lack of qualifications has been so obvious. When we finished the study we favored the Lehigh restraint test of Gilbert Doan, although the cost and inconvenience must weigh against a broad use. The various restraint tests (largely English) have a better record.

Travel by train and bus gave us the opportunity of becoming acquainted on a person-to-person basis. My neighbor was not a famous inventor of magnetic alloys, or an outstanding metals scientist, but a cultured and gentlemanly visitor from abroad. Our conversation would turn to his homeland, or to some aspect of life in the United States. That was friendship, on a first-name basis, with people from Norway, Sweden, Denmark, Germany, Holland, Belgium, France, the United Kingdom, Italy, Japan, Australia, and India.

I made it a point to converse with each member, while Tom's good spirits retained an informal atmosphere and Noel's managership kept the machinery in smooth operation. As hosts, we went all out to make this a memorable visit. We were all grateful to Bill Eisenman and others who made the First World Metallurgical Congress not only possible but a splendid success.

The Navy had problems that were suited to laboratory study; one came from cracks in submarine-hull welds. Clarence Sims and Maurice Banta, who had had experience with weld cracks, took charge of what sounded like a simple problem. Before they had finished, however, they had on their hands a rather obscure problem of banding, or of dendritic segregation of manganese. There was also the question of why some producers supplied crack-sensitive or banded plate and others insensitive plate, although of equivalent analysis. Banding cannot be eliminated economically by a high-temperature homogenizing treatment, which suggests that the steel-manufacturing practice at some mills reduced the segregation to an acceptable level. The solution lay in that direction; the rest was in the hands of the steel mills. Sims and Banta's paper was a major contribution on the weldability of the intermediate manganese steels in heavy sections.

Clarence is a top man on steelmaking and steel quality, with its various ramifications, and is well known as a careful and experienced experimenter. He is equally as good at telling stories. They usually have a typical twist, such as the one about the Scotsman who was about to start a round of golf. He was selecting a caddie and asked the prospect, "How are you on finding lost balls?" The caddie replied, "Oh, I'm very good at that, sir." "Very well, then," the Scot directed, "go find one and let's get started."

Speaking of stories, I thought up a riddle that I put to a group at lunch one time. "What is a musician's favorite pipe?" Of course no one had even a good guess so I told them it was the Three-B Pipe. And to "Why that?" I explained that it stood for the three big B's — Bach, Beethoven, and Brahms. Howard Russell spoke up and allowed that my riddle was highly obscure — it reminded him of the two Cockneys in Paris. They were in a butcher shop to pick up some meat. "Ah," said one, "Je suis." That was as dense as mine, so Howard scored and had to decode "I am." That was also obscure, so he said it meant "high ham" in Cockney, which may still be obscure.

Howard didn't seek publicity and may not have been as well known as many of lesser stature, but he was one of our strongest men behind the scenes. One needed only to ask, to receive sound advice and penetrating criticism. Much of his time was spent in the highly classified field of nuclear physics. He was in charge of that work at Battelle, which dated back to the time when the word uranium was classified and no one dared even to mention it. That reminds me of the time I was interviewing a German professor in 1945. I had asked him what work he had done with uranium. I was surprised at how startled he was as he protested that he had never touched the stuff.

Howard was always good company at lunch, especially as he discussed topics of the day with interesting elaborations of his own viewpoint. His early departure was a severe blow to Battelle and a great loss to its sponsors and to physics.

To get back to the Navy: We were asked to attempt to make anti-torpedo nets by welding. Welding would give an improved and cheaper net, but two of our major companies had tried and failed and Battelle was a last resort. The starting material was cold drawn high-carbon steel wire at ¼-in. diameter with a tensile strength of more than 200,000 psi. That was not a promising starting material for a product whose failure in service would render it useless. A good policy in research and development is to make an attempt with the feeling that hard work and a bit of ingenuity may get the job done. So we tackled it.

Flash welding is a standard method for joining metals and was to be used. The problem was one of securing a strong and ductile weld without producing a hard and brittle heat-affected zone and a weak zone adjacent to it. Straight flash welding would not avoid those deficiencies. Recalling my work at A. O. Smith on the "double bump," I decided that there was a good approach to a solution.

In brief, the effect of the weak zone could be avoided by keeping it thin. The hard, brittle zone posed a more difficult problem, yet it could be circumvented by causing the heat-affected zone to cool slowly, or at a retarded rate, as soon as its temperature had fallen below the critical point of the steel. To accomplish this we developed electrical timing that controlled the welding process and also applied a second current at a predetermined time to control the cooling cycle to give a strong yet ductile joint. The "hard work" phase brought this project to a successful conclusion. The Navy adopted our process and built a special plant to manufacture the nets.

Bob Bennett was the project engineer in charge. He attacked the work with his customary enthusiasm with no signs of ever being discouraged. Positive thinking is a good attitude in research and keeps one from wondering what would happen in case of a failure. Bob published our work in the Research Supplement to the *Journal of the American Welding Society* and showed that even an "unweldable" steel can be welded to give a serviceable joint — with hard work and a bit of ingenuity.

Speaking of flash welding reminds me of an important point of translating that was forcibly revealed when we rendered Hilger's book from German into English. The first attempt was by a translator in New York. When I received his copy I was horrified by its poor wording; it was so bad that some of it didn't make sense. I then sent both book and translation to a friend who had had some experience in the field. Even his translation was far from adequate. Time was running out, so I wished the job on myself.

The point here is (1) the first man knew German but did not know flash welding, (2) the second man knew flash welding but not German, and (3) I knew both. The translation was published by the American Welding Society as the only one in English on flash welding.

Many people came to Clyde for advice, which he gave in generous measure. One request that I recall was from the Association of American Railroads that he survey the industry on the question of establishing a research policy and how to implement it. As was his wont, he tackled the project with vision and enthusiasm. He asked me to help, which of course is the part I recall best.

Railroading has always fascinated me, so it was with great pleasure that I studied those phases of equipment and right-of-way that had metallurgical import. Needless to say, the railroads were replete with problems that I studied in eager anticipation of making worthwhile contributions.

Railroading was not known as a progressive industry, but a bright spot was the work of Ray McBrian, engineer of tests of the little road in the Colorado Mountains, the Denver and Rio Grande Western Railroad. Ray was the leader in his field, especially through his metallurgical approach to many railroad problems. One of his special achievements was his spectroscopic determination of the metal content of diesel lubricating oil, providing the measure of wear and hence the need of maintenance. That inspection is now used by truckers and others. I believe it must have been his work that helped to motivate the Association of American Railroads.

One project that stood out clearly was the substitution of HSLA steel for ordinary steel in sheet and sections and welding for riveting for freight cars. Those substitutions, on a straight strength basis, I found, would result in a ratio of loaded weight to empty weight that was above the handling capacity of the braking system. Solutions were elusive, although in this case, using the stronger steel in somewhat heavier sections to reduce the ratio would increase car life and reduce maintenance costs. Furthermore, a visit to the repair shop of a leading railroad revealed that welded joints would reduce or possibly eliminate leakage and local repairs. Research and development problems for locomotives and rails were readily uncovered.

Clyde presented his report at an Association of American Railroads meeting in Chicago. His discussion of research policy resulted in the decision to establish a research laboratory in Chicago, near the Illinois Institute of Technology.

It seems fitting to close this chapter on Battelle with a few general comments. A glance at the rosters of national societies and technical committees will show that Battelle staff members contribute to the activities of their professions as they enjoy the fruits of memberships. Many are invited to give papers and talks before national societies and their local chapters. These pursuits add to the individual's stature and to the Battelle image. It is a legitimate privilege that the research institute grants its staff: As the member grows he is worth more to the sponsors. On the other hand, the exposure of a bright young man to the outside world as a featured speaker creates at least a small hazard of losing him. In practice, the hazard is small because a major research institute is a highly desirable place to practice one's profession. Only a strong pull will attract a good man away, and some return.

The private, not-for-profit institute for sponsored research has a permanent place in the affairs of our country. In addition to Battelle, there are now several that are geographically distributed. Gordon Battelle would be proud of the enterprise that was born in his mind and initiated by his and his mother's endowment.

Chapter 7

The ALSOS Mission, 1945

Metallurgical specialist on supersecret atomic-bomb mission; German war work in metallurgy.

THE dust jacket of the book "ALSOS," by Dr. Samuel A. Goudsmit, scientific chief of the mission, states that "ALSOS was the code name for one of the strangest missions the United States Army ever created — a mission so secret that only the highest government and army officials knew of its existence. ALSOS was a brand new kind of intelligence outfit with scientists as supersleuths." In "The ALSOS Mission," Col. Boris T. Pash (ret.), military chief, relates the exciting military operations that found and captured numerous high-ranking German scientists. Both books are excellent reading. The mission ultimately reported to our leaders that Hitler would not have an atomic bomb.

ALSOS was formed at the request of Gen. Leslie R. Groves, chief of the Manhattan Project, which led directly to Hiroshima, to learn what progress Germany might have made on the atomic bomb. It was well known that the start had been made when the German physicist Otto Hahn produced fission of uranium, and also that a theoretical paper on a chain-reacting pile had been published. Word was also around that the Germans had ceased publishing in that field. It was inconceivable that we would be able to invade the mainland of Europe if they had the bomb. For our planning, we had to know what the Germans had accomplished.

Biological warfare, the proximity fuse, metallurgy, and a few other fields were included in the mission. The name ALSOS, incidentally, is Greek for grove, and its use caused some misgivings as did the lightning bolt of the insignia.

Dr. Sam Goudsmit was a Dutch physicist who came to this country and was well known for this work on electron spin. He knew Europe's leading physicists and spoke their language, so he was an excellent choice as the scientific chief. He was not in the Manhattan Project and thus could not disclose anything on our work if he were captured. I thoroughly enjoyed my association with him and found him to be sympathetic and helpful in my work.

Early in 1945, I received a message marked "SECRET" at Battelle. It came from Washington and informed me that I had been proposed as a metallurgical scientist who would be competent to serve on a mission in the European Theatre of Operations. Would I be able to accept the appointment? I had no idea of what was involved, but I replied that I would come to Washington to discuss details. However, an earlier incident gave me great concern about my being cleared for a secret assignment.

Early in the war, I was asked by Dean McQuigg, of Ohio State University, to head a team to investigate welding practice in plants that were handling classified contracts. That meant I would have to be cleared. Innocently enough, I agreed to serve and awaited clearance. A little later, the dean came to my office with the surprising information that it had been denied, with no explanation or further information. I asked him to go with me to the local FBI office to learn how to correct what I felt was a completely absurd action.

I explained to the FBI agent that I wished to take any steps needed to remove that stigma and get clearance for any duty for which I was qualified. Dean McQuigg assured him that I was well-known and respected and that I was a loyal citizen. The agent replied that the classification came "from the White House" and that he could do nothing for me. My assertion that I was as loyal and clearable as anyone in the White House availed nothing.

I left the FBI office greatly disillusioned and wondering what went on in the White House, "behind a curtain," as the Dean put it. My doubts about my ability to contribute to our war effort on classified projects proved to be unfounded, and later I had no difficulty in obtaining "top secret" and "Q" clearances. I can understand how a Communist can operate in our midst and be undetected, but my experience still baffles me. It was very clearly the kind of mistake that should never be allowed to occur. Several years later, in 1945, I was accepted for ALSOS, which had super "top secret" status.

After I arrived in Washington, I found myself with Allen T. Waterman of the Office of Field Service of the Office of Scientific Research and Development. During briefing sessions, they seemed to know a considerable amount about me. They were especially interested in my German associations and my ability to speak the language. I gathered that I would be used for work in Germany, but I was told I would not receive my instructions until I reported to headquarters in Paris.

The day finally arrived for me to report in Washington for last-minute briefing, medical checkup, procurement of my uniforms, and so on. Included was a session on safety measures at sea if our plane were forced down. We left for Paris on May 2, 1945, and arrived at Orly airport the next day. About the first thing I saw was a sign reading *Schweizraum,* or weld shop. Paris was completely blacked out, which gave us an eerie ride into town. I stayed the first night at one of those old hotels near the Tuileries, and transferred the next day to the very fine Royal Monceau near the Arc de Triomphe.

Dr. Allan A. Bates was my first and principal contact and the man I was to replace on the mission. Allan was a bundle of energy and had found numerous important people for me to contact (no mean feat, in view of their dispersion) and had gathered up many documents for my use. That information served to guide me, altough I was on my own to work out the operational details.

Allan had some good news for me, too. Bob Mehl was in Paris and would join me for a trip into Germany. He was stationed in London, in charge of our Technical Industrial Intelligence Committee, and wished to get a reprieve from paper work by going into the field for a change. He had arranged this appointment in ALSOS. Bob was buried in an old hotel of the Napoleonic era, so we rescued him and took him to our hotel. There, Allan briefed us, in extenso, on the work we were to do and regaled us with numerous adventures he had had under unbelievable conditions. Bob came quickly to life with prospects of action.

On the morning of May 6, we left with jeep and driver for Verdun, of World War I fame. I don't believe Bob knew much about World War I, but I recalled numerous places that had made the headlines between 1914 and 1918. History took me back a century and a half at Valmy Ridge, where an epoch-making battle had been fought and the might of revolutionary France first broke into the consciousness of Europe. One foggy morning, a young and fanatically inspired army forced the Prussians and their allies to abandon their attempt to restore the French monarchy. On that occasion a prophetic statement was uttered, the likes of which is hardly duplicated in all history. It was made by Goethe, who was there with the troops of the Grand Duke of Saxe-Weimar. Shortly after the retreat was in progress, he spoke to a group of officers: "*Von hier und heute fängt eine neue Epoche der*

Weltgeschichte an. Ihr könnt sagen Ihr seid dabei gewesen." ("From here and today a new epoch of world history starts. You can say you were there.") How true!

We also passed the famous Maginot Line, which still gave evidence of heavy fighting, and the Siegfried Line and its massive tiger teeth. I was told later that it would have been a formidable barrier if the Germans had had the capacity to defend it.

On arrival at Verdun, we had lunch at an officers' mess, our first contact with combat troops. Oh that I had had a tape recorder to catch all the tales we listened to on that and other occasions! Some of the men had come all the way from North Africa and Italy, but all had experienced the excitement and hardships of warfare, and had come through to tell about them. There were gripes, of course, and the principal one was that of replacing troops one by one instead of by units. The new men were too likely to end up as casualties before they were ready to do their share of the fighting. Furthermore, the veteran troops had no respite, short of getting wounded, or killed. That, they insisted, was a poor basis for the right psychological atmosphere. My young son, who went through the same experience in Korea, had the same gripe.

Then came Germany and Saarbrücken. There we got the first shock of war, or at least of the devastation of war. The Saarland is a prize of war and a natural target, so the city had largely become rubble. We could view the scene abstractly, as a drama on the stage, but the rubble and the faces of the few inhabitants we saw were evidence of the stark reality of warfare. I thought, "Sherman didn't know how right he was."

Soon we were rolling nicely along the Autobahn, through beautiful hilly country. Bob was in excellent spirits, figuratively sort of kicking up his heels, and thoroughly enjoying his outing. (I was too.) Then it happened! Our driver had pulled up alongside the road and was checking our jeep. The inspection dragged out quite a while and we suddenly heard rifle fire. "Good Lord, is this an ambush? We are too valuable to lose our lives here," we thought. "Just some GIs out after some birds for supper," we were told. By now the sun was getting lower and we were wondering what next, when a staff car came along. The officer was headed for his base at Mannheim, which is near Heidelberg, our destination. He told us to get in and that he would get us transportation from his base. That took care of our situation — or so we thought — and we were soon rolling and bumping along in the car.

Now we were in the hands of people who knew nothing of ALSOS, the supersecret mission, only we did not appreciate the significance at the time. Our spirits rose back up where they belonged. After more bumps the lieutenant delivered us to the officer in charge of transportation at the depot.

"Where is it you want to go?" he inquired. "To the ALSOS base in Heidelberg" was our innocent reply. "Well, where is that?" Then the full meaning of "supersecret" dawned on us. The only one around who knew was our driver back on the Autobahn. However, our new driver would take us to the G-2 base (intelligence) in Heidelberg, and maybe they could help.

By the time we reached G-2 it was pitch dark in the blackout. There was great reluctance to telling us where our base was, since we were total strangers. Finally, the officer must have become convinced that we were honest and he went to work over the field telephone to get our address.

There followed a wild-goose chase, abetted by the darkness and some misinformation from the local police. Then we tried a new approach on foot. That took us up a hill on a zigzag back road in pitch darkness. We were feeling our way along to keep on the road when a sharp shot rang out. "Good Lord, an ambush?" we wondered again, because SS troops were on the prowl. "No," explained our driver as he came up to us, "my revolver went off by mistake."

Farther along, up the hill, we saw an officer come out of a house to get into his car. He was Gen. E. L. Harrison, chief of intelligence, and he directed us back across the Neckar, about five miles upstream, and then several miles up into the mountains back of the castle. That didn't sound right at all, but Bob got a great idea as the general drove off. Going just two houses from where we stood, he knocked on the door. A young lieutenant appeared and invited us in with, "Where have you been? We expected you hours ago."

The next morning, after enjoying the exquisite view of the valley, the castle, and the city, we left for Göttingen, where our Northern Base was located. Our first shocker was Frankfurt am Main, the first large city we came to. We saw the devastating effects of bombing and artillery fire. The ruins were typical of every large German city I visited. I reflected on earlier visits to Frankfurt — the huge railway station, the excellent Frankfurterhof Hotel, the Goethe house — and what I saw for mile after mile made me wonder at the meaning of it all. Kassel was the same, only worse, and we were glad to reach Göttingen. It was serene and unscarred by war.

My thoughts turned naturally to Professor Tammann and my visit with him in 1912. Although he was gone, I made a mental note that Professor Masing was there to carry on the tradition of a great metallurgical school. But there were immediate things to be done and we hastened on past a long row of our tanks to our base.

We were greeted warmly by genial young Major Blake, a combat engineer who had been over since landing in North Africa. We were introduced to several officers who were attached to the base. One in particular was Lt. Gus Toepel from Madison, Wisconsin. He was a happy bachelor and was to be our driver. I was a colonel, and Bob, as head of a mission, was a one-star general, but to Gus we were strictly ersatz, and he frequently let us know it. I was sprouting a few gray hairs and Gus said it was a good thing, "to give you some dignity with these troops."

Bob and I had come up north to visit the Kaiser Wilhelm Institut für Eisenforschung in Clausthal-Zellerfeld, to which town the institute had been evacuated when the laboratory in Düsseldorf had been bombed out. The iron institute had quarters in the famous old Bergakadamie, or school of mines. Professor Anton Pomp was the acting director in the absence of Dr. Franz Wever, who, along with numerous other leading scientists, was being interrogated by our forces.

Professor Fritz Koerber had succumbed to the trials and tribulations of the war, but during his long and fruitful tenure as director he had built well. He had the confidence of his sponsors, the iron and steel industry, and had demonstrated the beneficial and strengthening influence that well-organized research can exert on its industry. His function was one of advancing knowledge of the metallurgical arts and sciences while leaving the application of that knowledge to industry.

My acquaintance with the institute made it clear that its work was of truly high quality, and I reasoned that the principal factor was the quality of the staff, including the supervisory management. Later, I asked a Krupp official if the institute metallurgists were truly outstanding, as compared to German metallurgists in general, and if so, why? He replied that they were truly superior and were *ausgesuchte Menschen*, or the pick of the crop.

Professor Pomp greeted us cordially and introduced us to the key staff members. They all seemed to appreciate our ability to discuss problems of mutual interest in German. Bob and I behaved like two visiting scientists and we maintained that atmosphere throughout our discussions. Furthermore, those people were not Nazis, although they did contribute to the German war effort, as a few examples will show.

Germany was plagued by shortages of key materials, such as brass for cartridge cases. The institute was asked to study steel as a substitute, a problem which is fraught with difficulties, as we had learned in the United States. They first determined the locations at which steel cases failed in firing. These were principally longitudinal splits, cross cracks on extraction, and stickers. They then studied each structural weakness that was responsible. That phase of the study gave the major point for solving the problem. In this way, they learned that stringers of carbide particles in the steel led to premature failure. The solution was a quick heating above the critical point to dissolve the carbides, and an accelerated cooling to disperse the carbides as an innocuous precipitate. This work permitted the early substitution of steel for brass. I saw evidence of this when I visited the battle site at the huge war memorial to the Battle of Nations of 1813, in Leipzig, and the famous Remagen Bridge over the Rhine.

Another contribution was the substitution of iron powder and low-carbon steel for wrought copper for rotating bands on projectiles. The former was used for small sizes and the latter for large sizes. Again they developed the basic principles to get the right start for the solution — a sufficiently low yield strength of the band to avoid too-high engraving pressures. The sintered iron bands were produced with the correct amount of porosity to serve the purpose. The wrought steel bands were produced with an excessively large grain size to lower the yield strength closely to that of copper and brass. That possibility was known in the 1920s from the scientific publications of the English metallurgist L. B. Pfeil. Actually, Charpy, Sauveur, and Hanemann had shown that a small deformation at room temperature and heating to an elevated temperature will produce the extra-large grain size. Koerber and Bardenheuer used that technique during World War I to produce rotating bands of extra-mild steel with a yield strength of about 15,000 psi and less. Firing tests indicated that the ersatz bands prolonged the barrel life, presumably because of the zinc coating and its lubricating effect on the rifling.

An ersatz steel was needed for machine-gun barrels to replace the chromium-molybdenum steel. The criterion was strength at 900 °F. They developed one with 0.10 to 0.12% carbon with eight times that amount of titanium, with a creep limit of more than 50,000 psi at 900 °F. It more than doubled the life on tests, especially when electropolished and chromium plated. The catch was making the steel, as we found out with Smith Alloy No. 10, and as others learned from type 321 stainless steel. Titanium is a problem alloy for steel.

Lightweight armor for aircraft was a problem with them as it was with us. The institute studied steels and light alloys by ballistic tests and high-speed photography, both theoretically and experimentally.

It was known that Germany had gone far in developing the gas turbine and that the institute had made a contribution. The efficiency is critically dependent on the permissible operating temperature of the blading; yet in 1945, the basic principles were but dimly understood and attention was focused on creep strength. Dr. Max Hempel, the fatigue and high-temperature man on the staff, had the results of laboratory investigations of combined creep plus fatigue, which control the behavior. The useful life depends on the way in which those effects combine as determined by the composition and structure of the blading alloy.

Dr. Hempel had good results on the effects of temperature on three different steels, and he had gone far enough to show that good design required knowledge of both effects for the material. After I returned to the United States, I lectured on Hempel's work before the Los Angeles chapter of AIME. The reception and discussion of his work indicated that it contained the kind of information our designers needed.

The powder metallurgy group also studied methods for making complex parts with powdered steel that had to be hard enough to withstand the effects of wear in service. I saw the results of this work when I visited the famous Metallwerk Plannsee in Reutte, Austria.

I had been interested in steel quality ever since my summer job at Midvale Steel, and since my work at A. O. Smith, where it played a dominant role in procurement. The Kaiser Wilhelm institute had two top steel metallurgists, Drs. Bardenheuer and Oelsen, with whom we had interesting and enlightening discussions on steelmaking. The German steel industry had a vested interest in steel quality and looked to the institute for basic studies in the field. Dr. Bardenheuer had many publications to his credit, but is no longer with us.

Willy Oelsen was interested in slags and in Thomas, or basic Bessemer, slags in particular. He told us, with unconcealed pride, of his verified prediction that two immiscible slags form. That is important work in Germany, because the fertilizer industry and the Thomas process are closely tied economically, one being a source of phosphorus and the other a user. Oelsen became rector of the school of mines at Clausthal, and then director of the iron institute at Düsseldorf under its new name of Max Planck Institut für Eisenforschung.

In the meantime, during our discussions indoors, Gus Toepel had been busy outdoors. He had discovered some playmates, and they him. When we came out we found him playing happily with some children, so we all had had a fine time.

For lunch, Gus took us up to the officers' mess, and we had instructive conversations with men who had recently been front-line troops. That was our daily experience during our visits to Clausthal. One day the Ninth Armored Division rattled noisily into town and we had its officers as table companions. A man-sized gripe was leveled at the order that stopped their advance on Berlin, well this side of their goal. They insisted that our army could have taken the place easily, and possibly with no loss of life on either side. Ninth Armored had had a sharp engagement east of the Elbe River and had dispersed the opposition. Then the order came to halt. Germans were surrendering in great numbers, so it would have been a march to Berlin rather than a campaign. We had no answer then for what was behind that order, but it came out later as a political decision.

One officer stated that the Germans could not understand how a man with the good German name of Eisenhower could be the supreme commander of the allied forces. Then he added that he got quite a kick out of taking a Nazi prisoner named Eisenhower. Just one of many amusing incidents that we heard. One was not amusing, though, and could have been tragic. A young officer with a small squad was out on a mission and asked for directions from an elderly lady. She gave them willingly enough, but so as to guide them into a trap. Our troops were alert, however, and the trap did not work. I asked what they did with the woman, and he replied, "Nothing."

All too soon, the time came for Bob Mehl to leave for duty back in London. I know he enjoyed the stimulating discussions at Clausthal and the daily adventures of being with ALSOS. We took a walk through town on the eve of his departure and summed up experiences and impressions of our work. One was the calm with which V-E Day was greeted. I was especially impressed because I remembered the celebrations of November 9 and 11, 1918. Another was the splendid discipline of the combat troops with whom we associated, including personnel of the armored division in Göttingen. Bob commented, "Not a nerve in a carload." I think he got that idea from some ad.

After adieus to Bob I looked up Professor Georg Masing at the university. He had been carrying on the Tammann tradition in fine style until the war interfered with his research and teaching. I had met him in Berlin in 1925, when he was with Siemens and Halske and involved with industrial problems. I recalled his work on beryllium as an alloying element. At Göttingen, he was the scientist, enlarging our knowledge of metal behavior, or metal physics, as evidenced by the copies of his publications he gave me. They had established him as a leader in his field.

One time, I visited the Masings in their home. Since we knew each other, our conversation turned easily to their political situation and life under the Nazis. Both he and Mrs. Masing told me about living in a dictatorship, with emphasis on self-preservation and how to achieve it. It was an account I was told frequently, usually with deep feeling, and sometimes with a great deal of emotion. Professor Masing told me that he had dared to discuss politics and express his true feelings only with his wife and family, and that he had had to conform or else.

It was also in Göttingen that I had a conversation with a youngster who had some stamps to sell, and one of those little daggers the Hitler Jugend used to wear on parades. I asked for his impressions of that Nazi activity. In reply he was plenty vocal, but I can sum up his attitude — "Pfui!" Was he clever enough to dissemble?

I left the institute and Göttingen, although with promises to return. By friendly jeep, I rode back to Heidelberg, where I met the fabulous Col. Boris T. Pash. He had been in charge of physical education at a Los Angeles high school, and had entered our intelligence service. He was quiet and composed in conversation, but our officers and drivers told me that he was a highly competent officer with all the fire and drive needed to accomplish a mission, come hell or high water. His principal concern, in my case, was to see that I had the facilities and service I needed.

Our Heidelberg base was my favorite place to work on reports and to attend to what business was at hand. Our house, as was the custom, had been the home of a ranking Nazi and was modern and comfortable. The weather and the view of the Neckar valley were terrific. One time, I arrived late in the day, ate dinner at our private restaurant, and then went to work on my notes in the living room. For a bit of relaxation, I stepped out on the porch. There I saw a couple of new arrivals, which was not unusual, and we introduced ourselves. Then one of them startled me by asking, "What river is that?" I told him, wondering how it could be that he didn't know where he was. He was an expert on Japanese culture, history, and language, and had spent his life in that field. Just two days before our meeting, he had been in Hawaii, ready to go to the Far East, where he would be put to good use. Then came an order to take the next plane to Germany to take up work in a country, and in a language, with which he was totally unfamiliar. He was keenly disappointed, and in the dark about the reason for his transfer.

My next trip was to Württemberg, another beautiful mountain region in western Germany. On the way down, I ran into a situation that could have prevented me from writing this. We took the Autobahn partway, skirted Pforzheim, and headed for Stuttgart. My driver stopped once to check for directions, whereupon I got out to examine a German fieldpiece that had been left along the road. It was not an unusual sight, but here was an opportunity to get a close look at one. As I got closer, I noticed that a shell was lodged in the muzzle and I contemplated getting it out so I could examine it. As I was reaching for it I recalled a Sunday-supplement article on booby traps. "Oh–oh," I thought, "better not touch it." I didn't. On my next trip along that road I saw, to my horror, that the shell was gone and the barrel was split wide open. So I can write about it.

My target, to use the military term, in Stuttgart was the Kaiser Wilhelm Institut für Metallforschung. It was the sister institute of the Institut für Eisenforschung and dealt with nonferrous metallurgy. The director was Professor Werner Köster. He had evacuated his office, library, and part of his laboratory to the little town of Urach, which was my destination.

Speaking of Urach reminds me of the time when Allan Bates was temporary mayor. He arrived there as my advance man, following the sweep of the Third Army across southern Germany. Urach was in the throes of the depredations of a lot of intoxicated displaced persons who had been released from German custody. With his two counterintelligence agents, Allan organized a company of French prisoners of war and armed them with confiscated weapons that had been in storage. With that force, he quelled the riot and set up a military government to keep the peace. As mayor pro tem, he handled civilian affairs until local authorities took over. That was just like Allan.

Professor Köster greeted me cordially when I introduced myself, and said he knew about me from my publications on Carboloy. He had just finished editing the book on hard metal carbides by Kiefer and Hotop of Metallwerk Plansee, which cited references to those publications.

I was first attracted to Köster, about 1920, by his pioneering work on the mysterious yield point of steel, with particular reference to nitrogen and its role in strain-aging embrittlement. Incidentally, Bob Archer in the United States had very similar ideas, but did not have the experimental evidence that Köster developed. I believe Professor Köster also associated the effect of aluminum in inhibiting aging with its reaction with nitrogen.

When the metals institute was moved from Dahlem to Stuttgart, in 1934, Professor Köster was appointed director. In its eleven years of operations the institute had many fine contributions, and was an outstanding research center and training ground for young metallurgists. My expectations of some worthwhile sessions with Professor Köster were amply fulfilled.

My first visit was brief, but sufficed to promise an interesting and instructive relationship in future visits. I did find that our troops were exerting the American influence, especially on the youngsters. As I was getting into my jeep, Professor Köster asked if I would do him a favor, which of course I would. Would I bring him some "Kautabak" for his young son who was not as fortunate as other youngsters in receiving gifts from GIs. I concealed my surprise at his request, but assured him I wouid bring some on my next visit. True to my word, I had a pocketful of "Kaugummi." You see, Kautabak is chewing tobacco, while Kaugummi is chewing gum. We also had occasion to discuss general subjects such as the old German Empire, the horrors of inflation, Nazis, Communists, dictatorships, and the war.

Then it was back to Heidelberg and on to Wiesbaden, on a special mission for the OSS, and up to Göttingen. Incidentally, that last name was really difficult to pronounce. It was in the English sector, and an English driver stopped me one time and asked the way to Go-e-*tinge*-en. After a little decoding, I was able to help him get there.

On my way to Braunschweig, I called on Professor Pomp to check on some reports that were being prepared. There I ran into Mr. Foulkes, steel-wire specialist of Bethlehem Steel, who was with the Technical Industrial Intelligence Committee. He was there to see Professor Pomp, who, he said, had made the best contributions to the metallurgy of steel wire and had established himself as the top authority. I remembered that when revising my "Metal Data" and included Pomp's data.

My mission in Braunschweig was a simple one, yet it turned out to be a wild-goose chase. I was trying to find three freight cars containing equipment of the iron institute that had gone astray. Locating them was not as easy to do as it may sound, because freight cars were scattered all over the landscape. I got in contact with the colonel who was transportation officer for that region, and he assured me he would find them and notify me. That ended my search.

My driver and I visited the base where Goering had trained his Luftwaffe. The occasion was a large and impressive military ceremony at which several Russian officers had medals pinned on their chests. It was very formal, except for the antics of our Air Force buzzing the parade grounds.

My next stop was in Wiesbaden, where I was to call on CIOS, a special intelligence group. I recalled the city as a beautiful spa with park, Kursal, and the inevitable small shops. I saw that it had been hit, although not badly. Wiesbaden had been selected as the headquarters of Bradley's Twelfth Army Group, and was off limits for bombers. Someone had goofed, however, and some buildings had been damaged. That did not interfere with my enjoyment of a bath in the special water of the resort.

The next morning, I left by jeep for Paris, and arrived at the Royal Monceau twelve hours later. One of the first men I met was Dr. Rettaliata, whom I had met at Allis-Chalmers in Milwaukee, and who later became president of the Illinois Institute of Technology. He was an expert on gas turbines, and had been sent over to check on a German submarine that had the new propulsion of Professor Walter. He left shortly afterward for Kiel to investigate it. I met him again, after he had returned, and he was highly enthusiastic about what he had seen and experienced. The Germans had made hydrogen peroxide in a much higher concentration than had been thought possible, and were using it as an improved fuel. He learned this during a ride in the submarine. The performance, both on and beneath the surface, had been an eye-opener.

Professor Walter had developed jet and rocket propulsion for planes and bombs. He received his title of professor from Hitler, as a high distinction, and it was reported that he received $3-million from Japan for use of his patents.

I also met Henry Allen, fellow student of my Columbia days, who had been chief metallurgist of Disston Saw and later became head of the Franklin Institute. He told me of his interview of Dr. Osenberg, with the help of Martin Fleishman of Timken as his interpreter.

We enjoyed considerable camaraderie among the various groups at the Royal Monceau, but there were serious occasions too. I had a good discussion with Sam Goudsmit; we talked about my work, and about the fine work of Professor Pomp and his staff at Clausthal. I commented that they were just twiddling their thumbs when they should be advancing the metallurgical arts and sciences. Finally, he gave me permission to evacuate the iron institute if I felt that was in order. I was working on plans to start up both Kaiser Wilhelm institutes, later in the summer, when orders to return home brought it to a close.

I had arranged with Bob Mehl to go to London to confer with some of his people. I telephoned Bob and then took the first boat train from Dieppe. The harbor was a wreck, yet we were able to cross to England safely by staying in the de-mined lane.

On the boat, an officer of our Seventh Army explained how they operated in cities and at the Siegfried Line. When the Germans were well supplied and still had the will to resist, they were skillful and tough defenders. Our superiority in firepower, thanks to our ability to supply the implements of warfare in great abundance, enabled us to overcome even dogged resistance. Prisoners expressed surprise that we could maintain gasoline supplies so far from the coast. Happily, the English had the answer to that one — a welded pipeline. I saw two of them far out in eastern France. The line was laid across the English Channel by winding up a long length on a huge reel on a boat and reeling it out the way a cable is laid. I never heard what became of that fabulous project.

On the train to London, I was seated with an Air Force colonel, a few other officers, and a nurse. She was on her first leave after months of ministering to the needs of our wounded and was going to London to recuperate.

In London, I met friends from the United States who had come over for the Technical Industrial Intelligence Committee — Jim Gill of Vanadium Steel; Alex Field of Rustless Iron and Steel; Rudy Schatzel, the copper expert from Rome, New York; as well as Bob Mehl. Gregory Comstock of Firth-Sterling had run across some interesting hot-press operations in Germany and had had one of the presses shipped back to the United States. Most of my time was spent on matters of mutual interest with these people, and also on getting untangled from red tape.

One day, Bob and I went shopping for books at secondhand stores and then ended up at Claridge's, where we relaxed and had dinner to soft music from behind a screen. Another diversion was a cocktail party at the English Speaking Union for visiting Americans. I also saw some of the bombed sections of the city, although nothing compared to what I had seen in Germany.

That visit was relatively short, and I was soon riding through the rural districts, heading for the Bovingdon airport and Paris. On the way, I was told Clark Gable had been stationed nearby. He had made a fine impression on the English with his manners and deportment. He was just another soldier and not a glamorous movie star.

During a short delay in Paris, on my way to Heidelberg, a paratroop officer of the 101st Airborne Division recounted his experiences at Normandy, Arnhem, and Bastogne, and on foot with the Third Army to Innsbruck. He had interviewed German field marshals who said they knew they were defeated after our Normandy invasion. I learned later that Hitler had commanded his western forces to defend their positions to the last man — a harsh lesson in futility. Incidentally, Battelle had a young man named Black with the 101st. He had been all through its campaigns, and after the war told me he had not been touched, not even a sprained ankle.

One day in Heidelberg, we were shown the boundary between East and West Germany — the "Iron Curtain" — to which the U. S. had agreed. I could hardly believe the map that was spread before us, and I haven't yet recovered from the shock. Of course we had nothing to do with political matters, but we had targets in the region that would soon be turned over to the Russians free, plus all the innocent victims. On that account, ALSOS was given information before the map was revealed to the public. We were told we should go there at once, while our army was still in control. On my way I stopped in Göttingen and paid Professor Pomp another visit. Then I headed east in my jeep.

My first stop was a "target of opportunity" — the mountain factory at Nordhausen where the buzz bombs and rockets were made to destroy London. A huge tunnel went through the mountain. At the entrance were the raw materials, inside were the workshops, and at the exit a few finished bombs. It was left just as it had been when in production. I confess to a bit of scrounging in the shape of a caliper micrometer. Another bit of scrounging provided the best eraser I ever used. It was about one inch square by one-eighth of an inch thick. This reminds me of some scrounging that Bob and I did not do. We were admiring some drafting and calculating equipment at the iron institute. Bob commented on how easy it would be to take it with us, but we didn't engage in that kind of scrounging.

During my visit at Urach, Professor Köster had told me about the work of his physics group under Dr. Fritz Förster. That included the Kawimeter, a highly sensitive device for detecting surface flaws of metals. (The Kawi stands for Kaiser Wilhelm Institute.) He gave me the name of the Lorenz Vacuum Tube plant at Mülhausen where the device had been installed. It was used to test each length of sealed-in wire of molybdenum as it dropped through a tube. Lengths that had die marks or other harmful imperfections were automatically cut out into a scrap box. The plant superintendent told me that the use of the Kawimeter had reduced tube leakers to practically nil. As an old lamp metallurgist, I held that to be an excellent contribution.

My next stop was Bitterfeld, where I. G. Farben had its chemical and magnesium plant for producing the well-known alloy Elektron. I had visited this plant in 1925, when investigating the aluminum-from-clay process. In 1945 it was still, with no signs of life or activity. The occasion for my visit was a tip on magnesium that I had received in London.

A few executives and key personnel were there, and they were pleased to have someone to talk to and to show around. There was nothing new on magnesium, but they did have a prize possession — a gigantic 30,000-ton press. I obtained a few photographs of it as souvenirs. It had been very effective in aircraft production, providing a quick and easy way to turn out propellers and structural members. I was told that when they lost a propeller plant they didn't bother to replace it because their press could easily turn out all that were needed.

Upon discussing aluminum alloys I was told that extruded bars of the 4.5% zinc and 3.5% magnesium alloy ran 70,000-psi tensile strength. That was the alloy that Dr. Guertler disclosed to me in Berlin in 1925, and has been more recently developed in an important series of high-strength alloys.

In Leipzig, my destination, I stayed at the CIOS house. It was a fine, large house near a beautiful park, and had been the residence of a Nazi. He had shot his wife, his daughter, and then himself. One of our men acquired a P-38 revolver and a fine camera. That reminds me of a story that went the rounds. A soldier was looking for a house for temporary quarters in a town that had just been taken. Approaching one house, he heard a shot, and as he reached the entrance a frantic maid appeared and asked, "My master has just shot himself. What shall I do?" Our quick-witted soldier replied, "Get me his revolver."

In London, I was told that a Leipzig foundry had developed an excellent technique for making steel castings. My informant suggested that I, a metallurgist, should look into it. "If we could turn out castings like those, we could use them in airplanes." Not one to let skepticism deter me, I looked up the foundry in question — those "steel" castings were made of magnesium.

I was also told in London that a Leipzig firm was doing exceptionally fine work in X-ray technique. So I went there and, after announcing myself, entered the office. Upon looking around, I saw a couple of Fräuleins at their desks, but no menfolk. In somewhat naive fashion I asked, "*Ist niemand da?*" (Is there no one here?) One of the Fräuleins answered quickly, "*Ja wir sind hier.*" (Well, we are here.) They did tell me about their work, but I saw no one on X-ray technique.

The principal purpose of my visit was to check on gas-turbine-blade manufacture at a Junkers plant. I was accompanied by the plant superintendent and the secretary of the general manager. Between them, they were acquainted with all the details. The plant was of medium size and had many machine tools for individual operations. Although turbine blades are of different sizes and of odd shapes, I figured that a relatively large work force was used. They prepared for bombing by erecting brick walls that served as revetments around the machines.

I inquired about the high-temperature alloys that they used, in view of their alloy shortage. The highest alloy they were permitted to use was 35% nickel and 20% chromium with titanium additions. Also, the blades were hollow, to provide access for cooling. I selected a small blade for an example, and found out later that it was 3% chromium steel. While I was there, the announcement came over the radio that the Russians were to take over Leipzig. Along with disbelief that our President had agreed, there was consternation.

On one of the trips around Leipzig, we drove past the huge granite monument that commemorates the allied victory over Napoleon after his disastrous retreat from Moscow. A much smaller battle had been fought there in 1945, and the technical reason for a visit to the scene was to secure samples of German steel cartridge cases. They confirmed what I had been told.

My way back to base took me through Jena and Weimar. The famous Carl Zeiss plant at Jena was my immediate interest. The town and the Zeiss factory had suffered some shell damage, but I easily recognized the university town where I had lived thirty-four years previously. It was still a little medieval town.

At the Zeiss plant, I was told that the man I wished to see was out and would return in a half hour or so. I learned, too, that my conversationalist of 1911, Fräulein Johanna Sondhaus, was still there, but time did not permit me to look her up. Instead, I made another visit that turned out to be a highlight of my whole trip.

The Schiller summer house is not far from Zeiss, so I decided to visit it. I found it easily enough — somewhat run down and with a large shell crater in the yard. I had no idea whether anyone would be there, but I rapped on the door and waited. The door opened and a puzzled, middle-aged woman appeared. Her expression seemed to say, "What does a man in an American uniform want?" I explained that I had some time on my hands and would like to visit the Schiller house. That did not get me an invitation, so I told her about my student days in Jena, in 1911, and of my interest in Goethe and Schiller, and that I knew about the times they had spent together on this spot, and of the significance of the stone bench out there in the corner of the yard. I talked on at some length to persuade her that I was sincere in my desire to visit Schiller's house. Then she invited me in.

My guide and hostess told me about the objects that were still there and of others that had been placed in secure storage. There were the bust of Goethe, among others, on the first floor, Lotte Schiller's bedroom on the second, Schiller's bedroom on the third, and his study in the attic. It was a simple ménage for so famous a person. She also showed me the outdoor kitchen in one corner of the yard, and, in another, the stone bench where Goethe and Schiller used to sit and converse. Part of Eckermann's quotation of Goethe, on the occasion of his visit in 1827, was on a tablet nearby: "In this arbor, upon this bench . . . we have often sat at this old stone table, and have exchanged many good and great words." Those two had coupled two of the greatest minds of modern times on that little bench.

By this time, she had warmed up to the subject of our conversation and had accepted me as one to whom she could unburden her pent-up feelings. We were in the home of Schiller, the great *Freiheitsdichter*, or poet of freedom, who had kindled the flame of freedom in the German people when under the yoke of Napoleon. "What a contrast," she declaimed with great emotion, "with Hitler and his unsavory dictatorship, which deprived us of our fundamental rights. It was by the evil use of force and fear that he brought his countrymen to heel. After luring us into his trap by scourging the country of the Communist cancer, and with the bait of social welfare and national prestige, came the day of *Machtübernehmung* [usurpation of power], after which he ruled with impunity. Schiller preached and lived in recognition of the great truths of freedom and human dignity." Then she continued, "Hitler tried to make us believe that those were impractical and vain figures of speech, but," and here she folded her hands emotionally over her heart, "I know here that Schiller spoke truly for us Germans, while Hitler betrayed us."

Her powerful outburst made a lasting impression on me. The setting was perfect and the timing was opportune. I have often thought that those few days must have been but a brief interlude, when she was free to express her true feelings and show her profound appreciation of what Schiller meant to her and to his countrymen. I confess that I left with a strange feeling of having had a glimpse deep into the German soul. Her parting words were, "Why must we now go under the Communist yoke?" I walked slowly back to the Carl Zeiss plant with her words ringing in my ears and contemplating the great good fortune that is ours.

My discussion with my man at Zeiss showed that they had produced only military equipment before and during the war, with no new developments. Later I learned that the best of the staff and equipment had been evacuated to our zone. Since then, Zeiss has again become a leader in optics.

A glance at the map showed that our highway took us through Weimar, the seat of the greatest epoch in German literary history. It was the home of

Goethe for over a half century. Instead of attempting to visit his residence, we turned off at the bridge over the Ilm, drove down and under it to a little park on the riverbank. We stopped there for our lunch of K rations. Those who have been there would recognize this as the site of Goethe's Gartenhaus.

There it was, just as I had photographed it many years before. Well, almost as it was. A poster in front read "OFF LIMITS," and although I was only an ersatz colonel, I knew what that meant. The yard had been untended, the grass had grown tall and unkempt, and the wooden fence needed repairs.

As my driver, Gerry Beatson, and I broke out our K rations I told him about the man whose retreat it was, considered by many to be the greatest genius who ever lived. Gerry was interested, prompting me to relate one incident that gave an example of Goethe's virtuosity.

Members of a small group of intellectuals at dinner were reciting their literary works, or would entertain with some personal contribution. The point here is that the poet Wieland was well established and that his works were quite familiar to the group. When Goethe arose, he extemporaneously composed poetry that was recognized as a take-off of Wieland in style and content. Everyone gasped in astonishment at this display of talent. An outside guest, who heard Goethe for the first time, asked his neighbor, "Is that Goethe or is it the devil?" His friend replied, "I say both."

Our repast over, we were soon on our way back to Göttingen and Heidelberg. There followed a few days on ALSOS reports, with but little time off to enjoy the sights of western Germany. One time, I browsed around a bookstore in Heidelberg and casually inquired for books about Goethe. The proprietor smiled and asked, "You mean the books we burned for Hitler? We have those over on that shelf." I selected a couple of books, which were one of the few items we were permitted to buy in German stores.

On my next trip I asked for Gerry Beatson again, but was told that I would have Lt. Brown. I learned again that Colonel Pash selected his men with skill. Since I was going far afield, I was given two drivers, and both were armed. This visit was to Munich to see Professor H. Borchers, the son of Geheimrat Borchers, and head of the metallurgical department of the Institute of Technology. His quarters in Munich had been destroyed, but I finally found him at Weihenstephan, a little town nearby. He was in charge of a pilot-plant operation on the carbon reduction of alumina by using oxygen. The idea was to make aluminum with a minimum use of electric power.

Professor Borchers had also developed a coated fabric for protection against radar detection. He had sent samples to Professor Wesch in

Heidelberg for testing, but had received no report on the results. When I interviewed Professor Wesch later, it was in the local jail. I learned that larger samples were needed before a definitive test could be run, and that ended it.

I also had a conference with Dr. W. Schmidt in Munich and was to see him again at Murnau on his work on light-metal alloys. On the way, we stopped at Bad Tölz, General Patton's Third Army headquarters. Incidentally, Munich had been badly hit, and the only buildings that were still sound were the city hall and the well-known Frauenkirche, the symbol of Munich.

One was not in Bad Tölz long before he realized he was in Third Army country, as the following incident will show. Brownie and I were strolling along the main street admiring the paintings on the buildings, and the Gutenberg house, when an MP approached, gave us a snappy salute and informed me that I was "out of uniform."

That was my first brush with the law, but I did realize that my trousers did not fit properly into my boots. I had never had any difficulty before, but ground rules are ground rules. Brownie was clever in such situations, and the MP permitted me to proceed on the promise that we would leave town. "But," he cautioned "if my major down the street stops you, be sure to tell him that I called you on it."

The next day we drove to Murnau, but Dr. Schmidt was not known at the address he had given me! However, in looking around, I ran into another of those experiences that one has under postwar conditions. I became aware of truckloads of people being driven northward along the highway. I asked a man who happened to be standing next to me what was going on. He told me they were Latvians and displaced persons being taken back home. He added that the Russians had just absorbed the three Baltic republics, and that the German Nazis were telling the displaced persons that they had to go home with the Russians. That was an attempt to get rid of them as informers, since they could identify Nazis for our military government. My informant commented on the evils of the Communist dictatorship. He was no slave and had refused to go because the Nazis could show no order to do so from our military government. Speaking of our military government, we got a laugh out of a column by Dorothy Thompson, a leading columnist of the day, to the effect that it had no plan or policy. If she had experienced the disturbed and unnatural conditions of those times she would have known that her gratuitous criticism was out of order. Things were just that chaotic.

My main target was the Metallwerk Plannsee at Reutte. Our route took us through Garmish-Partenkirchen, past the famous Zugspitze, the highest mountain in Bavaria, and into the snow-capped Tirolean Alps. The plant is well known in the powder metallurgy industry and was then in the charge of Dr. R. Kiefer, with Dr. W. Hotop in charge of production. Dr. P. Schwartzkopf, the founder, had gone to the United States in the 1930s to establish an American branch. I did not meet him until I returned in 1960.

It will be recalled that I learned about powdered iron and powdered steel parts when at the iron institute in Clausthal. Plannsee had been in full-scale production on rotating bands and machine-gun parts, as well as on other items. The former saved copper and the latter saved machining time, because even complex shapes were easily produced by powder metallurgy. Gregory Comstock, my friend of Carboloy days, also visited Plannsee and reported on his visit in detail. I brought back samples of the parts as examples of the new technique.

For the night, Dr. Hotop suggested a summer resort in Obladis, not far away, toward Innsbruck. That "Ob" means "above," or on a mountain top. Climbing up was a bit of a job for our jeep but the view was well worth it. The resort was then a refuge for youngsters who had no other home. The lady in charge welcomed us and fixed us up very nicely for the night. The next morning, we enjoyed a good breakfast, had another look at the scenery, and paid for the hospitality with a large assortment of candy bars and gum. The children had plenty of food, but candy was a very special treat.

As we reached the outpost of the district on our way back to Plannsee, we were stopped by a guard. His captain wanted to talk to us. After a telephone call, the captain of the guard arrived and wanted to know what our strange jeep was doing in his territory. (We had not checked in upon entering.) He added that if they had been able to start the engine, we would have had to walk down. Our driver, demonstrating the value of experience, had removed a part of the ignition system after he had parked the jeep. Brownie went to work again and we were soon on our way.

Back at Plannsee, I had another session with Dr. Hotop on the production of iron and steel powders. At lunch we joined an Air Force captain, who was in command of a guard detail at the plant. Being naturally curious, I asked him about his experiences as a bomber pilot. I then heard about the German jet and rocket fighter planes. He said we had no defense against the latter because they flew too fast to be hit, and that it was good that the Germans had produced only a few. They could make three passes at our squadrons with impunity before running out of fuel.

I commented that I had heard that a bridge is a difficult target to knock out. He agreed. He said he was glad that it was, too, and added, cryptically, that he deserved the Congressional Medal of Honor for failing to knock out one bridge. When I pressed for an explanation, he said he had tried to destroy the Remagen Bridge just a few days before we captured it. Furthermore, he had knocked out just one bridge and had lived to regret it. At my surprised look he added that that bridge was over a river between Reutte and his base at Kempten. Twice a day, he had to make a mean detour that would have been avoided if he had left the bridge alone.

Now back to Professor Köster and a visit with him. I found him greatly disturbed. The French had just taken away all his laboratory equipment, and now they had come for his library. The French officer was in his office at that very moment. I had other plans for the institute, so I decided to stop the removal of the library.

The French officer knew no German, but with my French and his English I got my message across — Professor Köster and his people were working for our army, and the library must not be removed. Apparently he respected my "authority," because he left barehanded.

A more critical situation developed with Fritz Förster. Professor Köster told me I should visit him in his physics laboratory in Eningen, so we jeeped over. There I saw what was really the start of his electronics plant, "Institut Dr. Förster," which later became prosperous and widely known.

I have mentioned the Kawimeter. Dr. Förster had another device that was sensitive to the earth's magnetic field. When I asked about it, he mentioned that it could be used to destroy a tank from an airplane. It could be adjusted to fire a projectile from a cannon as it sensed the change in the magnetic field near the tank. All the pilot had to do was to make a run over the tank. It had worked satisfactorily in a test — but was developed too late, for the Germans. That was one kind of proximity fuse, and I was the only one in the mission to discover one.

Dr. Förster also had a device for inducing a current in a metal or alloy that, by determining the resistivity, could identify the metal and get its hardness. That was the basis of the eddy-current method used in nondestructive inspection.

There was an aftermath. The next day, the French took him away for a special assignment. The Germans, as they retreated from France, had planted plastic mines that had to be located and removed. Ordinary mine detectors were of no use, but perhaps Fritz could devise a detector that would work.

I sympathized with the French, but I felt that they should use some other person because I was working on an idea for getting the two metallurgical institutes back in operation. Dr. Förster was important in my plan, so I discussed getting him back with Dr. Goudsmit and Colonel Pash, when I returned to Heidelberg. I received the necessary approval, and then did some serious thinking. Colonel Pash assured me he would "rescue" Dr. Förster if I gave the word, but I began to feel that the situation did not warrant so drastic a solution and, in addition, I kept thinking of those deadly mines. Fritz was established in a laboratory at Lake Constance with comfortable living quarters, and was well treated by the French.

I have mentioned my idea of reestablishing the operations of the two research institutes, and that I saw advantages in combining them. I had a discussion with both directors and they favored the plan. I hated to see their staffs marking time while they could be advancing metallurgical knowledge.

Like other wars, World War II had stimulated research, invention, and technology, the results of which presaged a new era in metallurgy and materials. I felt that the new technology would have to be supported by studies in basic and applied research. The two institutes had made significant contributions, so the longer they remained idle the greater would be the loss. As my knowledge of military government increased, however, I learned that it had ideas of its own.

Professor Köster had business to handle in Stuttgart, so he rode in with me from Urach on the last day I was with him. I recall standing on a street corner as we said Auf Wiedersehen, and looking out over the rubble that was Stuttgart. He had discussed the alternatives of restoring a city, or of moving to a new location and building anew. Then he commented, "When I look out there I just don't know what to think."

I had another trip down south, but it started with Hanau and the famous Heraeus plants. The vacuum-melting and nickel-alloy plant had been evacuated to a soap factory in Sterbfritz, a little town nearby. When I arrived there, the plant was being dismantled in response to the wishes of the citizenry. It was in the charge of Dr. Rohn, son of the founder. The citizens simply wanted their soap again.

The precious-metal plant, and ten years of laboratory records, had been evacuated to Onstmettingen in Württemberg; $2.5-million worth of precious metal had also been dispatched, but no one knew where it was. Hanau was in the U. S. Zone, so I tried to find the shipment. I was told later that the French had taken it.

Heraeus had worked on rhodium plating for mirrors for searchlights. That project was discontinued (too-low reflectivity), but it led to work on an infrared filter that would have the same absorption as water vapor in air. The filters had to be made to very high precision or they were not usable. The glass-plate surface and the salt film were controlled for smoothness and thickness to a few micro-inches with special Zeiss equipment. The work was continued in Onstmettingen under Dr. Auwärter. He let me have some samples.

A few years after the war, Dr. Auwärter came to the United States on a business trip that brought him to Battelle. We greeted each other like old friends and recalled the events of 1945 when we first met. Seems that our visit to Ontsmettingen had been an event. At lunchtime, Brownie and I purchased a bottle of milk apiece and ate our lunch out on a hillside. The citizens were incredulous that two adults would drink milk that way — a drink that in Germany was reserved for babies.

On our way back to our Heidelberg base, we almost met the crown prince, who was in residence in Huckingen at the Hohenzollern Castle. As we were approaching Huckingen, a Fräulein in a car ahead of us was waving to us. It turned out (not as you may be thinking) that she was the daughter of the keeper of the castle and was offering us the key to the gate. After crossing the drawbridge, we unlocked the huge portal gate and entered. We were greeted almost at once by a youthful voice from high on the wall surrounding the castle proper. "The crown prince sends his greeting but begs to be excused from giving an audience because he is having an afternoon nap. Please come in, however, and visit the public rooms." Thus we missed meeting a historical character, the son of Emperor Wilhelm II and pretender to the throne. I had seen him in Berlin, during my student days, when he used to ride in the Tiergarten.

In Heidelberg, we became aware of a flurry of excitement — Serge Jaroff and his Don Cossacks were performing for us that evening, out of friendship to Colonel Pash. I had heard them before, but never to better advantage, and we let them know that we greatly appreciated their visit. Colonel Pash, son of the Russian archbishop in San Francisco, participated too, and sang in both English and Russian. We were especially proud of him as he kept pace with the professionals.

The fiery little leader appeared to be truly inspired, and closed the performance with a toast. At the end, he worked up to a high pitch, quaffed his drink, and with a sudden gesture, dashed his glass to pieces on the floor. A little later, I commented to one of the Cossacks on the fervor with which his leader had toasted us. "Yes," was the reply, "he is a very emotional person and gives vent to his feelings." Still later, as one of our men was brushing up the broken glassware, I mentioned that Serge Jaroff had become extremely emotional. "Yeah," was the cool rejoinder, "he asked us ahead of time if we would object if he did that."

One time, in Heidelberg, I was asked to go to the city jail. There was a German prisoner there, Professor Wesch, who wanted to talk with an American officer who could speak German. I was selected. This man was described in "ALSOS," by Sam Goudsmit, as a "first-rate windbag and a second-rate physicist," and also as an example of a favored Nazi. Wesch tried to persuade me that he was a "good German," and a Nazi in name only. He claimed that he was an electronics expert, and that he could help Sir Robert Watson-Watt, the inventor of radar, if given the opportunity.

My conversation with him was in stark contrast with those I had with the German metallurgists.

While working on the reestablishment of the two institutes, I went to Frankfurt to discuss procedure with Colonel Zornig in the huge I. G. Farben building, and he told me some of the facts of life of military government. One institute was in the English Zone, and the other in the French Zone. The only rights we had were those of interrogation. Furthermore, we had no provision for handling what I had in mind, to combine them and set them up in our zone. Each institute would have to submit a plan to the military government of its zone, which would then accept or reject it. I had never felt more hopeless when confronted with a problem, yet I realized that the military had many difficult problems of a pressing nature. They would not look approvingly on one that could easily be dismissed.

After a brief visit to Wiesbaden for the OSS, I headed north again. When we traveled in the country, we would stop at a Gasthof (a few were open) and get hot water for soup and coffee. We were always made welcome, and I found it easy to engage the proprietress in conversation. For about the first time, Germans were free to speak openly, and it never took long to get around to atheistic dictatorships and how they failed to satisfy the needs of the people. One woman declaimed, "A country without religion — that can't be." Not one (man or woman) who mentioned it could understand how we could have Communist Russia as an ally and give in to so many of their demands. One man quoted Clausewitz, I believe it was, that war is an extension of politics, and said we were naive enough to let the Communists play the game to our discredit. "You should realize," he concluded, "that we're not all bad Germans and that we hate Hitler as much as you do." I came away from conversations of this kind with the feeling that the mutual antagonism between Nazis and Communists resembled that between the North Chicago Gang and the South Chicago Gang of our late Prohibition days — that the two were poured in the same mold.

One of our men, Dr. G. P. Kuiper, the well-known astronomer from Yerkes Observatory, returned one day from Jena with a special mission for me to handle. A Dr. König had been the top man on astronomical instruments for Carl Zeiss and had recently been released from custody by our troops. He had returned to Jena and wished to get word to his wife to join him. Frau Leonore König was in hiding in the home of Professor Zeigler in Osterode, and I was asked to deliver the message. I stopped at the house when I passed through Osterode on my way to Clausthal and Essen. Frau Zeigler, who came to the door, appraised me with searching eyes as I gave my message. I had to do a lot of explaining before she would admit to a total stranger that she harbored the refugee. Finally, Frau König came out and regarded me with a haunting and unbelieving look. I was face to face with a miserable human being, a refugee from Nazism. I went all out to assure her that her husband was alive and wanted her to join him, and that she was free to return to Jena. She came partly to life as this sank in, but she had given me a glimpse of Hitlerism. I wondered later if they were not jumping from the frying pan into a slow-burning fire under communism.

In Essen, I had an interesting session with Dr. Brühl, the Krupp quality man. Grain-size control had entered steelmaking, but they had problems with customers who specified just one grain size instead of allowing a range. Alloys for steel were a problem, but vanadium was recovered from blast-furnace slags and it became an important alloy. They also worked on columbium steel and titanium steel for machine-gun barrels as their best way to secure increased strength at operating temperatures. I had learned about this at the Kaiser Wilhelm iron institute. Bearing steel was a critical item. Dr. Brühl told me they found they could make it in the open hearth, although doubtless with some sacrifice of quality. I was a little surprised when he described flash welding steel spars for aircraft. I was introduced to the man in charge of steelmaking, and I had an interesting discussion of open-hearth practice with him. I was able to confirm some of the experience I had had at A. O. Smith on the importance of the silico-manganese block and on ladle rephosphorization.

Krupp, of course, was the major producer of heavy ordnance. They used acid open-hearth steel to secure adequate properties in the tangential direction, which also avoided flakes. That was the practice I had learned at Midvale, in 1916. Formerly, they had used explosive tests of prolongations of the gun barrels, but had switched to small hollow test bars from the wall. Split test coupons were a cause for rejection. They had also used notched-bar tests at low temperatures under the (mistaken) impression that the results simulated those of explosion or high-strain-rate tests at ambient temperatures.

Dr. Wiester, the associate director of research at Krupp, had been a student of Professor Hanemann. After getting acquainted through reminiscences about Charlottenburg, he explained their work on martensite as the aftermath of Hanemann's earlier studies. By means of high-speed cinematography, they had found that martensite forms suddenly, and not as a result of a metastable phase change. For example, martensite needles would be seen in one frame, although the preceding frame would show none. Professor Hanemann had discarded his conception of martensite as a phase in a metastable equilibrium — which speaks in forceful terms of his objectivity.

Dr. Wiester took me to the laboratory, where we discussed welding armor plate and bridges. They were surprised to find that they had an electrode that would not crack armor. Dr. Wiester ascribed it to the use of Krupp's Weicheisen, which is similar to Armco iron. I learned later from Toré Noren of ESAB in Sweden that it was their basic rod. Actually, it was a low-hydrogen rod.

Hitler had to have bridges for his Autobahn, which prompted the use of basic Bessemer steel and welding. That also resulted in numerous cases of brittle fracture, as happened to our ships. In that connection, they had not yet learned of our unhappy experience and, had they been told, probably would have discounted the report. Their failures occurred in sections that were one inch or more in thickness. Furthermore, their designers learned where to use riveted joints, but that was not a sound solution of brittle fracture. As I observed one time, a rivet hole initiated a brittle fracture when a bridge girder was hit by a projectile, although it produced only a ductile dent a short distance away where the projectile hit. (This again demonstrated the difference between the notch effect and the impact effect.)

Dortmund was not far away, so I decided to interview my friend of Charlottenburg days, Dr. E. H. Schulz. His house in the city was a complete wreck, and I found him and Mrs. Schulz in a simple room in temporary quarters. They were both in good spirits in spite of all they had been through. He had been chief metallurgist at United Steel Works, which led to a good discussion of steel quality versus welding. He was strong for making steel in the furnace and not in the ladle. During the thirty-two years since Charlottenburg, he had had a fine career in the steel industry and was one of Germany's top metallurgists.

I enjoyed meeting friends of earlier times and I benefited greatly from discussions of technical and other subjects. Duisburg-Huckingen was also nearby, so I dropped in on Dr. Werner Knapp of A. O. Smith days. He was assistant works manager of the big Mannesmann plant. He, like all the rest, was just twiddling his thumbs, although he had been ordered to start a small production for the British.

The need for reducing the nitrogen content of steel had been recognized, and he was developing a side-blow converter process. It was later taken up by U. S. Steel on a pilot-plant basis. On a project at Battelle, Clarence Sims had shown that the process was an improvement over the Bessemer process in that respect, although Dürer's oxygen process soon took charge in the field.

Werner made a cryptic comment one time. He said, "I thought Eisenhower wanted a bridge over the Rhine." I told him that I thought so too, but what was on his mind? He then told me that there was a perfectly good, and unused, bridge at Huckingen. Some troops had come partway across and then had turned around and gone back. Presumably, they had thought that the bridge was strongly defended, but the Germans had only a few Landsturm troops (militia) of very little value in defending that bridge. I had never heard of that incident.

I was soon on my way back to Paris via Cologne and the Rhine. Cologne was really a wreck, having sustained terrific bombing and shellfire. The cathedral, the only building not badly damaged, stood out even more prominently than before the war. It paid silent tribute to the marksmanship of our bombers and gunners. A bomber pilot told me he had a red circle around the cathedral on his map plus strict orders to avoid hitting it. The railroad station, however, had been badly damaged by bomb and shell fragments, and it was onlv a short distance away. Incidentally, I feel that a serious mistake was made when the city was restored. That cathedral deserves to have a beautiful park in front of it, to set it off properly, yet the view has been, and is now, obstructed by buildings that are too close.

At Bad Godesberg, a youngster pointed across the Rhine to the hotel "where Chamberlain was in 1938." He had a wry smile on his face, and I did not pursue the subject.

Speaking of Rhine bridges, we visited the famous Remagen Bridge. It had become a wreck by then, although there was evidence of fighting. I have read since that Hitler sent some SS men to execute the officers who were in command of the defending troops.

After spending the night in Verdun, I arrived in Paris the next morning to close out my affairs with the army and our Paris office. While there, I heard of an amusing incident that involved Colonel Lindbergh. He had flown over on a special mission for the Army Air Force and was with us for a few days. During the flight he engaged the radio operator in conversation and was asked, "Is this your first flight across?" "No," came the calm reply, "I crossed first in 1927, then again in 1936 and now this time." That passed quickly around the Royal Monceau.

One day I had lunch with a naval officer, a lawyer and an assistant of Mr. Justice Jackson, who was in charge of our war-crimes prosecution. He had just interrogated several SS men who had murdered an American aviator. They were cold-blooded killers of the gangster type and he described all of the gruesome details. They would get a fair trial, but were doomed.

While in Paris, I received a letter from Clyde Williams about a special mission for me in England. This time I flew there. The day was clear, and as we crossed Paris the city had the appearance of a scale model, an interesting sight. The highlight of the flight was the charming countryside of England. In London, I was fortunate enough to secure a room in the Cumberland Hotel at the Marble Arch. During my first stay in London, I had had to stay at the Lincoln Hotel. It was an old, old one and Cockney English had made communication difficult.

I spent some time with U. S. government officials passing along information I had gathered. Bob Mehl had returned to the United States, but I had good sessions with Professor W. B. Plank and Bradley Stoughton, who had taken over the London office of the Technical Industrial Intelligence Committee. I had known Bradley since World War I days, but I became much better acquainted with him in London. He invited me to his apartment for the evening and we discussed our work and points of mutual interest. He was an excellent companion. I also had a few missions of my own, one of which was to look up Dr. Desch to have lunch with him at our embassy. I had met him in Sheffield, in 1925, and again when he came over to the United States, and those meetings were always stimulating and educational. He was a truly admirable person and his knowledge of metallurgy was encyclopedic. Dr. Desch was one of our most productive Men of Metals, and was universally loved.

Some business with the Ministry of Supply involved a meeting with Dr. Hubert Sutton, arranged to pass information between us and the English. One subject was the Nimonic superalloy of Mond Nickel that had become important for gas turbines because of its excellent high-temperature properties. I also had lunch with Drs. Griffis and Pfeil of Mond Nickel for a special discussion of Nimonic and work to be done on it to increase its useful service temperature.

After going through channels successfully, with Hubert's help, I visited the National Physical Laboratory to review the work of Tapsell on combined stresses. By means of ingenious methods, and test specimens of suitable shapes, he tested Nimonic and other alloys when subjected to combined creep and fatigue at elevated temperatures. Although the technique differed, his investigation paralleled that of Hempel at the iron institute in Clausthal.

This work impressed me strongly as the basis for improving jet-engine performance. As soon as I got back to Battelle, I started to have Howard Cross, our high-temperature man, go over to look into it. I had in mind that Tapsell had told me that our piston engines would be obsolete as soon as the operating temperature of jet engines could be increased somewhat. That is precisely what happened — a fine example of the hidden role of metallurgical research. Then we ran into red tape that delayed Howard's trip to England for many months. When permission finally came, it did not extend to Germany, with their method of cooling the blading, nor to consultation with Max Hempel. Seems that one must reckon with red tape and adverse decisions!

Howard had an interesting trip, and came home with an amusing anecdote to tell. I had recommended the Cumberland Hotel and he secured a comfortable room there. The weather was raw and cold, so he closed his windows on arising the next morning, as Americans would do. When he returned in the afternoon, however, the windows were wide open and his room was powerfully inhospitable. The next day, he told the maid to leave his windows closed. That afternoon, his windows were again wide open and his room chilly. The following day, he tried strategy, strong strategy. He called the maid in and showed her candy bars and cigarettes. Then he asked her if she liked them, and upon receiving the expected affirmative answer, he told her she could have them if his windows remained closed. That did it; like a good scientist, he solved his problem.

The letter from Clyde called for me to go to Coventry to look into the production of tungsten carbide bullet, or projectile, cores and any differences between their practice and ours. It also brought out the effects of small changes in process details. Then it was back to London, and final preparations for leaving for home the next day.

We flew first by a British plane to Shannon, Ireland. A bus picked us up at the airport to transport us around the bay to Foynes, where we were to take a clipper ship, or seaplane (flying Pullman sleeper) for the crossing. By then it was time for dinner at the seaport landing. The table partner at my left was a Hungarian refugee named Biro, the man who made writing history. When we were having coffee, he took a pen out of his pocket and proceeded to write on a menu. He said it was a "ball point pen" and he was on his way to the United States to introduce it. He came from Argentina, and had been in Europe to sell his new pen. Finally, he put a piece of paper at the bottom of his water glass and demonstrated that his pen could write under water. He showed us the tiny ball at the end of the pen, and commented on the long life of one filling of ink. He was very optimistic about the future of his pen. The reaction of one woman, however, was all too typical of a common reaction to anything new — "But who wants to write under water?"

On the flight across the Atlantic, we slept in berths very much as though we were on a cloud. The next morning we landed (watered?) at Botwood, in Newfoundland, for breakfast. The flight to New York was beautiful, but uneventful except for the final ten miles down on the water of Long Island Sound. That was my fastest boat ride and was exciting, to say the least.

After a few days in Washington on reports, and closing out government business, I arrived home in Columbus.

My tour of duty with ALSOS was both interesting and rewarding, although it lacked the intense war excitement that Allan Bates experienced. My scientific and technical findings were all preserved in Washington, along with many thousands of other reports from other missions and intelligence agencies that had been abroad. I never heard of any use to which they may have been put, but the informed reader may feel that some of those I have mentioned would have been of great value to us. There can be no doubt, however, that our mission performed a necessary and highly useful function. Sam Goudsmit, Colonel Pash, and others make that abundantly clear in their books.

Before I close this account of ALSOS, I would like to make some general comments on our bombing accuracy, and also on education under the Third Reich. One time, in Wiesbaden, I had lunch with a member of our bombing survey team. I told him I had been favorably impressed with our bombing accuracy in hitting desirable targets (marshaling yards, etc.) and avoiding others (the Cologne cathedral, Frauenkirche, etc.), although I had learned that heavy machinery escaped serious damage. The bomber pilot at Plannsee was well pleased with our bombing now that he was on the ground and could see the results. The strategic bombing survey, however, did not substantiate my impression of our accuracy.

Germany has been proud of its educational system since it was given new direction and composition by Wilhelm von Humboldt, in 1810. In my experience, I felt that the strength of the system lay in the high level of competence of the faculties. Not the least factor in attracting leading engineers and scientists to the teaching profession was the title of Professor, one that was highly respected and eagerly sought. An additional attraction was the opportunity for professional activities outside the classroom. There is a subtle difference between Germany and the United States in the way we use titles. When I was teaching, I was both Professor and Doctor, but I, like others, preferred Doctor, whereas Germans prefer Professor — subtle, yet significant. In Germany I am usually Professor.

My curiosity was aroused about education under the Third Reich through comments by both earlier and more recent graduates. The latter complained that education had taken second place to being a good Nazi, at least outwardly. Parades, celebrations, and meetings interrupted classes, but the student had to attend them if he were to retain a good standing at the university or technical school. I learned the effects of this practice from old-timers who told me that it took a couple of years of additional training, in plant or office, for recent graduates to complete their education and become useful employees.

Eighteen years later, I was able to attend an ALSOS reunion in Washington, and to renew my acquaintance with mission friends. Sam Goudsmit and Colonel Pash were prominent among the relatively few who made it. Both of them reminisced about ALSOS, and General Groves and Dr. Frederick Seitz were our principal speakers. One of my drivers, Gerry Beatson, had become a successful broker. Captain Wallace Ryan was an executive with Lea and Perrins, the Worcestershire Sauce company, and doing very well. He told me that although others had tried to duplicate the original sauce, none had so far discovered the secret.

ALSOS remains a fond memory. My experience with and in our army left little to be desired, in spite of the apparent confusion about my security clearance that had been generated during the previous year. We had comfortable quarters and good meals, excellent transportation in jeeps with friendly drivers, as well as beautiful country. Our ration cards took care of desirable and useful extras. I even enjoyed C rations and K rations, especially when eaten alongside a stream, or in a city park, or in a Gasthof.

The interviews with German metallurgists were always interesting, and were conducted in a congenial atmosphere. I noticed little difference in discussions with them and with our own people at home.

A comparison of German metallurgy with ours of those days would no longer be useful. However, at the risk of overemphasis of significant differences, I felt that their development of rotating bands of powdered and coarse-grain iron, powdered steel for machine-gun parts, and steel cartridge cases was especially important. Other intelligence agencies could discover what the Germans were doing, but I was able to report how it was done.

Although a country is advanced technologically, it pays to keep in touch with other countries whose top people make equally significant contributions to technology.

Chapter 8

Battelle International Division, 1952

In Europe to appraise the prospects for laboratories in Geneva and Frankfurt.

A few years after World War II, Clyde Williams began his plans to expand into Europe. As a preliminary step he established the Battelle International Division (BID) with offices in Geneva, Paris, and London. He commissioned me to survey metallurgy and welding in the industrial countries of western Europe prior to establishing Battelle laboratories. My mission was to explain the Battelle program and to inquire into attitudes toward, and the prospects for, sponsored industrial research.

On this tour, I interviewed management and technical personnel during visits to plants and laboratories in the metal industries. Thanks to the World Metallurgical Congress, I also met friends who paid back hospitality they had received the year before. Fortunately, I was able to have Edyth, my wife, with me, and to introduce her to Europe. She tells about her experiences in her book "When Spring Comes Home to Europe and the Middle East." While I was busy during the day, she made good use of her opportunity to browse around many interesting cities in Europe. Evenings and weekends, we had free time for dining with local friends, or to make excursions into the country.

On March 11, 1952, we embarked on the *Liberté* and arrived in Paris on March 17. After a pseudo-French dinner, we left by sleeper for Geneva, where our principal European office was located. The next morning, we went to the Hôtel de la Paix, which became our Geneva home.

After indoctrination in Geneva we went to Zurich, where we stayed at the beautiful Hôtel Baur au Lac. My first interview was with Max Huber, chief metallurgist at the Franz Fischer Foundry in Schaffhausen. I had heard glowing accounts of the Fischer Method at the Lebanon Steel Foundry, and it was the reason for my visit. Max took kindly to the possibility of studying his problems on a laboratory or pilot-plant scale.

While in Zurich, I spent a day at Sulzer Bros. in nearby Winterthur. My host was Werner Felix, chief of the materials testing laboratory and a participant in the World Metallurgical Congress. His company is an important producer of metal goods. The subject of brittle fracture came up readily, especially because he was studying the use of tests for various applications. By trying different notches, he had shown that a test bar with an effective notch in it will break with a brittle fracture, if broken at a sufficiently low temperature. The appropriate notch to use would depend on the application, or on the object of the test. He had also included the controversial Schnad test bar, which, as I recall, had not demonstrated any advantage over other test bars of more conventional design.

Felix was interested in the amount of plastic deformation that precedes or accompanies brittle fracture, if any does. There had been conflicting evidence on this point. He took X-ray-diffraction patterns of the fracture surfaces of test bars that had been broken at different temperatures, to detect the presence or absence of deformed metal. When the energy to rupture was at the low level, brittle fracture occurred with no discernible effect on the diffraction pattern.

Leaving Zurich, we traveled overland on a mountain railroad to Interlaken to see the beautiful Alps. We had planned to stay at the National Hotel, but it was closed and we went into town to the Jura Hotel. This resulted in an interesting incident.

As we were enjoying dinner, a charming young lady approached our table. "Would you mind my joining you? I see you are Americans." She oozed personality and her voice was nicely modulated, and of course we would be pleased. Our table talk was quite informal, but she looked around frequently as if she were expecting someone. She explained that she had just come from Rome and was to meet her husband from the United States at the National. They were getting in a skiing vacation together. She had come to the Jura hoping that he would find her. He did, too. After dinner Edyth asked me if I knew who she was. Of course I had no idea, and then she told me, "Ingrid Bergman, that was who she was." I retorted, "Yes, of course, and I am Clark Gable, remember?" Edyth insisted and told me later that she had checked with the desk clerk. (Ingrid, were you actually there?)

Our next stop was Milan, where we stayed at a hotel adjacent to the imposing cathedral. That night, we heard an ominous vocal rumbling that came up from the street. It was a multitude of angry Communists. On walls, we read their slogan, "Ami, go home."

I was in Milan principally to see Dr. A. Reggiori, director of the Breda Institute. He had been in my tour group during the World Metallurgical Congress, and had taken graduate work at Carnegie Tech under Bob Mehl and Vsevolod Krivobok. I also contacted Mario Signora of Falck. They knew what was needed for research and production, and also that new practice should be preceded by laboratory and development experience. There was no difficulty in finding worthwhile projects to improve the technical and economic position of the Italian steel industry, but the necessary financing would not be easy to provide. On the favorable side, they welcomed the idea of our cooperation.

While I was busy with Reggiori, Edyth enjoyed the beauty of the cathedral and explored the shops in the vicinity. All was peaceful for a while, but later she was suddenly and unexpectedly in the midst of a milling mob on the plaza. They were mostly Communists on a wild demonstration. She was in danger amid the club swinging, but managed to escape to the hotel, safe but ruffled.

From Milan, we rode to Florence on one of the new Rapido trains, forerunners, I believe, of the luxurious express trains that now ply between major European cities. Florence without some sightseeing is no visit, and we did not miss the pleasure of seeing the cathedral with its adjoining basilica, the Pitti Palace, the Ponte Vecchio, etc. Upon retreating, the Germans had blown up all bridges except the Ponte Vecchio. The little shops, the statue by Benvenuto Cellini, and the secret overhead passage were left intact. The Medicis used that passage to go across the bridge safely from one palace to another.

My contact was Dr. Paolo Lombardo of the Italian Metallurgical Co., the principal Italian producer of aluminum. The concept of a private research institute was not new to him — the Breda Institute was similar to Battelle — which led to a discussion of its role in industry in general, and in Italy in particular. Our talk developed the need to make a substantial capital investment; I felt we would have to keep in mind, also, a corresponding commitment to a staff. Italian industry was not large, according to our standards, but it was fairly well diversified. Similar diversification would be required of a research organization confronted by such a situation if it were to handle a satisfactory volume of work. That situation and its consequences will come up again. On the other hand, I felt that financial and technical support would be useful to Italy in her industrial evolution.

Dr. Lombardo demonstrated that he was a good host by taking us to dinner at one of those gourmet restaurants in a cellar, the Bocca Lape. The aroma of a huge roast beef greeted us, immediately stirring the gastric juices. As we finished our excellent dinner, about ten o'clock, he asked if we would care to go to his office. I thought it a little odd at that late hour, but, of course, we would like to. After driving around on a personally conducted tour, we finally drove up to a Medici palace. I was still surprised as we entered, but we soon saw a curious mixture of the old and the new. The walls and staircases had paintings, sculpture, and decorations that had been there in the Middle Ages. In the center of the rooms were the files, desks, and office paraphernalia of a modern business office. The government owned the palace, but rented out office space to Dr. Lombardo's company.

Our next stop was Rome, with the Colosseum, the Forum, the Palatine Hill, St. Peter's. All came up to our high expectations. An unexpected treat, however, was a marble statue in the Borghese Museum. Rembrandt painted with such skill and precision that his people, and especially his fabrics, looked real. It was the same with that Bernini statue, a human figure frozen in marble. I wondered what might be the dimensional difference between that statue and an ordinary sculpture if measured in mils, or even in micro-inches. That mood, however, did not interfere with our enjoyment of the exquisite artistry of the sculptor.

I visited with Messrs. Colari and Vergillo, of the University of Rome and the Italian State Railways, respectively. As usual, these visits were to discuss the general research and development situation, and specific programs and problems.

Then, by train, up the beautiful west coast of Italy, past Pisa with its problem tower and on to Genoa, where Columbus is honored. One of my contacts was Dr. A. Scortecci, director of Finsider (the steel combine that is partly owned by the government) and in charge of research and technology. He had been in my tour group. In the evening we were guests in his apartment, which overlooks the bay. The next day, I visited his laboratory at the Genoa plant. His laboratory did not measure up to modern requirements for supporting an important steel company, but apparently that was traceable to a shortage of funds. For technical information, Dr. Scortecci explained that he had to rely heavily on his library, where a vast amount of literature had been collected. Each article was abstracted for convenient reference, and the more important ones were translated for staff use. For retrieval, Dr. Scortecci was one of the first to devise a system to match the huge volume of literature. He was proud of his metallographic work, and rightfully so. His exceptionally fine technique was clearly evidenced by the collection of his slides, projected on a screen for my benefit.

In Milan, I had been referred to Dr. Umberto Guererra, director of the Italian Marine Welding Bureau and founder of the Italian Welding Society. His laboratory was in Genoa. From his comments, I found that his problems were similar to ours. He was posted on the work of our committee on ship steel, and seemed to be favorably impressed.

Dr. Guererra told me of an interesting problem that he had just solved. A welded structure near Rome had failed from a weld crack of unexplained origin. He had first reproduced the crack in a laboratory study, then had diagnosed the cause, and followed through by finding how the joint could be welded without cracking. The answer was discovered by changing the welding technique to alter the stress system under which the weld metal solidifies and cools. I remembered that later, during my consulting practice, when confronted with a crack in a penstock weld.

A few days later, we arrived in Paris. We found it to be gray, rainy, and chilly — winter still lingered. Taking a cue from Browning, I exclaimed, "Oh that spring were here now that we are in Paris." Of course, inclement weather does not keep one indoors, especially on a first visit, as this was for Edyth, to Paris. We had spare time to go places and see things, including dinner at the Café de la Paix, and Berlioz's *Damnation of Faust*.

My first order of business was a visit with my friend of the tour, Dr. Georges Blanc. I have already mentioned him as a live wire, and I daresay no one took back from the WMC more information, useful information, than he did, nor left a better impression on his hosts. Georges took me to the headquarters of the government's Foundry Technical Organization and then to his large research laboratory in Sèvres, a suburb of Paris. The central office counseled on all facets of the foundry industry, while the research laboratory advanced the science and technology of foundry practice. Although the latter was in the growing stage, it was so well staffed and equipped that I doubted if Battelle would be commissioned to aid the foundry industry.

Georges announced that we would visit his museum to see an exhibit of the work that got his laboratory off to a good start with the foundry industry. It turned out to be a collection of stoves of the kind used to heat rooms. The iron castings are thin-walled and of relatively large surface area. There is more art than meets the eye in the making of those stoves. Georges had developed the necessary technology. While in Paris, I also had the pleasure of a visit with Georges's brother, M. L. Blanc, chief metallurgist at Citroën.

That Saturday, Edyth and I went to St. Germain-en-Làye for luncheon and a pleasant visit with Georges and his family. After only brief acquaintance with his mother, it became abundantly clear that Georges came by his enthusiasm and ability naturally. In the afternoon, we visited the palace of Francis I and SHAEF headquarters, traveling on the new four-lane superhighway. On this occasion, Edyth had a session on the Bible with Georges and his friends.

A live subject in those days in Europe was the Low Shaft Blast Furnace Project. The committee in charge was holding a meeting in Paris that Clarence Lorig, the Battelle representative, was supposed to attend. Since he could not make it, I was asked to represent him. The design of the furnace was discussed in detail, because this furnace was supposed to fill the need for small operations, and possibly be availble for experiments on blast-furnace practice. I met Dr. P. Coheur, and arranged to meet him in Liége.

IRSID, also in St. Germain, was next on my schedule. It is a large and well equipped research institute for the French steel industry, but is so well known that it is hardly necessary to relate the details of my visit. I also visited the president, M. Melcor, in the Paris office. The idea of sponsored industrial research was new in France, yet I had no doubt that it would be truly beneficial if conducted there as it was in Columbus. IRSID covered research and development for industry broadly, but I had in mind the numerous small firms that had special and local problems yet lacked the necessary funds to support a research and development program. Not long after I returned home, a small steel mill sponsored a project of that type at Battelle-Geneva.

With three days' free time at Easter, Edyth and I took a tour of Normandy, one of many intersting tours that are available for regional sightseeing. At Caen, where the English sector of the Normandy landing beach had been, we visited the ruins of the castle of William the Conqueror. A certain Gil le Sage was an ancestor of Edyth's (née Sage) and a leader in the army that invaded England in 1066, so we had a personal interest in the castle. We next drove through the bocage country, whose ridges of dirt with closely entwined tree roots had provided protection for the defending German troops and had made our advance in 1944 extremely difficult.

That night we stopped at Avranches on the south coast of the Normandy peninsula. Nearby is the centuries-old monastery and castle, Mont-Saint-Michel. It surmounts a massive rock formation, just offshore, and is a truly impressive sight. A visit to it is well worthwhile to all who appreciate these gifts from early times. Moving on, we were soon in the Loire Valley, the château country of France, with romantic old castles, fruit trees in bloom, ancient cities, and wineries.

Photographs of Blois in travel literature had always intrigued me, and now I saw it. Farther up the valley we paid a special visit to Château Anjou to see the famous tapestries depicting the Apocalypse of St. John. We left the tour group at Tours to visit a friend of Edyth's at Château Condé.

Some years ago, two brothers Bedaux came to Grand Rapids, Michigan, to bring order and prosperity to the furniture industry there. There they met two sisters, married them, and took them back to France. It was Charles Bedaux, the head of an international consulting firm, who acquired the château. Upon the death of her husband, Fern Bedaux, who was a good friend of Edyth's, took it upon herself to rehabilitate the firm after the war, and continued to live part time at Condé. The Prince of Condé had been a famous general during the Thirty Years War and, later, under Louis XIV. His generalship made it possible for him to have one of the most beautiful palaces in France. A visit to the château was a singular treat; it is so romantically equipped, and contains so many relics of historical and cultural interest in the great hall, living quarters, library, and dining room.

I thought that Mme. Bedaux's organization and Battelle might have mutual interests, and took advantage of this visit to explain our work to her. She arranged a meeting for me with her president, whom I saw in Paris. Incidentally, the Duke and Duchess of Windsor were married at Condé in 1937, and posed for photographers on a porch that overlooks the grounds and forest. I settled for a photograph of the porch.

Next, we were off to Brussels. Like France, Belgium had a government agency to assist its industry and agriculture. It goes by its initials, IRSIA. After checking in at IRSIA headquarters in Brussels, I went to Liége to see Professor Coheur, who was in charge of one of the two government steel laboratories. Almost immediately, I got into a dispute with their man in charge of gas and foreign-inclusion analyses over the validity of the fractional vacuum fusion method. He said it was inaccurate, but he could not explain away the results that Merrill Scheil and I published for the standard oxygen samples of the Bureau of Standards. That settles one point, but I know it doesn't settle all.

Dr. Coheur was full of life and ideas, reminding me of Georges Blanc. The steel industry had awakened to the need of some scientific support, and to the realization that it could no longer rely solely on craftmanship and experience if it were to grow and be healthy. Coheur was a principal in the new movement, and doing quite well too. From the industry standpoint, however, he was limited to industrywide problems, very much as IRSID and similar laboratories were limited. Individual units of the industry have problems or plans they wish to work on that they wish to keep confidential, or that are solely of concern to them. I kept that situation in mind, because Battelle fills that gap nicely.

Belgium, like Germany, was largely dependent on the Thomas, or basic Bessemer, process for tonnage steel. With the advent of welded construction (Belgium's brittle bridges and Germany's Autobahn bridges) the conventional process was running out of gas, so IRSIA and Coheur were doing something about it: working to improve the notch toughness of commercial steel. However, that is another story, and it comes up in a later trip to Europe.

Professor Coheur had to attend a meeting in Brussels, so he drove me back in his car. It was a Ford, a so-called large car, which prompted me to ask, why not a small car? He replied, "Those little things have no power," and I soon learned what he had in mind. It was "full speed ahead and damn the torpedoes," by which I mean narrow roads, curves, and even the towns we went through. I had visions of chickens getting hastily out of the way and cackling lustily as they did. On a later occasion, I also went into Brussels with him, but this time we rode the train. He confided that he liked the quiet ride better — less excitement.

Back in Brussels, I visited Dr. St. Cyr, at the Belgian Welding Society, for a discussion of bridge failures. He told me to contact Professor Soete at the University of Ghent. I did that the next day, but I went first to the Forges de Clabecq, at Tobize, for a discussion of steelmaking with Jacques Houbaer. He had attended the World Metallurgical Congress and was enthusiastic about his experience. At noon, Edyth met me at the railway station to go to Ghent. At the university, Professor Soete had his secretary take Edyth for a tour of Ghent and Bruges.

After preliminaries in his office, Professor Soete took me to his laboratory to see the work in progress. Dr. De Graff, of the Royal Netherlands Iron and Steel Co., had developed a notch test for steel that had the virtue of testing the full section thickness. It had attracted attention in Europe, as Noah Kahn's Navy tear test had in the United States. Professor Soete was studying a modification of it that gave more uniform results. Whether it was simply less discriminating is a good question. One can conceivably get variable results even with a perfect test, because of variations in the steel. Van der Toorn, et al., have shown this to be correct for commercial mild steel plate by analyzing the results of Charpy V-notch test bars.

I was quite favorably impressed by Professor Soete's work on the brittle fracture of welded joints under three-dimensional stress systems. I was equally impressed by the design and simplicity of his testing machines. He has carried on in this field, and has developed into a leading authority on the performance of welded structures.

While at the university, I had a pleasant visit with Professor De Sy, who was head of the department of metallurgy.

We ended our Ghent visit, as guests of Professor and Mrs. Soete, with a drive around town and dinner at a restaurant that had a view of trees, greensward, and a little brook. Back in our hotel, the next morning, we had *a* strawberry for breakfast. Customarily, one has strawberries (plural), but in Belgium one has a strawberry (singular). They grow them three inches in diameter. I photographed one in a display window to show to skeptics. After a sightseeing tour to the Cave of the Han, where the river flows right through a mountain, and along the beautiful Meuse Valley, Holland was next on our itinerary.

We expected to see windmills everywhere, but they are now few and far between. It is cheaper to use electric pumps these days, although there are still a few windmills operating so the tourist can take a picture to show back home. Furthermore, these are dewatering pumps, which pump water from the land into bays and canals. Another sight is the tulip beds that make you grab your camera.

My first contact was Jan Palm, who had been on my tour the year before. Fortunately, I had a weekend free, so I rented a car for the Palms and the Hoyts to take a sightseeing tour. Jan was an excellent guide because he is well posted on the history of his country, with emphasis on architecture and genuine antiques. I was especially interested in the massive dykes that are highly publicized. One time, when we were riding along the top of one, I asked Jan, "Would you show me where the hole in the dyke was that the little boy plugged with his finger and saved you from being flooded?" All I got in reply was a broad grin.

After a visit with the Palms in Beverwijk, and dinner at our hotel, Jan posted me on research and metallurgy in Holland. As one man whom I wished to contact, I mentioned Dr. Geerling, of Royal Dutch Shell Oil, in my best Dutch. Jan's face was blank. After a supreme effort by both of us, Jan's face lit up as he said, "Oh, you mean Dr. *Geerling*," only in real Dutch.

Jan made an appointment for me to visit him. Shell, like other oil companies, was interested in brittle fracture and weldability. Geerling, like many others, was doing intelligent work, but different in its details. It is unfortunate that there is not more uniformity in the concept of what needs to be tested, in how tests should be conducted, and in how the results should be interpreted. This situation exists in spite of national and international conclaves, and individuals and committees who attend to such matters.

My next stop was Delft, to visit the metallurgical laboratotry of TNO, the research organization for Dutch industry. After my introduction to Professor Bratsma, head of TNO, I discussed their work in metallurgy and welding. Actually, I was lucky to get there, because I had gone astray in the rain. The administration office in town had given me directions. I was to go just so many canals one way and then so many bridges another way, but after thinking I had done that I was still not there. Happily, a baker boy on a three-wheel cart came to my rescue. I sat on his box and he drove me along his route until we were within pointing distance of the university building where Professor Bratsma had his office. In the meantime, Edyth had been exploring museums and china shops, and had come away with some prize possessions, as usual. We had lots to tell each other when we met at the railroad station, also as usual. Back in Amsterdam, we packed and set out for Geneva, where I wrote my inevitable reports.

We were soon on our way to Sweden. In view of my known associations with Germany, and also of its industrial importance, it is strange that I did no more than pass through that country. Bert Thomas, in charge of BID, told me he did not want to risk endangering Battelle's possibilities in Germany. He had the mistaken notion that I was compromised because of my work with the ALSOS Mission. I knew he was wrong, but he was too far from the scene for a sound decision.

We decided to go by train to Düsseldorf, for some weekend sightseeing, and then by air to Sweden. We found Düsseldorf partly restored, but with the results of World War II bombing still in evidence. We did have comfortable quarters in a new commercial hotel.

The next day we took a train for Bonn, with a brief stop in Cologne, to view the beautiful cathedral. At Bonn, we crossed the Rhine by boat to take the cog road up to the famous castle Drachenfels, high above the river. That was the same trip I took in 1912 with my friend of Jena days. After taking some pictures, and lunch in the castle restaurant, we returned to Düsseldorf feeling that we had had a very pleasant outing.

Toré Noren met us at the Göteborg airport, with flowers for Edyth. He was chief metallurgist of ESAB and a well-known leader in the welding industry. He had been on my tour. Toré and his boss, Bjorn Kjellberg, the chief engineer, were excellent hosts and were helpful to me. ESAB was a leader in the welding industry with solid achievements to their credit. One was their "basic electrode," which permitted the welding of armor plate without cracking it, although they didn't know why. It was the first of our low-hydrogen electrodes.Bjorn and Toré had us as dinner guests that evening; the centerpiece on the table was a United States flag.

To continue my discussions with ESAB, Toré flew to Stockholm with us. After we were settled in the Grand Hotel, he took me to Jernkontoret to meet Sven Fornander, technical secretary of the Swedish Ironmasters Association, which functions in the interests of the steel industry and assists in the export of high-grade Swedish steel, sponge iron, and iron ore. Starting with J. A. Brinell, of hardness-testing fame, it has operated a technical bureau that fosters scientific and technical work. Mr. Fornander was very helpful in giving me good advice on how to spend my time. We were pleased to be the guests of Mr. and Mrs. Fornander for our final dinner in Stockholm.

I called on Professor Axel Hultgren at the Technical University, a modern-day leader in Swedish metallurgy. He had been a student at Charlottenburg under Dr. Hanemann just after I left, so our talk included reminiscences of those days as well as discussions of metallurgy. He had had a distinguished career, which closed fittingly with the leading chair of metallurgy in Sweden. A major contribution of his, and a boon to all who process hardenable steels in thick sections, was his method of controlled cooling to prevent the development of hydrogen fissures. The producers of steel rails used his method to make sound rails, thereby greatly reducing the incidence of rail failures. Vacuum pouring is the modern method for handling hydrogen.

Edyth and I were his guests at dinner one night, along with other friends. There we became acquainted with a national dish, cold spinach soup with a cold poached egg in the center. We also learned something about toasting at dinner parties. Each guest must toast the hostess, so she must use some special technique when returning the toasts, or she will pass out from overindulgence.

The Metallografiska Institutet had become famous under Professor Carl Benediks, but was in the charge of Professor Erik Rudberg at the time of my visit. He deserves a high rank among eminent Men of Metals, and I can heartily recommend a visit to his institute, even though it is off the beaten path. It fills the need for a more profound understanding of metals and their behavior, yet leaves for others the necessary studies of applied or technical metallurgy.

Mr. Fornander recommended a visit to the Bofors Co. at Karlskoga, where Nobel had set up his explosives works and steel plant. It is one of the better high-grade steel mills and produces ordnance materiel. My contact was Mr. Amberg, who was in charge of the laboratory and who later became professor of metallurgy at the Technical University when Professor Hultgren retired.

I saw at firsthand the care and skill with which Mr. Amberg handled the problems of a quality steel mill. To use a cliché, he put science into the art of steelmaking. One of his major studies was the old one of foreign inclusions, with emphasis on their identification as an aid to their control or elimination in steelmaking practice. One of his techniques was a spectacular method for examining steel for foreign inclusions. After nicking a specimen, he broke it open to provide a (ductile) fracture face. He then blued it with a temper color and examined it in a binocular microscope, which revealed the inclusions on a brilliant background. A brittle fracture was useless for providing this observation.

Mr. Amberg had a problem that he had thought he might want Battelle to work on, and asked me to discuss it with his assistant, Dr. Åke Henz, a younger man and a keen student of metallurgy. From his exposition, it was clear to me that Battelle would be a logical choice for a sponsorship. That summer, Åke came to Columbus to get the project started, and returned again toward the completion. While in Columbus, we had him for a party, with Battelle friends. The Swedes are very formal socially, as Edyth learned in her visits with the ladies, but he seemed to enjoy our informality in dress and behavior. He even removed his coat in the presence of the ladies, as we did.

When discussing steelmaking with their chief metallurgist, Mr. Wahlberg, I was impressed by their generous addition of Swedish sponge iron. That is the charge material the Sheffield steel companies used when making their high-grade tool and cutlery steel. Bofors used two induction furnaces of twelve-ton capacity for some of their production. They were the world's largest at that time, he told me with pride.

Mr. Wahlberg had some friends in for dinner and a musical evening, and included Edyth and me. His excellent record library provided an interesting treat. It was then that we met Dr. and Mrs. Weibull. Although drinking was very minimal, I noticed that the guests all returned home by taxi. The Swedish law on driving after drinking is strict.

Dr. Waloddi Weibull was science advisor of research at Bofors, and had an excellent laboratory for explosives and steel behavior under high strain rates. He is very well-known for his work on fatigue of metals, but is also an oceanographer, a violinist, and a gracious host. Mrs. Weibull invited Edyth to come to their home to speak to a party of local women who were interested in the Bible and biblical archaeology. On the day we left, Mrs. Weibull and her two daughters dressed themselves in their native costume to bid us goodbye, providing photogenic material for my camera. After a plant tour of a factory inside a mountain, we bade Bofors goodbye.

Our next stop was at Borlänge to visit the Domnarvet Steel Mill, the principal producer of commercial steel grades. My contact was Åke Josefeson, chief metallurgist, whom we met at dinner, along with other personnel. Count Bo Kalling joined us briefly to extend greetings, as he was leaving for Stockholm. As is well known, Professor Kalling was conducting interesting investigations in steelmaking. This work came up the next day when I was with Dr. Josefeson, discussing it and other points of mutual interest.

Edyth was taken on a tour to Falen and the famous Stora Kopparberget copper mine. It is the oldest company that has been continuously in business, and was the producer of revenues that financed Gustavus Adolphus for his incursions in Europe during the Thirty Years War. Edyth also saw 300-year-old log cabins whose exteriors had turned to stone from red paint made from copper slag. A highlight was her visit to the home of Selma Lagerlof.

I thought that the contact that held the most promise for BID was the Sveriges Mekanförbund, the association of the Swedish machinery industry. Mr. A. Björklund, managing director, was very favorable in his attitude toward what Battelle had to offer, and stated that they needed our type of industrial research. Apparently, they found but little application of basic research in their work. The idea of our studying the operations of a company to set up a research program appealed strongly to him.

I regretted the lack of time that prevented a similar interview with Mr. S. Venner-Carlsson, managing director of Jern och Metallmanufakturförening. It is the organization of the fabricating industries — stamping, forming, pressing, welding, enameling, etc., and I had been told by informed people that this branch of the Swedish industry was in need of applied research.

I had been warned to be cautious on one point. The Swedes were very proud of their Metallografiska Institutet, and any suggestion that BID would compete with it would be regarded unfavorably. Actually, I had nothing of that sort in mind. Furthermore, during my visit with Professor Rudberg, he looked favorably on the Battelle program as a method of taking advantage of our experience and know-how.

Our final night in Sweden coincided with "Welcome to Spring," the celebration of the coming of May as the harbinger of sunny days, after months of limited daylight. The college graduates paraded and sang their songs, while citizens danced among blazing bonfires in Skansen Park.

We visited Norway and Denmark, but our time for them was limited. Oslo has a small steel mill that uses cheap electric power for melting. That was the only study of Norwegian metallurgy that I made firsthand. I had to forego a visit with Professor A. Winterbottom at the Norwegian Institute of Technology at Trondheim. He had been on my tour, and had read a paper at the congress that reviewed their situation in metallurgy, research, and education.

I do have a word of caution, should you ever travel from Stockholm to Oslo by sleeper — get a good night's rest before you go. We were told that we would not be disturbed at the border. Yet, in the wee small hours, our compartment door was opened and the lights were turned on. Just someone wanting to know how much Swedish money we had. I guess we hadn't reached the border yet, because we had gone back to sleep before we were awakened again the same way. A man insisted on seeing our passports. I say insisted, because his intrusion had started an argument. Back to sleep again, whereupon the customs inspector went through his routine. The

Norwegians repeated the performance on their side of the border, and we were fit to be tied. Finally, a waitress opened the door, turned on the lights, greeted us with a smile, and cautiously asked if we would like an early cup of coffee. She had doubtless had trouble before. I presume the best way to go is to fly.

In Copenhagen, I met two friends who had been on my tour and now were to post me on metallurgy and research in Denmark. I came away convinced that our type of industrial research would help them, technically and economically, to advance faster than would be possible if they were left to their own resources. They had able people to work with, but no organization that was staffed and equipped like Battelle.

In spite of limited time, we were able to see something of Copenhagen, thanks to Professor Borge Lunn of the University of Denmark. After a visit to his laboratory, he took us to Tivoli Park, which is rightfully famous for its beauty and diverse attractions. We left in time to witness the changing of the guard at the royal palace. Their formation in marching was especially interesting and differed from anything I had seen before. During a visit to an old castle, we learned why the staircase was built as a close right-hand spiral going up. This gave the advantage to a defender, who would be higher up than the attacker. The former could use his sword effectively in his right hand, which would not be true for the attacker. They didn't tell me how a left-handed person would fare in either position.

While I was occupied with metallurgical matters, Edyth was out on the town window-shopping. She stopped at the Hôtel d'Angleterre, having heard so much about it. As she was leaving, she stopped to buy postcards. The clerk told her, "Your General Eisenhower is coming in." Edyth turned, and Ike was just three feet away and saying to his companion, "Now take your steel industry . . ." (He must have known something about such matters. Anyway, we came away with this Ike anecdote.)

That evening, we were entertained at dinner by the Bent Christiansens. He was chief metallurgist of the Danish Steelworks.

Time being always of the essence, we flew to Geneva, where I wrote reports and closed my business with our office there. That Sunday, we went on a side trip to Chamonix, a charming nearby town in the French Alps. One can start from there for exciting rides up into the mountains. Of course, Mount Blanc, the highest in Europe, is the central attraction. We went on one of the easier rides by téléferrique to a nearby mountaintop. Our concierge at the Hôtel de la Paix had recommended a ride from where we were, high over a valley to another mountaintop that was much higher. After one look at a car that was going over and up, we decided that our mountain was high enough. We did enjoy the grandeur of that scenery and wished that we could stay longer. But Edyth had a lecture engagement in

London, and I had business in Geneva and Paris, which called us back to business on Monday.

Edyth flew to London, and I took a night sleeper to Paris. I had hoped to meet Professor Portevin, a distinguished French metallurgist, but my time was short and he was not available. He had published on foreign inclusions in steel, but I felt he would benefit by using the polarized-light technique that Merrill Scheil and I had published. My principal contact was M. Grolleau, president of the International Bedaux Co., for a discussion of possible mutual interests with Battelle. I went then to London, by way of Calais and Dover, and joined Edyth at the Mount Royal Hotel, near the Marble Arch.

At that time (1952), England was on an austerity program. I encountered it head-on the first morning at breakfast. I saw an English couple eating, and apparently enjoying, some good-looking sausages. I ordered some, too, and prepared mentally to enjoy a good English breakfast after those continental imitations. One bite, and I was sadly disillusioned with whatever had been substituted for meat.

One day, I visited the British Welding Research Association on Marylebone St. (Some of the London names arouse one's curiosity. This one came from "Marie la Bonne.") Many of us were seriously interested in the characteristics that steel should have for welding. There was still great confusion in this field, which added spice to our discussions of underbead cracking of hardenable steels and brittle fracture of structural steels. I thought that they had the latter under better control, or understood it better, than the former. Thus, in Europe, I found them ready to make more use of fully killed and fine-grain steel in the normalized condition than we were. The same held for finishing hot rolling at lower temperatures to end up with the steel having a finer grain structure.

Our discussions of the mechanism of underbead cracking of hardenable steels was less satisfactory. As may be suspected, that was because our views were divergent. It is easy to say simply that hydrogen is the cause, and, indeed, using low-hydrogen electrodes eliminates those cracks. If we are to understand underbead cracking, not just how to avoid it, we must go further. Of two different steels of the same analysis, tensile properties, and hardenability, one will crack because of hydrogen, while the other will not, although exposed to it in the same way. The same is true for the same steel having two different microstructures. Simply saying that hydrogen causes underbead cracking falls far short of accounting for these differences. Nor does it account for crack-free welding with the atomic-hydrogen arc, which produces large volumes of hydrogen in its most active state. Thousands of aircraft parts were successfully welded with hydrogen-producing electrodes by the simple expedient of controlling the microstructure.

I was fortunate to find Louis Reeve in London for lunch and reminiscences of our days at A. O. Smith. The weldability test that he developed at Appleby-Frodingham was popular because it bore directly on the ability to make a sound joint under restraint. It is used, in principle, as the Controlled Severity Test (CST). Welding literature discloses a variety of tests of steel and weld metal that impose different degrees of restraint, cooling rates, and so forth, but there was more chaos than order in this field.

I also visited the National Physical Laboratory, and went over the work that Dr. Reese had presented to the World Metallurgical Congress the year before. I believe his work sheds light on one problem, the effect of adding another element or an impurity to iron or steel. He avoids the confusion, whether recognized or not, that must occur when the base metal already contains impurities. When the Charpy bar with V-notch is used to test commercial steel, the impact values change gradually over a range of the testing temperature. Dr. Reese showed that his highly pure iron changed abruptly in one degree Celsius from a high to a low impact value, although he used the V-notch. Inasmuch as a 2% manganese addition still gave an abrupt transition, although the transition temperature was lowered by 250°F (at 0.05% carbon), one would very likely assume that impurities were responsible somehow for the gradual change. He made a significant observation: The problem of controlling the notch behavior of steel "may resolve itself into one of controlling the effects of very small amounts of such elements as oxygen, nitrogen, phosphorus, and carbon, which have low solubilities in alpha iron."

During this week in London, I had very little time for the pleasures of sightseeing — Edyth took care of that — but I was able to contact a number of people for discussions of metallurgy and research. Among them were Hubert Sutton; Harry Gordon, our representative in England; Col. Charles Guillan, secretary of the Institute of Metals; H. G. Taylor; C. L. M. Cottrell; P. T. Houldcroft; N. P. Allen, director of the Metallurgical Division of the National Physical Laboratory; and Edward Epremian of our Naval Research Office.

On our first Sunday, we took a boat ride down the Thames, past the Tower and the Bridge, to Greenwich, and back. Later, we visited Westminster Abbey, where I ran across a statue of an ancestor, the Duke of Argyll. Then on a Saturday, another day off, we joined a tour to Oxford and Stratford-on-Avon. It is a treat for an American to visit places he has known previously through reading, as it is educational to realize as never before the meaning of our English heritage.

In Oxford, Edyth noted a geographical connection with our home in Columbus. We lived on Blenheim Road, which is a few miles from the center of town at Broad and High. Oxford also has an intersection of The

Broad and The High, which is about the same distance from Blenheim Castle, the residence of the Duke of Wellington and birthplace of Winston Churchill. Very few people would be aware of this coincidence, which maybe it was.

The highlight of this tour was Shakespeare's *The Tempest* at the memorial theater in Stratford. The English have done very well by their Bard, not just with the theater and performances by outstanding Shakespearean actors, but also with the park on the Avon River, Anne Hathaway's cottage, and the memorial building with dining room. I recalled what a thrill it was when, as a high-school student in Minneapolis, our English teacher told the class that a Shakespeare performance was coming to town, and then prepared us for it.

The next day, we were guests of the Hubert Suttons, starting in Windsor. They were excellent guides as we walked through the public rooms of the castle and around the grounds. The English know their history, and the stories of famous people and historical events that the Suttons told us brought us close to them. We also saw the playing fields of Eton, where the Battle of Waterloo was won.

After a look at Nell Gwyn's house, roast beef and Yorkshire pudding for Sunday dinner, and a ride through the country, we arrived at the Suttons' in Farnsborough. It was peaceful place, surrounded by terraced lawn and flower gardens. That part of the Sutton establishment provided Hubert with a hobby that I am sure was an acceptable diversion from his demanding work at the Royal Air Force establishment nearby. That night, we were on a sleeper traveling to Edinburgh.

Time was running out, and Scotland was a must. My mother was a Campbell of the Argyll clan. Being half Scot, I had always wanted to visit the land of my ancestors. The next morning, we were in Edinburgh, which is pronounced Edinbro, although I have never learned why.

As one leaves the railroad station, he is greeted by an impressive view — the central valley that divides the city in two, beautiful gardens, and monuments to famous Scots, such as Sir Walter Scott, Gladstone, Livingstone, and Hume. After getting oriented, we took a tour of that interesting city.

The Castle, high up on a towering hill, dominates the city. It is larger than it appears from down below, with plazas, courtyards, memorial halls, and chapels. Our guide gave us its history, at times in dramatic fashion. After passing through the main portal, he stopped and pointed back up to a small barred window. "That," he said, "is where they kept the Duke of Argyll prisoner before they took him out and — swish — off with his head." Later I asked why. "Oh, the Argylls were Protestant Covenanters. The Catholic McGregors were in power, and they got rid of all Covenanters they could capture. Several hundred of Argyll's followers were boarded up in a stockade with no exits and simply left there." The Roman Catholics and the Protestants played rough with each other in those days. The Campbell clan was not gentle, either.

We stopped in the chambers where Mary Queen of Scots had lived. We looked far down the steep wall from which her little son was lowered in a basket to freedom. That little baby became King James I of England, who is well known because of the King James Bible. One could have an interesting time just browsing around the Edinburgh Castle.

Edinburgh has a "Royal Mile," with the Castle at the top. On our way down, we passed the house in which had lived a man who was a respected councilman by day and a bad robber by night. His double life was the inspiration of Stevenson's "The Strange Case of Dr. Jekyll and Mr. Hyde." After viewing the cathedral and other landmarks, we arrived at the bottom, at Holyrood House, a favorite palace of royalty. The Queen's flag was up, so we could not visit it.

Back in town, I took time off to procure some shoelaces. I complained to Edyth afterward that I had to settle for short laces because they didn't carry my long kind. "What is so wrong with short laces?" she wanted to know. I admitted that they were all right, but that I preferred the long ones because when one broke I could still use it. She just guffawed, "You're Scotch for sure, but this is Scotland — remember?"

Our next stop was Glasgow, close to Argyll country. When we were in London, we had had lunch with Herbert Harris, chief metallurgist of the Glasgow plant of Babcock and Wilcox, and a friend from A. O. Smith days. I had told him that we planned to visit Glasgow and the surrounding country, which prompted him to place a car and a driver at our disposal.

Our car picked us up the next day, shortly after breakfast, and we were off for the country, Argyll country. We visited Loch Lomond, the Trossachs, and surrounding hill country with green in abundance and trees and gardens in bloom. We stopped for lunch at the Buchanan Club. While we were there, our path crossed that of another motion-picture star, Katharine Hepburn. She was appearing in Shaw's "The Millionairess," in Glasgow. She and her retinue were seated in a corner of the dining room. It was Katharine, all right, so when she left I gave her a smile of recognition as she was getting into her car. Back came a clever little hand wiggle, à la Hollywood, I presume. Edyth insisted that it was for both of us, and I let her have it that way.

The night train to London, a visit to Lloyd's, and after final arrangements, the boat train to Southampton to board the *New Amsterdam* and head for home. But first, London. We were fortunate in having Herbert join us for lunch, so we could express our thanks in person for our excursion of the day before.

My acquaintance with leading members of the profession abroad has made me feel, rather strongly, that we should invite more of them to deliver our annual and memorial lectures. In that connection, I had Rex Shepheard in mind for the Adams Lecture of the American Welding Society. He was top technical man in Lloyd's, and a leading proponent of welded ships, and of sound practice in fabricating them. He was out of town when I called, and I learned later that he was in the process of stepping up to the top technical position in English shipping. Instead, I saw Mr. G. M. Boyd, his principal assistant and a well-known figure in marine architecture. I was unable to accomplish anything definite, but I did leave the message that I intended to promote Mr. Shepheard for the Adams Lecture. I wished to secure at least his reaction to an invitation, should one be tendered by the American Welding Society. (That point is history. He did give the lecture.)

My session with Mr. Boyd developed an interesting point. Instead of using a universal specification for ship plate, they visited interested suppliers to examine the details of steel manufacture, the equipment in use, the results of tests, and service performance. Once they decided that a plant is capable of producing satisfactory plate, they write up a procedure for that plant to follow. They do not insist on a common procedure for all mills, but adapt one to a particular mill, which does not disrupt its regular practice any more than is necessary.

Once aboard the boat train, your fate is sealed — you are going home. My emotions at that time were mixed — with thoughts of home intertwined with those of the interesting days so recently experienced. Fortunately, the mechanics of getting from hotel in London to stateroom aboard ship are compelling and distracting; by the time of that first meal aboard, when seated with new traveling companions, a routine has started.

The return to the United States by ship gives one an opportunity to go over recent experiences in Europe, before becoming engulfed in problems and duties at home. I reviewed my mission in terms of what we should do, and how we should do it, to establish a successful operation in Europe. I had become aware of both positive and negative elements, but the advantages to Europe appeared to outweigh any difficulties we might encounter. I also felt that Clyde, on an earlier trip, had correctly visualized the conditions for a permanent operation.

The question of the confidential status of sponsored work would require correct treatment. European industrialists are much more secretive than we in the United States are. I discovered that personally in 1925, in England and especially in Germany. For example, before starting on our investigation of the aluminum-from-clay process, Metallbank required Chris Dantsizen and me to agree in writing that we would not disclose anything relating to the process to unauthorized persons without its written consent. (I was released from that obligation about twenty-five years later.) The Electric Bond and Share Co., on the other hand, just relied on us not to disclose their plans for Priest River power.

I ran into that attitude toward secrecy again in 1952, when discussing Battelle with managers. They looked favorably on the possibility of having work done that would not require them to make staff and equipment commitments. The conversation would proceed along that line, but would inevitably turn to the question — how were they to know that the results of their sponsored projects would not become known to their competitors, or even get back to unauthorized concerns in the United States? To put it bluntly — were we a listening post for our people back home?

There is a good answer to the question of secrecy, and, in fact, there must be; otherwise, our not-for-profit research institutes would find it difficult to secure sponsors. I would explain that our staff in Columbus was trained to maintain a project as confidentially as the sponsor desired. I also added that we consigned patent rights to the sponsor, and had to be careful to avoid leaks. This practice made group participation practical, and we were able to bring in two or more disciplines to work on a project. I would assure them that we had no problem of satisfying sponsors on secrecy. Otherwise, we could not have grown into an important segment of the research and development industry. They grasped the point readily, and I felt they were satisfied that we operated on an ethical plane.

The second point was the lone worker versus teamwork. I emphasized the latter as an advantage of the Battelle multi-discipline method of handling complex problems. In Germany, at least, a man in research wished to be assigned a problem or project and then be left alone while he worked on it. He liked the challenge, and welcomed the opportunity to demonstrate his ability to produce. I had a good example of that at A. O. Smith, with a moral.

I had a young German metallurgist on my staff who was a good example of the lone worker. He had written an excellent Ph.D. dissertation, showing that he was a promising research worker. One time, I assigned to him a project in which Walter Jominy was interested. It was not an easy problem to solve, and Walter needed the results, so we both followed the progress at the start. Apparently, our close supervision was not to his liking, because he asked me to allow him to work in his own way. I wished to have him agree willingly to do things our way, so I agreed to his request, with the thought that if he needed help, we would find it out and then act.

One day, Walter came to me and said that he had to have the results and that we should review the program. In brief, it had not progressed as it should have, but Walter and I brought order to the somewhat chaotic situation. Incidentally, we showed the young man that experience is of great help in research, as it is in other fields. I have already told about another project he was handling, and of how willingly he later accepted my guidance in converting a brittle alloy into a ductile alloy.

More recently, I have learned that the group effort by teamwork is practiced in Germany, greatly to the advantage of all concerned. I explained this phase of the Battelle operation to management personnel, along with its advantages when working on complex problems of modern technology. I would also point out that only a large and multi-discipline institution can operate satisfactorily in this way, and that only a few companies can afford a similar practice. The concept of a Battelle type of institution for sponsored research was new to Europe, but I felt that it was regarded with favor once it was understood.

Battelle established branch laboratories in Geneva and Frankfurt; management and staff were recruited locally, so we may say that one is a Swiss institute and the other a German institute, although both retain close ties with Columbus and operate according to the principles that have been proven to be proper and successful by the original institute.

I regretted the absence of Germany from my itinerary. It was a natural for me, and was the country for which I was best prepared. There was also abundant evidence that German industry was ready for, and in need of, the assistance Battelle could provide.

Professor Henry C. Wallich, of Yale University, tells the story of those postwar years, in the broad terms of economics, in his book "Mainsprings of the German Revival." The role of science and technology is buried in statistics and generalities, but to a technical man, who uses different tools to probe industry and its future, there could be no doubt that closing the technology gap was a major problem. Let us take a quick look at postwar Germany from this viewpoint.

Germany had been cut off from the outside world, as I learned personally in 1945, and its leaders were under interrogation, as I learned from correspondence. Available engineers and scientists were largely immobilized, or under strict Allied control, and their laboratories were either inactive, destroyed, or removed as reparations. Important segments of German industry had been cartelized into a few major groups, but after the war they were split up into numerous independent firms to generate competition. That was the situation inside Germany. Outside, new technologies had developed that were certain to guide the future of industry.

All these circumstances placed Germany in need of our assistance, more so than other countries that I visited. I do not suggest that Battelle would have been able to solve all the technical problems of German industry, or to close the technological gap. I do suggest that the ground was fertile, and in excellent condition for cultivation. Battelle would have a market of exceptional size and availability for its services, while, in providing them, it would benefit its sponsors directly, and industry indirectly.

It will be well to take a closer look at research in Germany to define a major field of an independent and not-for-profit research institute. There is no need here to consider conventional university research. Federal and state governments finance research in Max Planck institutes, and at universities, under the aegis of research councils. The projects are general or basic in character, and the results are available to all. Private firms are left to their own devices to carry out projects in which they have a unique or proprietary interest. Many companies in this category could use the staff and facilities of a large research institute to advantage. They would thus be relieved of commitments for staff and equipment, yet would be assured of expert handling of their projects. Furthermore, a Battelle institute would be broadly diversified, and more so than most companies could afford to be. Of course, the institute, once it had established itself, should attract sponsors from additional categories, such as governments, industry groups, and foreign firms.

I was well enough acquainted with German thought to feel that a local institute would have to be German from top to bottom, if it were to be accepted. That was also Clyde's intention. When it came to assembling a staff, the privilege of working under favorable conditions in association with a major United States research institute would attract highly competent personnel. That would be an advantage, because private firms in Germany, as well as in the United States, may have to rely on less than competent personnel, and frequently do.

The situation with a new research institute in Germany, however, presents a problem of acceptance, and therefore a problem for business. It can be likened, in some ways, to the situation of a new family that has just moved into the neighborhood. Germany had fostered research, and had accorded it a significant role in science and industry as it had enjoyed its fruits. Obviously, a new institute would be appraised with a competent and critical eye, and be evaluated in comparison with well-established institutes. That is one good reason for assembling a strong staff — acceptance by the local scientific and technical community, or prestige. Another reason comes into play when the staff communicates orally or in writing with sponsors and prospective sponsors.

My interest centered around metallurgy and welding, each of which had advanced substantially during and after the war. Primarily on this account, and also from my knowledge of the country, I felt that a German Battelle could and should readily develop sponsorships in both fields, as we had in the United States.

Germany would also be convenient for the Scandinavian countries. Bofors, in fact, had already sponsored a project at Columbus, under the supervision of Åke Henz. As I have noted, Sweden was especially promising for sponsorships in metallurgy.

The Battelle laboratories in Geneva and Frankfurt have established sponsored industrial research on a firm basis in Europe. My only criticism, although I cannot claim pure objectivity, is that the role of metallurgy and welding in their operations is relatively small in contrast to the leading role that I had visualized.

Chapter 9

Department of Defense, Frankfurt, 1959-60

Staff representative for Department of Defense in NATO countries; studies of metallurgical research and technology.

IN 1959, I was commissioned by the Department of Defense for a tour of duty in the NATO countries. My home office was the Research and Engineering Division, under Dr. Herbert York, and I reported to the Materials Section under J. R. Townsend, my liaison. My particular interest was research, development, and technology of the materials industries, although my terms of reference included special assignments. My anticipation of an interesting tour of duty was fully realized — a mild expression for a highly gratifying experience. And doubly true, because Edyth was included in my travel orders. At this time we were still living in Columbus, Ohio. When we returned to the United States, we moved to California, to our new home in Berkeley. This DOD mission also terminated my official connection with Battelle.

After the preliminaries had been completed, Bob Townsend asked me to report to his office for final briefing and sessions with Personnel. While I was with him, he received word that Deputy Secretary of Defense Donald Quarles had had a fatal heart attack in his sleep. That was a close call for me — an indefinite delay had been avoided, since Mr. Quarles had signed the necessary papers just the previous afternoon. That was a sad day for our country, and especially for DOD, where he was a dedicated and effective executive and an able administrator of his department.

It is not an object of this discourse to go into all the preparations for our stay abroad. DOD had no office in Germany, so I was assigned to the Army's European Research Office in Frankfurt am Main for administrative and logistical support. With travel orders for Edyth and myself, we loaded up the family car and headed for New York and the good ship *America*. My car had to be at dockside a day ahead of embarkation, and that gave us a day in the big city, and an evening at "The Music Man." Edyth had lecture engagements to fulfill and could not accompany me, so we waved farewells as the *America* was pulling away from its pier. The voyage over was enjoyable but uneventful. My table companions were a cotton merchant and his wife, from Bremen, which gave me an opportunity to brush up on German.

My first concern, on landing at Bremerhaven, was to recover my car and be on my way to Frankfurt. In other words, I thought of myself as a civilian tourist, but that illusion was soon shattered — I was in the Army now. I first became suspicious that something was wrong when I noticed questioning glances coming my way. Then a Navy officer approached me with the information that I had to have my car "processed," and provided with Army plates. That could not be done until the next day, Monday, and I could stay at the Army Dependents Hotel in the meantime. I began to get the message.

My new Navy friend drove me to the hotel, and I was soon in the hands of people who knew just what I was to do. I also learned something about rank. My rating of DEF-2 meant nothing to the Army personnel. They thought it must be a very low rating, judging by the GS system for civilians, but I explained that the DOD rating went in reverse order, and that I rated above a three-star general. "Then you are a VIP and you will get a suite," I was told.

Before getting my "U.S. Forces in Germany" plates, my car was inspected for mechanical condition, and I had to pass a physical examination and demonstrate knowledge of road signs and rules of safe driving. There are about a hundred signs and seventy-five rules, so I had to bone up for the tests. Also, before securing my Army driver's license, I was required to secure adequate insurance, with a Green Card as evidence of it. That cost me a hundred dollars more than equivalent coverage would have cost at home. (I wonder now why I didn't include that insurance on my travel voucher.) The Green Card is required by most European countries, and must be shown before crossing borders. Otherwise, you stay put until you have one.

Finally on my way to Frankfurt, my next experience was with German traffic. I found it to be quite different, with bicycles, motorcycles, pedestrians, tractor-drawn farm vehicles, oxen-drawn farm vehicles, lone tractors, automobiles, and trucks, including speeders. Then came small towns, Bremen, lunch at a Gasthof and conversation with a Fräulein, and, finally, the Autobahn familiar from ALSOS days, and Frankfurt. With only a minor indiscretion or two, I reached my temporary quarters in the Army hotel, Ambassador Arms. It is a conveniently located, motel-like hotel that the Army maintains for authorized transients. The staff was German, as at other military hotels in Germany. I transferred after a few days to the Bachelor Officers Quarters (BOQ) to await Edyth's arrival. Meals were both good and inexpensive at the officers' club in the I. G. Farben casino.

The next order of business was one of being absorbed into NACOM, the Northern Area Command of our armed forces. That meant more forms to complete in Personnel, financial affairs to arrange, getting my precious ID card, and becoming acquainted with local facilities, such as the PX, the commissary (supermarket), garage, filling station, post office, and laundry. For permanent living quarters, Personnel wanted to put us in a house, but we much preferred to be in one of their nicely furnished apartments, and had been advised by friends to secure one if possible. We were most fortunate in having one in the "State Department" complex on the plot of the Army sector. John J. McCloy, our high commissioner after the war, had had a group of apartment buildings erected for his people when we thought that Frankfurt would be the capital of West Germany. When Bonn was selected instead, those quarters were made available for such as me. And we enjoyed Ernst Schwendlerstrasse 9 immensely, as well as our view of the Taunus range in the distance.

Then came the day in July when a TWA plane from New York descended on the Rhine-Main airport, and Edyth arrived at last. Need I say that we enjoyed this experience to the hilt? If I do, it is to point up the fact that we metallurgists have glorious opportunities to enjoy life as we pursue our chosen profession — a continuing theme of this account. Except when she was lecturing in Stockholm, Malmö, Paris, London, Frankfurt, Stuttgart, Berlin, and Berchtesgaden, Edyth was able to join me on my trips — at no expense to the government, let me add. While I was busy with my contacts, she visited places of local interest, while weekends gave us time to go farther afield. But I had a job to do, and I plunged into it with real élan.

At the European Research Office, I was greeted by Col. Monroe ("Monty") Freeman, brother of Orville, the former governor of Minnesota. Orville Freeman was then serving as Secretary of Agriculture, and Monty was later to be with the Smithsonian Institution. I was given a desk alongside Lt. Col. L. ("Lou") G. Klinker, their materials man and an ordnance officer. They helped me arrange my program of visits, and that was a huge task. Bob Townsend sent over various lists of projects, and other material, to guide me. I was personally responsible for selecting and scheduling my visits, and for reports to the Pentagon. Lou Klinker was a pillar of strength in support of my mission, while our Master Sergeant Gordon handled many necessary details. He ran the office and handled transportation and expense vouchers, while keeping me in line with the regulations. Before completing my tour of duty, I had reported on one hundred visits that dealt with work with which the Defense Department was concerned.

A few words may be in order at this point on the activities of the Army's European Research Office, the Navy's office in London, and the Air Force's office in Brussels. Our military services support unclassified research in Europe through these offices. Two conditions must be met before a research project can be financed — a qualified investigator must request support, and his proposal must be of interest to some military agency back home that has funds available. After that, a contract is signed, and is administered by one of those overseas offices. The investigator reports on his work, and is free to publish his findings. An example, from many, was the preparation of a compendium of ternary-alloy systems by Professor Guertler, who had been my teacher in Charlottenburg. I doubt if this would have been done, however valuable to metallurgy, had not our Air Force supported it. Professor Guertler had had many years of compilation experience, and was an excellent choice. He died before I could visit him. Mrs. Guertler carried on his work. If someone were to review our contracted projects, I believe he would find that our armed forces had supported many able investigators, whose contributions were beneficial and worthwhile.

Sometimes, new developments were considered by one or another of the offices. Lou Klinker seemed to be investigating one or more most of the time. An example was the Wankel rotary engine, which later reached the production stage. Another was a new detergent that cleans even greasy work clothes in cold water. Dainty linens can be added with them and they will come out clean!

Our three offices did not act independently, as if in competition with one another. By way of maintaining teamwork and coordination, each office advised the two others of its programs, and was always ready with help or advice. In addition, tri-service coordination meetings were held every three months to discuss mutual problems.

In the field, one of my first contacts was Professor Werner Köster, director of the former Kaiser Wilhelm (and now the Max Planck) Institut für Metallforschung. We had last said Auf Wiedersehen as we stood viewing the rubble that was Stuttgart, during my ALSOS mission. We were happy to meet again, under distinctly improved conditions. He, and his assistants and students, had rebuilt their buildings and laboratories, and the metal institute was again in business — a big change from Urach, Reutlingen, and Eningen in 1945. The entire central city was well along with its rebuilding program, with new buildings, streets, parks, and public utilities — evidence of hard and constructive labor. A notable feature was the tall television tower surmounted by a rotating restaurant.

Professor Köster had assembled a staff of promising young men who were working under older department heads, such as Professors A. Seeger, E. Gebhardt and U. Dehlinger. A handy little brochure showed that the institute had an impressive list of publications on physical metallurgy and metal physics.

I had the good fortune to be there in the fall of 1959, attending the institute's twenty-fifth anniversary in Stuttgart, after its transfer from Berlin-Dahlem. That was occasion for a celebration and for dedication of their Special Metals Institute. Edyth and I were invited to attend, so we were able to observe a gala performance, staged by a top German organization.

The formal sessions were held in the new concert hall, and it was interesting to note that Stuttgart, as was also true of London, Berlin and other cities, had given priority to that sort of structure in their rebuilding programs. Invited guests and staff members lectured on their fields, contributing to current metallurgical knowledge. A truly exciting presentation revealed the possibilities of the new thermionic-emission microscope, which originated at the University of Tübingen under Professor Möllenstedt. Dr. H. Düker was in charge of the work at the metal institute. This instrument has less resolving power than that of the electron microscope, but it has the distinct advantage of direct observation of the specimen, even at elevated temperatures. Thus, we were shown the structural transformation of alpha iron into gamma iron and back into alpha iron. I believe that was the first filming of that phase change. Needless to say, I emphasized this technique in my report.

Germany was in the process of adapting nuclear energy to peacetime uses and recognized the vital role of materials. The metal institute had been selected to house work in this field at the new Special Metals Institute that had been dedicated during the celebrations. The traditional ceremony of passing the key constituted the dedication. It was the key to the front door of the new building, and started in the hand of the architect. After describing his work, with appropriate remarks, he passed the key to Professor Otto Hahn, president of the Max Planck Gesellschaft. Otto Hahn is well known as the first to split the atom. He was decidedly not the ivory-tower scientist or stuffy physics professor as he discussed the role of the new laboratory in a delightfully humorous fashion. He then pressed the key into the hand of Professor Köster, who explained technical features of the institute. Then turning to the director, he said, "Now, dear friend Gebhardt, I pass the key to you and declare the Special Metals Institute dedicated. It is in competent hands, and I wish you success." Dr. Gebhardt then told more of the plans for supporting the new industry, and explained that his staff was assembled and ready to move in.

Of course, there was a banquet. It was held on Friday night in the Kursal in Bad Cannstatt, a spa and a suburb of Stuttgart. We were seated with the Franz Wevers and Erich Schmids. Professor Wever had turned the direction of the iron institute over to Willy Oelsen, and was now a senator of the Planck Society. His new duties took him to various branches, as well as to outside institutes with interests in common. I had missed him in 1945, when he had been in custody because of his position as director of the iron institute. He assured me that he had been well treated and that he had appreciated the opportunity to speak English. I have had similar reports from other Germans who had been in custody.

I had met Professor Schmid in Berlin in 1925, when he was a specialist in crystal structure and a student of deformation structures. He, like other members of that group, attained international recognition in later years. In 1959, he was on the other side of the Iron Curtain — a professor at the University of Magdeburg in East Germany. I regretted the conditions that precluded acceptance of an invitation to visit. He would have been a good man for one of our research contracts.

Good food, colorful background music, and interesting table talk — not in any particular order — insured our having an enjoyable evening. Yet we were to have some frosting on the cake, even two frostings. As we became aware of a change in the music to the lilt of an early Bach minuet, we turned to the entrance of the hall. There, dressed in court costumes of the eighteenth century, we saw Professor Köster and his charming daughter leading a group in that graceful dance. The applause afterwards was so insistent that we were treated to a repeat performance. It was truly beautiful and in marked contrast to the zigzag of modern jazz dances.

A twenty-fifth anniversary calls for a retrospective view. That was presented by Dr. Fritz Förster in his inimitable style, with lantern slides and commentary. He was in good form, and to all who are acquainted with his exuberance, that will bring to mind the fun that we had, too.

One might wonder how Edyth fared in that German atmosphere. The answer was simple — Professor Wever had another opportunity to speak English.

The Kösters invited us to take coffee with them on Saturday afternoon, in their restored home. I took advantage of this occasion to ask a special question. Professor Köster had favored me with a replica of the crystal that is the metal institute's symbol. On one side is a stress-strain curve of mild steel with pronounced yield point. It symbolizes his early work as the first effective step in understanding that unique elusive behavior. On another side is, fittingly enough, the representation of a dislocation, which also suggests modern work in metal physics. In response to my question, Professor Köster explained that the idea of the crystal came from Dürer's "Melencolia." The etching shows the crystal among various scientific instruments of the day, although Dürer left no explanation of the thoughts of the philosopher as he contemplates the tools of science. Edyth procured a copy of "Melencolia" for home display, along with the crystal.

On another visit to Stuttgart, the Kösters took us to dinner at a favorite restaurant on the Rotenberg, one of the hills surrounding the city. It was serenely quiet and peaceful, and all was well with the world. I was suddenly inspired to recite Goethe's little stanza that he wrote in the cabin in the woods above Ilmenau. It begins "*Über allen Gipfeln ist Ruh,*" (roughly, "Peace bedecks the mountain tops."). It demonstrates the poet's rare ability to express his thoughts in beautiful and forceful words.

We engaged in a bit of reminiscing of ALSOS days, during which Professor Köster told us about his "imprisonment." I asked him to tell it in English so that Edyth could hear it firsthand. His rank in science had made it necessary to follow a directive from Washington that he be placed in custody. Just north of Stuttgart is the beautiful Ludwigsburg palace, a former resident of Württemberg rulers. That was his prison. However, since the first persons he saw when he entered the palace were three Nobel laureates, he felt better at once. He spent the time pleasantly, and in good company. It is surely to the credit of our military that they carried out the directive humanely.

I couldn't help wondering why the top scientists were confined. As a member of the ALSOS Mission I had become acquainted with many of them, and I could conceive no sound reason for it. We had turned down the Morgenthau proposal, and since we were presumably committed to revitalizing German industry, the directive could only delay the work. Insofar as I know, scientists who knew Professor Köster and the metal institute were not consulted on the utility of the order, at least I was not. From other information, it seems clear that the directive allowed no discretion on the part of our military. However, custody of top German scientists proved to be no hardship on them, judging by my information.

Leaving Stuttgart for one of my trips, we drove to Munich, the "town of the Little Monks." In 1945, only the Frauenkirche (the symbol of Munich) and the city hall were whole; in 1959 there was a new city, as in Stuttgart.

The metallurgical department of the Munich Institute of Technology had moved into town from the brewery in Weihanstephan to which it had been evacuated in 1945. Professor H. Borchers was in charge, as before. After greetings and reminiscences, he took me through his laboratories, but he explained that he would have new and modern quarters in a new building. Bavaria is not a metal-producing state, so his field was product metallurgy. He was no longer working on the carbon reduction of alumina, his war work, in spite of recent interest in that process.

I was especially interested in the sonic method (acoustic emission) of Dr. J. Kaiser, which produces signals when a test bar is first loaded elastically. Upon release of the load and reloading a second time, no signals were observed until the load exceeded the prior load — the Kaiser effect. Professor Borchers described an examination of a failed pressure vessel by this method. The stress in the wall at failure was determined, as well as the origin.

This called to my mind the early work on our welded Liberty ships, which brought steel to the attention of marine architects as the principal problem of the failures. They were surprised to learn that ships were breaking up at the very low stress of about 10,000 psi. That finding took a couple of years in time, and cost a huge sum of money, yet a few test bars from the ship *Schenectady* should have produced the same evidence. I had reported on this method back home, but learned later that Lessels and Associates were already studying it. It is only fair to say that acoustic emission has become an active field of research and investigation.

As an example of prematurity, I had attempted something of this kind when at the GE research laboratory in Schenectady. I was aware of "tin cry," and hoped that some sophisticated equipment might be available that would work for steel. To that end, I asked Dr. Langmuir to suggest how I might proceed. After a discussion, he told me to check with Dr. Hull, from whom I received a superheterodyne, a pickup, and circuitry. Our tests were not favorable and we stopped work and turned to more pressing matters.

While I was at the institute of technology, Edyth was at the New Pinakothek across the street, enjoying that famous picture gallery. By some quirk of the visiting hours of that day, they made her leave prematurely and unprotected in a rainstorm. When I joined her later, still in the rain, I heard the sad story, too complicated by a mix-up of names, and too sad to relate here.

Professor H. Neuber is also at the institute. He is the author of our wartime bible on the notch effect. It was translated and distributed to those who were concerned with the brittle-ship problem. He was then working on a new edition of his book. Professor Neuber, under an Air Force contract, had developed a modification of the mechanics of the notch effect, which brought theory and experiment into closer agreement. I thought it was unfortunate that he was working with round notches and did not include the more dangerous crack-like notch. Professor Neuber was also measuring room-temperature creep of strong steels in the elastic range. He used a very high magnification of the deformation and long testing periods. From his work, and the finding of Dr. Kaiser on internal effects in the elastic range, it appears that the metal physicist would have a fruitful field to study, possibly to include something on dislocations.

We went to lunch at a nearby restaurant and, as usual, our conversation centered on points of mutual interest — namely, the notch effect. I commented on the tardy acceptance of notch brittleness as a real effect, but said that his book was read with real interest. I then asked him to tell me how he happened to write it. He smiled as he told of the circumstance that made it possible. It was back in the 1930s, when Hitler told every German to go to the polls and vote *Ja*. He was so preoccupied with his work that he failed to show up. Then one day, there was a knock on his door, and a couple of SS men took him to court for a jail sentence. While under lock and key, he called for writing materials and came out later with his manuscript.

At the university, I visited Dr. Weigel, whose field is rare earths and actinides. At the time of my visit, he was setting up equipment for a full work program. He had spent a year on the Berkeley campus in preparation. After we had moved to Berkeley, he attended a symposium on the rare earths, and we had the pleasure of his company for dinner.

During my mission, studies of reactor graphite were being actively pursued in the United States, England, and Germany. This prompted several visits to producers and research laboratories in Germany and Switzerland. Natural graphite had become interesting, according to advice that I received from the United States. Upon checking into sources, I learned that natural graphite was mined in Bavaria by the Graphitwerk Kropfmühl AG. The home office was in Munich. On my visit there, I was posted on their high-purity, high-density grades. The values cited for density were impressive, especially the uniformity within a block and from block to block. They had joined with the Ringsdorfwerke, in Mehlem, near Bad Godesberg, to form a research and development subsidiary to study applications of the natural graphite. I visited this plant later, and learned that although they had an active program, they were still in the study stage.

Siemens-Plania, in Meitingen, was working on the development of graphite-sphere fuel elements for reactors. This change in current practice stemmed from Krupp, and would eliminate the need to encase the fuel elements. I was told that the problem at that time was one of securing adequate strength to resist the impacts during operation. This appeared to herald a major improvement in reactor design, and I became quite excited over the prospects, so much so that I decided to check further on graphite. I read recently, with considerable interest, that this fuel-element design has been tested in a "pebble bed" reactor by the Arbeitsgemeinschaft Versuchsreaktor in Düsseldorf.

The coal interests of the Ruhr have a large research institute at Essen-Krey, where a team is unraveling the mysteries of that seemingly simple substance. Their comment to me that there were large gaps in their knowledge showed that they were off to a good start. I also visited the Battelle Institute in Geneva, where physicists were correlating the structures of pyrolitic and other forms of graphite with their properties.

After these trips I would return to Frankfurt and write reports, but that duty will be skipped here, as I wished it might have been at the time, to turn to northern Germany and Düsseldorf. The name comes from the little stream, Düssel, that flows, unseen, through the city and into the Rhine. Düsseldorf is the financial and business center of Westphalia, and is also the home of various metallurgical organizations. Those of prime interest to me were the Verein Deutscher Eisenhüttenleute (the German Iron and Steel Society) and, of course, the Max Planck Institut für Eisenforschung, what I have been calling the iron institute. On the consuming side is the Forschungsgesellschaft Blechverarbeitung organization, which is devoted to the problems of the sheet-metal industries. A former assistant of mine at A. O. Smith, Dr. Hans Groebler, was in charge of the technical activities.

We found a delightful hotel for temporary quarters, the Park Hotel. The former royal palace has a wooded park, with lagoons for waterfowl. The Park Hotel is at one entrance to the park, and close to the center of the city, the opera house, and Königsalle or Royal Boulevard. During another visit to Düsseldorf, we stayed in our first German motel, the Hoch Garage Motel, a very satisfactory lodging.

The first time I checked out of the Pa·k Hotel, I had a session with the cashier — a friendly man, but all business in money matters. I offered to pay either by a check on my Columbus bank or by one on the American Express Bank that headquarters had set up in the I. G. Farben building. I asked which he would prefer. On selecting the former, he commented that Germans are accustomed to transferring money by postal money orders, and that they have inadequate laws for personal checks. He knew that we used checks, and that our laws governing forgery gave severe penalties. So I signed my Columbus check and handed it to him. When he asked for identification, I handed him my Army ID card with my photograph and signature. He ignored the picture, but studied the two signatures. I asked why. He replied that signatures gave a more positive identification. I observed that the Park Hotel telephone operators already had the push-button method of dialing — or is it pushing?

The Max Planck iron institute was a mecca for visiting steel metallurgists. Professor Wever had turned over the direction to my friend of Clausthal days, Professor Dr. Willy Oelsen. He greeted me warmly when I called, and offered me the run of the place. I was pleased when he expressed appreciation for the help we had given Germany during the rejuvenation period.

We first had a session with the principal staff members, and Willy recounted his work on basic Bessemer, or Thomas, slags which proved the existence of two immiscible slags. When I inquired for my friend and tutor of Charlottenburg days, now Dr. (honorary) Angelica Schrader, I referred to her as the "Queen of Metallographists." He assured me she would come directly, and added that I must also meet the "Crown Prince" who was being groomed to take her place. He was a keen young man and was receiving the best of instruction. He surely had some man-size shoes fo fill, although that is merely a figurative allusion. Far from being an Amazon, Fräulein Schrader is really petite. Even so, that change must have been like grooming a replacement for Babe Ruth.

A friend of Clausthal days, Dr. Max Hempel, has a large and well-equipped fatigue- and creep-testing laboratory. In fatigue, the formation of an extrusion at the location of the ultimate fracture was well known. That was the limit of the light microscope for observation of the mechanism of fatigue initiation. The electron microscope identified the actual crack at the center of the extrusion and thereby provided the basis for coordinating a theory of fracture mechanics with experiment, at least for the material that was studied. Dr. Hempel was also participating in a large-scale, multi-laboratory cooperative program on creep testing, to secure significant data on important steels and alloys.

The iron institute has come back after the war as a well-staffed and well-equipped laboratory for the steel industry. Both metallurgical institutes of the Max Planck Society offer prestige positions for young scientists, a healthy condition for the continuation of excellent work.

At Phoenix-Rheinrohr in Düsseldorf, Dr. Baerlecken, chief metallurgist, discussed his work for a major producer of commerical steels. That included some basic studies, but was principally on controls and inspection for product quality. His laboratory was cooperating on creep tests with German and other laboratories, to determine the most probable value and the spread in actual tests to set standards for selected steels.

Krefeld is not far from Düsseldorf, which made it convenient to visit a specialty-steel producer, the Edelstahlwerke. I felt they had good programs on problems in their field, but were shy on doing new things — unless, that is, they were not ready to discuss them. However, they had become interested in our superstrength steels and had a program going on them.

One evening, Edyth and I invited Fräulein Schrader to take dinner with us at the Park Hotel. She gave us many impressions of life in Germany since 1913. We could touch only the highlights, but they showed what can befall people as the fortunes of war and peace rise and fall. Insofar as they went, her accounts coincided quite well with other impressions I had gained. (I have long felt that if one has friends in a foreign country, he can gain a better understanding of that country in conversation than from most written accounts, and he can better detect propaganda or biased criticism.)

I was very happy to learn that her work had received the recognition it so richly deserved when an honorary doctorate was conferred upon her by the Charlottenburg Institute of Technology. She was basically the same (as she said I was) as in the reasonably stable days when we first met. However, it was inevitable that intervening events had saddened her, and she felt especially sympathatic toward those who were imprisoned behind the Iron Curtain. We finally drove her to her home; our Auf Wiedersehen may have been our last.

From Düsseldorf, it was easy to visit the main Krupp plant in Essen. There I was turned over to Dr. H. Kreiner, director of research, who briefed me on his activities, and especially on the graphite-sphere program. (His brother was chief metallurgist at Gebr. Böhler.) They still had the problem of breakage as the spheres were impacted, as they would be in operation. A successful solution promised to provide a great improvement over encapsulation for fuel elements. For that reason, I had visited several graphite operations, and since then I have read of the German reactor at Jülich that uses this new form *(Kugelhaufen)* of fuel element.

After lunch, I visited the research laboratory with Dr. O. Rüdiger, the director. I was especially interested in their project on foreign inclusions in steel, which included identification of inclusions at various stages of manufacture and also of any that changed in the steel or even left the steel. Experiments with electromachining showed twice the rate of metal removal compared with other methods whose results were available.

I closed this visit at the Widia cemented-carbide plant by reminiscing about the early days of Widia and Carboloy with Dr. J. Hinüber, the manager. Under the full pressure of Hitler, Widia was important in Germany's rearmament, without which, in the terms of Senator Bone, Hitler could not have prepared for war as quickly and as effectively as he did.

My time was short for Widia, but from what I could observe, it was a fine, modern powder-metallurgy operation. I had not seen it when I visited Krupp in 1945.

It was always with pleasurable anticipation that we started a trip that would take us along the Rhine. On our way to Brussels one time, we drove to Bad Godesberg, the seat of our Bonn embassy. I stopped there to meet our science attaché, Professor Audrieth, who had been a friend on the Picatinny Arsenal scientific advisory council.

We met him and his family, and enjoyed a meal at the Embassy Club on the bank of the Rhine. From there we stopped at Cologne to visit the Krautkrämer plant for ultrasonic-testing equipment, and to have lunch with Dr. J. Krautkrämer. He and his brother were professors at the University of Cologne before becoming industrialists, and successful industrialists, I can add.

Leaving the Rhine, we drove on to Brussels by way of Aachen. In Brussels, I became acquainted with our Air Force research office, the better to coordinate my program with theirs.

While in Brussels, I visited Professor Soete at the University of Ghent, where Edyth and I had had a most interesting visit in 1952. He was an outstanding student of welding, and had contributed constructively to a clearer understanding of the effects of design and applied stress fields on the behavior of welded joints. This included biaxial and triaxial stress systems, section thickness, and general problems of brittle fracture. His test specimens were full-scale, as contrasted to the conventional small laboratory test specimens. His forte in experimental technique was devising relatively simple means for proper "interrogation," as someone has put it.

On our return journey from Brussels, we stopped at the Kurhotel Quellenhof in Aachen, where I was to attend a meeting of the materials and structures panel of the Advisory Group for Aeronautical Research and Development. AGARD was initiated by Teodor von Karman of the California Institute of Technology, as an advisory group to NATO countries. It was a deliberative rather than an executive body. I had been invited to attend this meeting by Dr. F. L. Wattendorf, its secretary. We were represented on the panel by Nate Promisel of the Navy, and arrangements for the meeting were in the charge of Professor F. Bollenrath of the Technical University at Aachen.

In spite of the limitations that are imposed on an advisory group, I thought their meetings served a useful purpose; they provided the opportunity for open discussion of important problems by representatives of several countries. The meeting dealt principally with the refractory-metals group — one rich in promise for air and space technology, and also in problems.

Aachen is a famous old German city, and was one of the seats of the Holy Roman Empire. There are relics of the days when Charlemagne was crowned, and also of the still earlier times of Roman occupation. The AGARD meeting was not all business, however, because Professor Bollenrath had arranged an interesting evening for members and their wives at Monchau.

Monchau is a little town in a narrow, wooded valley south of Aachen. The scenery is beautiful, and the food was good at the local restaurant. Our master of ceremonies was Professor Bollenrath, and he invited Edyth to take a seat next to him. He speaks English easily, so she had no language problem. She told me later of his account of Monchau during the winter of 1944–45. It will be recalled that there was heavy fighting in that region, when we were held up for a considerable period. Our troops were on one side of the little stream, and the Germans were on the opposite side. Though the armies were but a few yards apart, not a shot was fired. Professor Bollenrath explained that "there was no hatred in their hearts."

The Technische Hochschule at Aachen has had its fair share of Men of Metals. The names of Wüst, Borchers (the elder), Oberhofer, Goerens, Pomp, Houdremont, and Schenck (the elder) come readily to mind. They say that people make the institution, and that must be true because of the outstanding position of its metallurgical department. As a student, I followed the work of Wüst, Goerens, and Borchers, and later I followed that of Oberhofer, who made the role of oxygen in steel his special field. Goerens and Houdremont, both Luxembourgers, graduated to Krupp, and rose high in the direction of the company. Professor Hermann Schenck (the son), who was the current head of the department, followed rightfully in their footsteps as a top expert in steelmaking and president of the German Iron and Steel Society.

During a later visit, I met Professors Lücke and Schmidtmann of the metallurgical staff. Professor Lücke is what we would call the brains of the department: His field was basic research in metal physics. Even so, he is not up on cloud nine, but concentrates his work in areas that are technically important. He gave me a lucid lecture on his program, and I thought I detected a blending of German and American pedagogy. That was probably correct, because he spent a year at Brown University. There is good evidence, however, that he had extracurricular interests, because he brought back a charming New Englander who is now Frau Professor Lücke. Edyth and I met her at a social hour at their home.

Metallurgy had progressed a long way since I visited Goerens in 1912. That occurred to me as Professor Schmidtmann, head of the steel section, took me through his department. After a general discussion of what's new in steel metallurgy, he brought up explosive forming as a field of special interest. He had a test station in a discarded bunker for the forming operations. His work was more basic in character, and laboratory studies were planned that would parallel the bunker tests. I hoped that he would be able to predetermine compositions and structures that will perform satisfactorily at high strain rates.

To go back somewhat in time, shortly after my arrival at the European Research Office, Lou Klinker suggested that I go to London with him for a meeting to discuss contract proposals on materials problems. This trip introduced me to the modus operandi of traveling when under the Army. We flew over, checked in at the Air Force's Columbia Hotel in Bayswater, near the Marble Arch, and then proceeded to the Navy research office. That visit was included to acquaint me with Capt. B. F. Bennett, the commanding officer, and his staff. I also met the librarian, who was George Sachs's daughter.

Our meeting was at the Ministry of Supply, and I found that the group was headed by my friend Hubert Sutton, chief of the Materials Section for Air. (One of his outstanding jobs was the investigation of two Comet aircraft that had exploded in the air.) A joint meeting of this kind is scheduled when our two governments are considering bilateral sponsorship of a research and development project. The details are discussed first, and then a program is set up at the working level, prior to final action and financing at a higher echelon. The number of worthwhile projects was so large that each one had to be closely examined, and its priority established.

While in London, Lou also took me to our embassy to meet our science attaché, and the military attachés, and not solely to conform to protocol. Those sessions provided the background information necessary to keep a neophyte informed, and they also helped me to arrange my program for certain areas to supplement, rather than duplicate, the work of others.

England has a fine tradition in metallurgy that is being upheld by organizations, private and public, in research and technological developments. I was able to visit several of them. Actually, I felt I should limit the number in England, where other Americans would have no language problem, so that I could concentrate on contacts where familiarity with German would facilitate dialogs. (I've been trying to get that word in someplace.)

Sheffield is a steel center that enjoys worldwide renown. In addition to the metallurgical department of the university, whose members include Sorby, Arnold, and Desch, there are research institutes for steel and cutlery, industrial research laboratories, and, of course, steel mills. The whole city is steel oriented. For example, when I wished to visit the British Iron and Steel Research Association, I simply told the taxi driver, "Take me to BISRA."

Steel quality has been uppermost in my mind for a long time; when I think of quality, cutlery steel comes quickly to mind. The cutlery industry operates its own laboratory to check on the quality of the product and, incidentally, on the quality of the steel that is used. As one would expect, both keenness and durability of the cutting edge are determined. There are nearly six hundred companies in Sheffield alone, ranging in size from family affairs to larger firms; several may have space in the same building on a rental basis. I was a little disappointed in not finding more work on the metallurgical requirements of cutlery steel and a closer relationship to the steel producers, although the industry has technical surveillance of the product, which helps to maintain Sheffield quality.

At the University of Sheffield, I found that the metallurgical department had added staff and facilities for modern instruction and research since my visit with Desch in 1925. I was greeted there by Professor R. W. K. Honeycombe. He was heavily engaged with electron microscopy for studying metallurgical problems. I was especially interested in his work on the austenitic stainless steels, and the structural difference between the sensitive and insensitive conditions. We knew that heating the steel to temperatures around 900°F caused chromium carbide to precipitate at grain boundaries and, presumably, to impoverish the adjacent grains in chromium and permit chemical solutions to dissolve the low-chromium envelopes. That was the "sensitive" condition that led to failures, as proposed by Ed Bain. That was the current explanation, but I knew from my work at A. O. Smith that ordinary metallographic examination could not identify it positively. Professor Honeycombe examined these steels under the electron microscope and identified a thin rim around each grain as the impoverished layer. It is always gratifying to have positive support of an assumption based on indirect evidence, and that was the first time I had seen it for this case.

In another study of grain boundaries, he explained the effect of platelets and envelopes of sulfide impurities in steel. In superstrength steels these intergranular impurities have recently acquired considerable importance, so these studies took on a significance that had been ignored in the past, especially when a high sulfur content of 0.04% is permitted by specifications.

Professor Honeycombe also described a new instrument for examining submicroscopic structures for correlation with metal properties and behavior. It was the scanning electron-probe X-ray microanalyzer of the Cambridge Instrument Co. The fine distribution of selected elements can be determined by this method.

I had never expected to visit Cambridge on metallurgical matters, but that venerable university had become an important center for materials research in the early 1920s under Heycock. Mainly because of the electron-microscopy work of Cottrell, Tipper, Nutting, and others, it still ranked high in this field. I was disappointed that Professor A. H. Cottrell was not available during my visit, because I had been told that his views on notch brittleness were similar to mine, which must be rated as a somewhat extraordinary coincidence. So one day we rode out to Cambridge from London.

A visit to Mrs. C. F. Tipper's laboratory revealed a wealth of information as she described her work. It became clear why she was widely recognized as an authority on brittle fracture. It also became clear that Cambridge had added studies of a practical or engineering type that one associates with a technical institute. I had to wonder what the early dons would think of this departure from conventional pedagogy.

That noon, Mrs. Tipper was our guest at the neat little Garden House Hotel, just off campus. She told us, with some evidence of sadness, that she was to retire soon. Would a person of her experience and proven ability be lost to the profession? I reminded her of the time I had first met her, in 1925, at the Royal School of Mines in London, when she was working with Professor Carpenter on aluminum single crystals. I also commented on her long and fruitful career, first in basic research, as Constance Elam, and then in her present field, as Mrs. Tipper. She would rate high on any list of Women of Metals, and, in fact, she has the distinction of gaining international recognition under both her maiden name and her married name, and in different fields.

That afternoon I visited Dr. Nutting, who was a substantial contributor in electron microscopy. His main interest, then, was with the newly devised method for producing extremely thin films of a metal sample and examining them by direct transmission instead of by replication, the conventional method.

One time, I went to Birmingham to visit the research laboratory of the Mond Nickel Co. I was greeted by Dr. G. I. J. Bailey, who took me for a tour of the various departments. I thought it was similar in staff composition and equipment to BISRA, although Mond Nickel is directly concerned with nickel and its various alloys. I went there, however, with what I thought was one of my most important discoveries in mind. It was their work on superpure steel.

In the past, in its pursuit of higher quality of structural grades, the steel industry has gone from the Bessemer to the open-hearth to the electric-furnace process. The possibility of improving the refining stage increases in the order given. More recently, vacuum refining and degassing, and vacuum-arc remelting, have been adopted to improve the quality still further. There is no quarrel with that, but Mond Nickel decided to eliminate impurities, a principal cause of lower quality. I had just recently wrestled with superstrength steels for Polaris-missile casings for a client, and was well aware of the limitations imposed by commercial electric-furnace steels. Their deficiencies caused higher fabrication costs and greater scrap losses, which disrupted the shipping schedules. That experience opened my eyes to the need for high-purity steel of improved uniformity for these special applications, to insure greater reliability in service.

After preliminaries, I asked permission to visit the steel and welding sections. There I found that my conception of superpure steel was confirmed. Grain-boundary films and grain-boundary segregation, even on a minute scale, must have a critically important bearing on steel that is used at high strength levels. In brief, the superpure steels retain higher notch toughness than commercial steels, they are more foolproof in welding, and the welds are more ductile. Incidentally, experience at Mond Nickel and Bristol-Aerojet showed that fundamental studies, as contrasted to technical investigations, should use superpure steels to avoid anomalous effects and faulty conclusions, if unsuspected impurities affect the results. Directly thereafter, this idea bore fruit.

The European Research Office was placing a research contract with the British Welding Research Association (BWRA) for a basic study of the heat-affected zone (HAZ) in strong steels. The project was to use a very pure steel as a base to which various impurity elements would be added, at first singly and then in combinations. This was in line with the finding that the effect of an element depends in part on the steel in which it occurs.

The BWRA project focused on the HAZ, and was regarded as a study of the effects of composition and impurities on the weldability of the parent metal. Many investigators consider a notched-bar test to be the proper test for the purpose, and they use test bars in which structures have been produced that simulate those of HAZs of actual welds. This gave me no little concern, and I had this situation in mind when discussing the program with Mr. H. F. Tremlet, who was in charge.

When a metallurgist contemplates the time-temperature cycles of the HAZ, and recalls that high temperatures (overheating) and rapid cooling (hardening) can produce structures that adversely affect steel properties, and especially the notch toughness, he becomes concerned about the integrity of the welded joint in hardenable steels. He is also concerned about defects and cracks. The great emphasis on "zero defects" and the detection of defects by nondestructive inspection speaks for the latter. It is a good question — whether to place the emphasis on notched-bar tests of simulated-HAZ microstructures, or on the incidence of cracks and defects, as a better check on weldability.

From examinations of service failures and the results of laboratory tests, I have observed, insofar as the parent metal is concerned, that the real hazards are underbead and toe cracks of the HAZ, or cracks that initiate at steel defects. I have never observed a failure of a sound joint initiated by notch brittleness of the HAZ, although toe cracks are an exception. After all, notch toughness (fracture strength) becomes important when we have notches or cracks.

One time, I was discussing this point with Connie Voldrich, then in charge of our welding division at Battelle. We were studying the 4130 aircraft steel, and I was saying that the hazard was not the hardness of the HAZ, but actual cracks that would grow and produce a failure. We knew, for example, that many engine mounts were in service in aircraft that had hard HAZs, and we also knew that many had been rejected as soon as cracks had been discovered in them. In fact, the latter was the major reason that induced Messrs. Sims, Banta, and me to study underbead cracks at Battelle. I then asked Connie to prepare a test specimen of 4130 steel with the weld running longitudinally in the center, and to pull it as a tensile-test bar. One might expect the hard HAZ to fail first, but actually, it was in the weld that the fracture initiated. The HAZ had no notches or cracks.

I wanted the results of the BWRA program to correlate with service, so I suggested that tests for HAZ cracking be included along with the effect of prior parent-metal structure. I anticipated some interesting and important results, but my mission ended too soon for me to stay in contact with the program.

At one time, Dr. R. Weck, director of BWRA, told me of his design of a simple and inexpensive fatigue-testing machine that uses the principle of induced vibrations in a novel way. I saw machines of this design in Berlin, and I will mention them again.

During my wanderings around England, I learned that Bristol-Aerojet was producing missile casings and, in fact, was the sole producer in the United Kingdom. My visit there turned out to be most interesting. They had the same welding and fabricating problems with superstrong steels that we had in this country but at that time I felt that they had gone further toward their solution. Naturally, I was interested in the reason. Later, I concluded that the answer lay in close control of all functions by competent engineers and production people. The most important difference was steel procurement, but extended to all shop operations and laboratory control.

These thoughts surged through my mind as Dr. C. I. M. Cottrell, chief metallurgist, escorted me through the plant and back to his laboratory. He had made extensive studies of steel and welding from which he had developed a test of steel shipments that correlated with their shop practice. He would then either reject steel that would give more trouble than it was worth, or else change the technique to accommodate variations in the steel.

Dr. Cottrell had experimented with superpure steel and was aware of its vitrues, which gave me the idea that a production run should be made. With their Mr. F. J. Williamson, assistant chief engineer, we devised a research and development program to appraise it for missile casings.

Again, my mission came to an end before we could initiate it. The very real interest in zero defects, and reliability in service, places superpure steel in a favored position.

I had an interesting visit at the Hawker-Siddeley Nuclear Power Co., where they were attempting to develop an organic heat-exchanger medium for ship propulsion by nuclear power. The work I saw appeared to be headed up a blind alley, yet I read more recently of a ship reactor that uses an organic coolant. The problem at Hawker-Siddeley was sludge formation.

So much for visits in England. One time, we decided to go to the opera at London's Covent Garden. It turned out to be *Lucia de Lammermoor,* and was sung by completely unfamiliar performers. It was not long before we became conscious of an outstanding performance by the singer in the role of Lucia, so we settled back to enjoy a treat. We awaited the Mad Scene with special interest; it is a severe test of the singing and acting of a coloratura soprano. Then, as Lucia slowly came down the staircase, clad in white robes, we were spellbound by the performance. The audience response was terrific, and reminded me of Caruso's triumphs at the old Met. We took a second look at the program and saw the name of Joan Sutherland. From one performance, one knew that she has a place among the supergreat of opera. She had been singing in Europe, and we had to wonder how long it would be before the Met would "discover" her.

Back in Germany, I had a visit with Professor E. Molvo of Erlangen, a top man on semiconductors and metal surfaces. With research and by training students, he was responding to the demands of the German electronics industry. At least that was the general idea, but he complained that too many of his graduates were emigrating to the United States. I thought his studies of surfaces were especially interesting.

One weekend, we left Munich to go by way of Nüremberg to Bayreuth, where we heard Wagner's *The Flying Dutchman.* I had read about the departure from conventional staging and choreography and was expecting it, but I could not quite decide if I approved. I believe I enjoyed the Dutchman in Stockholm, in 1912, better.

An interesting interlude came between acts, with promenading outdoors, delicacies, drinks, and conversation, among an intensely musically oriented audience. Edyth found an American girl who was studying in Germany, and I an elderly German who had had a rich experience in the musical world. Bayreuth is truly a magnet for lovers of Wagner. The present regime is carrying on the tradition, but in a modern version.

Our return home took us through the Franconian Alps, a district rich in local color and scenery, yet quite off the beaten path of tourists.

While I am talking about weekend extracurricular activities, I would like to mention Rothenburg-ob-der-Tauber, Nüremberg, Dinkelsbühl, and Goslar — interesting examples of medieval German cities. Of these, Rothenburg just naturally leads the list in interest, and in freedom from the effects of modernization. There are still the ancient walls with towers, the old buildings (although some had to be restored after the bombing of World War II), shops, churches, the city hall and city square, plus water fountains where the citizens drew their supplies, and cobblestone pavements. Hotels such as the Eisenhüt and the Goldener Hirsch offer modern accommodations.

We scheduled our visit to see the "Shepherds' Dance," which commemorates an event of the Thirty Years War when the city was saved from destruction by the guns of Tilly. That was achieved by fulfilling the requirement of Tilly that someone quaff a huge flagon of wine, nonstop. The Bürgermeister was the volunteer, and history records that he had his stomach pumped immediately thereafter. The feat is depicted each day at eleven o'clock by animated figures in the town clock.

The citizens were so relieved when the city was spared the fearful destruction visited upon other German cities during the Thirty Years War that they danced in the streets. The shepherds and shepherdesses came in from the countryside and joined in the celebration. It was the latter that led to the "Shepherds' Dance." It opens with the Tilly episode, and closes with the dance on the cobblestones of the square. The pavement actually slopes, and one must admire the dancers' ability to maintain a quick half step under adverse conditions. It was a beautiful performance.

As one walks around the quaint streets, one soon becomes aware of early wrought-iron signs on shops, hotels, and doorbells. It was one on a doorbell that attracted Edyth's attention, and through the medium of sign language, she found an escort to take her to the shop of Herr Pflüger, where they are made. Soon thereafter, I was also escorted there. From the gleam in Edyth's eyes, I quickly recognized it as a shopping trip, and we obtained two of his beautiful creations, a doorbell and a light fixture.

The workshop has the appearance of an engraving out of the Middle Ages, and indeed, the tools and methods remain the same as they were at that time. During a later visit, with a group, Herr Pflüger demonstrated his skill in making small parts, such as rose petals, for the doorbell. He is a friendly person and welcomes visitors, but he does not speak English. If you visit his shop, and especially if you wish to obtain an example of early hand-wrought ironwork, it will be well to have an interpreter.

I saw him for the last time in 1967, when his staff was reduced to two middle-aged helpers. He was still filling orders, but he complained that he should have at least six helpers to keep up, and to sustain the art. I couldn't help feeling that his art was dying out as a sacrifice to modern technology.

Turning to business, I made a special visit to the University of Göttingen, which has a memorable place in metallurgical history. Professor Tammann was gone, as was Professor Masing, who was the incumbent when I was there in 1945. Finally, the longtime associate, Professor Otto Vogel, took over, and when I say finally, I mean just that. The Tammann regime had come to an end, and I was there close to its demise.

The quarters were small in contrast to Tammann's, in 1912, and had two graduate students; an air of sadness permeated the institute, which I could understand. For more than half a century, the establishment of constitution diagrams was a major function, and Professor Vogel had made many valuable contributions. Those diagrams are a part of the tool kit of the practicing metallurgist, without which he would lack the guidance that intelligent treatment of alloys and slags demands. It would be comparable to the plight of a motorist without a road map in a strange land. Evidence of their importance, if needed when so many uncommon metals and their alloys are being considered for use, is the recent and extensive work on constitution diagrams. Our armed forces were also sponsoring the compendium of ternary systems by the Guertlers and the work on dilute alloys of iron by Hume-Rothery.

As Professor Vogel and I were discussing his work and viewing his laboratory, I confess that these thoughts passed through my mind. I wondered if his extensive experience would be allowed to lie dormant, or be put to use in the future. I inquired if Göttingen was dropping metallurgical research, since he had told me that he would retire at the end of the semester. He replied "No," that a new group had been formed. When I asked what they were working on, he replied with a wry smile — *Versetzungen* (dislocations). The authorities had decided to switch to fundamental research in metal physics. Actually, that is a more appropriate program for a German university, and leaves the applied fields to the technical universities. This visit was topped by coffee with Professor and Mrs. Vogel. She is a concert singer, and well acquainted with Berlin as a musical center, so I enjoyed an interesting musical interlude.

As a student of Tammann, Vogel was a contemporary of Matthewson and Guertler. He recalled my visit in 1912, and recounted early experiences of the time. Tammann could be a hard taskmaster and must have belonged to the no-nonsense school of professors, but his students were proud of being graduates of Göttingen.

From Göttingen it is but a short ride to Clausthal, where I visited Professor Gunter Wassermann at the famous Berg Akadamie, or school of mines. He was at Dahlem in 1925, and is well known for his studies in metal physics. His laboratory is oriented along that line, but the scope was soon to be extended to applied fields of physical metallurgy. He commented on work on grain boundaries, and the difference between the effects of small and large angle orientations. Afterward, I wished I had raised the point of grain-boundary segregations and how they were handled in basic studies.

I also visited a new department for ore treatment, smelting, blast-furnace practice, and steelmaking. I thought I recognized the hand of Willy Oelsen at work when he was rector. In any event, this development reflected the influence of the expanding steel industry. That finished my visit to Clausthal, except for lunch at the restaurant that had been an officers' mess in 1945.

Passing through the Harz Mountains, we arrived at Goslar. A highlight was a visit to Barbarossa's palace, which takes one vicariously back into the days of the Salinger and Stauffer emperors of the Holy Roman Empire, that odd concoction of the Middle Ages. Someone has said it was neither holy, nor Roman, nor an empire. It was German, but not until Bismarck did Germany acquire an empire.

We finished with Goslar all too soon, and continued north to Braunschweig, to discuss the welding research of Professor W. Hofmann. The transformations that occur in the zone adjacent to the deposited weld metal are critically important in steel weldments. To study this, Professor Hofmann instrumented test welds to record cooling rates and volume changes at various distances from the fusion line. It was interesting to compare some of his findings with our own on aircraft steel at Battelle. His instrumentation created some interest in the United States.

Professor Hofmann was also a leader in studying cold and friction welding. During the course of his work, he has uncovered various fine points on which the production of strong welds depends. I had never studied the process, so the marked effect of apparently minor variables on joint efficiency came as a surprise.

The department of technical physics at Braunschweig is under Professor E. Justi. This was a fairly large group, and some of the programs were of interest in metal physics — for example, cryogenics and superconductors, the Hall effect, superstrength magnetic fields, electrical conductivity and Matthiessen's law, zone melting, metal whiskers, adsorbed films, dilatometric determinations of TTT diagrams of steel, and fuel cells. Most work on superconductivity requires the use of liquid helium to provide a sufficiently low temperature. The technique for producing the strong magnetic fields, for example, would be greatly simplified if a conductor were available that became superconducting at the temperature of liquid hydrogen, or at about 20 K. A material with this characteristic was found — the alloy columbium nitride.

Professor Hofmann invited us to dinner that evening, at his home in the suburbs. A pleasant surprise was the presence of Angelica Schrader. The conversation was general, since most of it was in English for Edyth's benefit. Somehow, it turned to wartime experiences, mostly on humorous episodes, although perhaps they were not so humorous at the time. One that I recall was told by Fräulein Schrader. A group of friends was at dinner, and a large soup tureen was in the center of the table. Suddenly, a bomb fragment crashed through the window, landed in the soup, and splattered that course over everyone. We spent that night in an elegant Victorian villa that accommodates paying guests in fine style.

After a short ride, we arrived in Hannover, where I visited Professor A. Matting at his Institut für Metallkunde (physical metallurgy). He had a full program of sponsored research on about the same basis that we have at our universities. A program on fatigue followed the rate of crack propagation that has the important technical application of permitting the calculation of useful remaining life of a part that has developed a crack. In welding, they were studying aluminum, gray iron, spot welds, and weld metal that would give a low modulus of elasticity. Other programs were on thermal shock, journal bearings, galvanizing, corrosion, flame plating and bonding to the base metal, and ultrasonic inspection. These are recognized as old problems, yet the arts must be continually studied if they are to keep pace with current (and future) engineering technology.

After Hannover, we drove to Bad Pyrmont and the Kurhotel for the weekend. We were joined there by a young friend of our Columbus days, Marcia Manning, whose father is a top metallurgist at Battelle. She was an exchange student living with a German family nearby, and brought a son and daughter of the family with her. Marcia was enthusiastic about her experience and her German family. Later, their daughter spent a year in Columbus with the Mannings and attended Ohio State University. Many young people on both sides of the Atlantic have enjoyed, and benefited from, the exchange experience. I presume a book could be written on the subject, but surely the privilege of living with a family of a different culture, and the opportunity to learn another language, are among the benefits.

A stay at a spa is a delightful experience. The grounds and park are beautiful, the food is excellent and is served in a fashion to increase one's enjoyment of it, and all seems well with the world.

On Sunday, we witnessed the reenactment of *The Pied Piper of Hamelin*, at nearby Hamelin. After the piper had agreed with the town fathers on compensation, a lot of cute little rats followed him off the premises. He then returned for his pay, only to be laughed at by the citizens and told to get going — the rats were all gone, weren't they! Then followed retribution — all of the little folk followed the piper with his quaint and plaintive piping — a powerful lesson on the evils of welching. This fable is not to be confused with that of the Mouse Tower at Bingen on the Rhine. In the latter, the mice chase and devour a bad and unprincipled man.

My mission took me to Paris to visit French laboratories, and also to contact the Mutual Weapons Development Team; SHAEF; Dr. Frederick Seitz, the Science Advisor to NATO; and Dr. F. L. Wattendorf, the Secretary of AGARD.

The venerable Sorbonne is a prime attraction for a visiting scientist. Henri Le Chatelier was gone. Although a man of his caliber cannot be truly replaced, younger men were carrying on with the same spirit of scientific inquiry. I was attracted by the work of Professors Benard and Michel of the École Nationale Supérieure de Chemie. Professor Bernard was especially interested in metal surfaces and the phenomena of surface reactions and diffusion at elevated temperatures. This is a fine field for basic studies, and technically it has become important. Professor Michel had a large-scale project on carbides in steel, and his findings should shed valuable light on the response of steels to rapid heating and cooling as in welding and induction heating. Our work at Battelle has shown the significance of the composition and structural form of carbides on the incidence of underbead cracking, and later work on induction hardening has shown the same for successful case hardening. Insofar as I know, we have no detailed investigation of carbides for these important processes.

The École Polytechnique is an exception to our institutes of technology. It was founded by Napoleon and is, in effect, a prep school of university rank for the military service. The curriculum is designed for scientific and technical education, and only young Frenchmen of outstanding record and promise are accepted. I visited Professor I. Jacqué, a chemical engineer with extensive experience in the French oil industry. He places that experience at the disposal of industry by means of a private research and development firm with which he is associated. At the time of my visit, he was studying reactions between various metals and technically important gases.

One day, a black government limousine called for me at the Hôtel de Paris and I got into the front seat so I could talk to the driver. As we drove along a beautiful stretch of the Seine south of Paris to St. Germain-en-Laye, I was reminded of the engineering feat of laying a water main from the Seine, over the hills on the left bank, to Versailles. It was laid in the seventeenth century, and is said to be the first of its kind. My destination was IRSID, the large research center of the steel industry. I was greeted by M. Delbart, whom I had met on a previous visit. After introductory comments, during which I explained my mission, he turned me over to M. Borion, who was in charge of the section on mechanical testing.

IRSID has studied the problem of brittle fracture in great detail, and it was my special interest. As is customary with workers in the field, they have developed their own philosophy. Also, as is usual, while one worker will accept the experimental data of another, he does not accept the other's conclusions or concepts when they disagree with his own. Which is something like the blind men deciding what an elephant is like. What brings this up is the assumption at IRSID that temperature, strain rate, and notch geometry are mutually related as dependent variables, in the same way that pressure, volume and temperature are related by the gas laws. That concept started in this country during World War II, and has also been accepted in Germany and France. I have observed elsewhere that this cannot be true (*Metal Progress*, December 1961, page 103). Temperature and strain rate fix the critically important property of yield strength, and their variations change the yield strength independently and by different amounts for different steels. The notch geometry fixes the (initial) stress distribution but not the properties. Experimental evidence has been available that shows clearly that temperature and strain rate are *independent* variables.

These comments do not reflect the wide scope of this work at IRSID, and I believe that they are as well aware as anyone else of procedure for handling brittle fracture in practice.

The electron-probe microanalysis is used extensively at IRSID. Their work on segregation in grains appealed to me as being especially significant, although it is probably not yet sufficiently evaluated in practice. A steel with only a small amount of an impurity element in bulk was found to be present in a high concentration at grain boundaries. Professor Honeycombe found the opposite condition in stainless steel after carbide precipitation. Surely these conditions must affect the way metals respond to service conditions in case grain boundaries control the behavior.

Upon leaving, I reflected on the fine contributions of IRSID to steel metallurgy, and also on the use they made of the contributions of others. It focused on communication and the great value of information distribution. The great mass of this information is served by modern methods of retrieval, yet in spite of the ease with which the message is received by the eyes of the retriever, the weak link has been that of reaching his mind.

From St. Germain, we drove to Beauchamp for a visit with Professor Eudier, chief metallurgist of the powder-metallurgy plant, Metafram. They produce a variety of products from metal powders, such as continuous strip, extrusions, control rods for reactors, friction products, and uranium oxide by hot pressing. In his second role as professor at the University of Paris, he was developing a new method for making semiconductors of intermetallic compounds, such as Bi_2Te_3, that he said was far cheaper than our fusion method. However, I have been told by a specialist that those compounds are less satisfactory than pure elements, such as silicon.

On our return to Paris, I stopped at SHAEF headquarters to call on Dr. Black, head of the scientific staff, and his English, French, and German assistants. As on other occasions, I was impressed with the close association of scientific and technical personnel with the military.

Lt. Gen. T. B. Larkin (ret.) was in charge of the Mutual Weapons Development Team that handles bilateral contracts on weaponry with a foreign government. My function was one of assisting on problems of materials. Unhappily, I had too little time away from my major mission for this duty, but I was able to visit two French arsenals with United States congressmen.

One visit was by auto to Bourges, where the French have an arsenal and a proving ground. The subject matter was classified, and I do not know if it has been declassified, so I shall skip that part of the visit. The session was handled on our side by my friend from Picatinny Arsenal, Paul Tweed. There was always someone in the United States who followed the technical content of the project and who traveled to Europe for these meetings and, one may surmise, for a look around Paris. To indicate where Paul's interests lay, let me say that he introduced me to a cafeteria, just off the Champs Elysées, where we had a fine meal.

Incidentally, our hotel in Bourges was one in which Napoleon stayed when on his trip north from Elba to Paris to start his Hundred Days escapade that ended at Waterloo. Since then, I have stopped at other hotels with plaques to Napoleon, commemorating his journey.

One time at Bourges, I was surprised when I was told that the French do not use molybdenum to suppress (reduce) temper brittleness in alloy steels. I had first learned of this practice from a French inspector at Midvale, in 1916.

A visit to Paris means to me a visit to Dr. Georges Blanc, whose foundry research institute in Sèvres I have already described. He had installed new scientific equipment for more precise studies of foundry practice. In metallurgical circles, we frequently discuss the merits of castings versus forgings versus weldments, aside from economics. Generally speaking, a good casting is equal in quality to the other two, but the process controls must be taken into account. The equipment and controls that are available for forging and welding place those processes in a preferred position, so a major function at the Sèvres laboratory is a study of the many steps that must be controlled for the production of good castings. This requires considerable support by a competent staff, plus laboratory facilities that are beyond the reach of many foundries.

Switching now to wrought steels, I called on Col. J. Faguet, chief engineer in the French Armed Forces Ministry for Air and French representative on the materials and structures panel of AGARD. I had met him in Aachen, where he was chairman of the materials group. He was especially interested in superstrength steels, for which he had numerous test results that confirmed our own. He was well aware, however, of the need for a more basic approach to the problem of quality and reliability.

Travel to Paris on one trip proved to be quite interesting. We drove through Wiesbaden, across the Rhine, and down to the famed Lorelei rock. We stopped to enjoy the sights and the ingenious flag control of river traffic. The current is swift and the passage is narrow, and I found Heine's poem to be a special treat after viewing the scene. There is now a plan to blast a deeper channel at the Lorelei to ease the traffic problem.

Riding on down the Rhine, we turned off at Coblenz to go up the Moselle, past Burg Cochem, and then to Trier. After viewing the remains of a Roman outpost, which always excite one's wonder, we drove to Luxembourg for the night.

We had no time to spend there, because I was due shortly at the Air Force research office in Brussels. The highway took us along the route of Hitler's final thrust to Bastogne. There the central square reflects the famous "Nuts" episode, with a nuts pharmacy, a nuts souvenir store, and a nuts restaurant. Of course, General McAuliffe is duly honored by a prominent plaque. There is an interesting little sequel to this.

On a later visit to Berchtesgaden, we had our grandson, Michael Van Vleck, with us. Edyth and Mike had just joined me after a tour through the Bible Lands. One day, we took him up to the Eagles Nest that was built for Hitler on a mountaintop with a glorious view of the Bavarian Alps. We noticed a man with a family group on this trip, and I saw him again at the Berchtesgadener Hof Hotel. After parking my car, I returned to the lobby and found Mike in conversation with him and his wife. As I joined the group he introduced himself — "McAuliffe is the name." "We are the Hoyts," I responded, "and this young man is our grandson. You wouldn't be General McAuliffe, would you?" "Yes," he replied, "General McAuliffe." We had told Mike the Nuts story, and his eyes opened wide when he learned the identity of our companion. He quickly asked, "Not the Nuts man?" The general was highly pleased at this instant recognition by a youngster, and he confessed to his identity. Mike secured a prized autograph at a little conference the next day. General McAuliffe's performance at Bastogne was an incident to be proud of, in World War II.

The next morning, Edyth flew from Brussels to London for a lecture engagement, and I went to the Air Force office, under Col. N. L. Krisberg, to review their metallurgical contracts. That accomplished, I had to make a decision — I was due next in Paris, and I had to decide between using my car all the way (that is, fighting the Paris traffic) or leaving my car in the outskirts. My decision was instant and positive. I drove by the Waterloo battlefield and a large and well-kept United States burial ground in the Ardennes, and I arrived at Château Thierry, famous in World War I. I left my car in a garage there and rode serenely into Paris by train. Just as serenely, I rode back to Château Thierry, recovered my car, and drove home through quaint villages, battlefields, and the beautiful Saarland.

It was on one of these occasions that my boss, Dr. York, came to Paris for a session with MWDT. I had intended to seek his approval, as head of engineering, to develop more contracts for metallurgy and to improve our assistance on materials on MWDT contracts, but our session was too brief for any adequate discussion of policy. After General Larkin's gala reception for Dr. York, with the officers in full dress uniforms, I left for Frankfurt, somewhat puzzled.

Berlin! There is no need to mention here the role of Berlin in world affairs, and I skip it, but it was my home town before World War I, and a visit to that famous city was a special treat. I believe that our Army is keenly aware of the chameleon-like idiosyncracies of Communist personnel, because we have a special routine for government people who are "on travel" to Berlin. A well-identified train plies between Frankfurt and Berlin each night, and we were required to use it. My papers were carefully checked at the station prior to departure, to insure having them strictly correct in case of Communist inspection. However, our two trips were quite uneventful. A young officer met us the next morning and escorted us to the Dahlem Guest House, where we registered. We had our meals across the street at the Harnack House. It was named in honor of the first president of the Kaiser Wilhelm Gesellschaft, and had been the meeting place of many famous German and foreign scientists.

As compared with the days before World War I, Berlin had changed in outward appearance, but in spirit not at all. The city was being rebuilt in the modern way, an example being the Hilton Hotel. I overheard new names and phrases for whatever attracted the Berliner's interest, as an example of his spirit of earlier days. The new convention hall is "the oyster," and a nude rider on a horse that points the direction to the Olympic Stadium is "Lady Godiva's Brother." I did miss the old Berlin with its uniformly high buildings, horse-drawn droshkys, and early-vintage Opel taxicabs with circular hoods. Taxis there today have two different color schemes, one for West Berlin and the other for East Berlin. The Kurfürstendamm was gay and brilliantly lit by night, but not so attractive by day.

On a weekend, an English friend of Edyth's, who was on the staff of the British consulate, gave us a tour from the Brandenburg Gate through the denuded Tiergarten, past "the oyster," and on to the exposition grounds for lunch on top of the television tower. At one place, we passed a huge artificial hill formed from rubble. The salvageable portion was used to rebuild the city, and in winter, the citizens used the hill for skiing.

Our Army had daily bus tours into East Berlin, and we took one of them. The shock was not one of disillusionment, for we had been told what to expect, but one of surprise that the Communist regime would picture such a glowing account of itself on such a shabby basis. The few visible inhabitants, the few vehicles on the streets, the rundown buildings and shops, and the tardiness of reconstruction all gave mute evidence of the bleak pall that had fallen on East Berliners. This picture of gloom and despair was highlighted by two details that I will touch upon directly.

The Russian burial ground was laid out and maintained by the Germans as a well-kept park, and was the only beauty spot on the tour. At the center, an impressive monument was erected to the memory of the troops who fell in capturing the city. I recalled my conversation in 1945 with the tank commander who had just been recalled from the Elbe and was disillusioned by the orders that had prevented our army from taking Berlin.

An incredible sight was the slogans with which the city was adorned, if adorned is the correct word. Those banners prated brazenly that we in the imperialist-capitalist countries were slaves to our industrialists and warmongers, but that Communism brought freedom, peace, and prosperity to its people. They were textbook examples of deception by upside-down doubletalk that deceives no one, except the younger generation. Those banners were not there when I visited East Berlin in 1967.

Another exhibit of the same type is Stalin Allee, formerly the famous Unter den Linden. It is all facade, a few feet in depth, and would serve well on a movie set as a Russian street, for a front view only. There was nothing in the background except empty space.

We returned to West Berlin by the Brandenburg Gate, after passing the site of Hitler's bunker and the burned-out Reichstag building. Two policement were stationed on the western side, and I engaged them in conversation. I mentioned those funny slogans I had just seen and asked if anybody could believe them. Those policemen, ex-soldiers, were experts on slogans, so I felt I would get a well-informed opinion. Their reply was a shock. They agreed that adults simply ignored them, but said that the Communists were working on the children and were bringing them up in a world of make-believe. Young people who were intelligent enough to do their own thinking were not deceived, but they were the ones who were escaping to the West. We became well acquainted with one in Frankfurt whose mother gave him her blessing as he escaped to freedom. He has since then moved to the United States and is doing nicely.

In spite of the conditions that prevailed, the West Berliners were going about their daily business very much as of old. In visits and conversations, I noted but little change in the fields of education, science, technology, and industry, except in content.

The former Friedrich Wilhelm University is in East Berlin, but West Berlin now has its Free University in Dahlem. It is dominated by the attractive Ford Hall, as we would call it. One of the buildings dates back to the days of the Kaiser Wilhelm Institut für Physik and contains the laboratory where Otto Hahn first split the uranium atom and initiated a new era. At the Free University I visited Professor Ludwig, who was studying the basic physics of the plasma arc in support of a group that deals with experimental and practical problems.

Before my first visit to Berlin, I had written to Professor Max Pfender, president of the Bundesanstalt für Materialprüfung (BAM), in Dahlem, to request an appointment. I added that I once attended lectures of Geheimrat Martens. Along with an invitation, and much to my surprise, he wrote that he had been trying to find someone who had known the Geheimrat, and that I was the only one who was still alive. Would I be so kind as to write up my recollections of him for their archives? I did that later, when in an appropriate mood. I also wrote my recollections of Professor Heyn at the request of Dr. Werner.

The BAM, or federal institute for materials testing, is much like our National Bureau of Standards, although the latter also includes functions of the Physikalische-Technische Anstalt. As the heir to the Materialprüfungsamt of Martens and Heyn, the BAM carries on very much as before in its relations with government and industry. Special sections for physics, chemistry, and electrochemistry, and a strong metallographic department under Dr. Werner, investigate areas that are important to the applied sections. Nondestructive testing had become important in Germany as elsewhere. In support, BAM has a department that studies and uses the various methods, housed in the Geheimrat's old residence on the property. (Dr. O. Vaupel was business manager of the German Society for Nondestructive Testing.) The development of the materials sciences for the support of technology is still a guiding principle, in the spirit of Martens.

BAM is a public institution, yet it accepts projects from industry, and even from the police. A project from the diesel-engine manufacturers is a striking example of outside projects. This one was in the charge of Professor Bussmann of the fatigue laboratory, and related to the design and fabrication of crankshafts. They were supplied in full size and were tested on a fatigue-testing machine of a special and very ingenious design. It has been described by Dr. Weck of BWRA, and also by BAM in its publication of the results of the crankshaft investigation. BAM built its machine from odd pieces of steel that the Communists failed to liberate.

This project developed useful information for design, fabrication, surface finish, and inspection, to insure long service life. It was so successful that I am surprised that this testing machine and its use for industry and government have been ignored (at least, as far as I know) in the United States. For example, the locations at which defects were either dangerous or harmless were determined, thus contributing to reliability in service, or to lower fabrication costs and fewer rejections. An example of its value came from a ship crankshaft that had developed a fatigue crack at sea. In port, it was decided that the ship could safely make an additional cruise, instead of being held up while a new shaft was being made for installation. The amount that was saved would pay for a lot of research and testing.

The old Institute of Technology was back in business, although under the new name of the Technical University Berlin (TUB). I knew that it was strong in metallurgy, although somewhat limited in scope, and it was on my list for a visit. On the campus I found new buildings, but the old and imposing main building was being restored.

I was in the hands of Professors Dahl and Detert for a review of their activities and a visit to the institute. The laboratories were not as completely equipped as I had come to expect, although the work in progress was being properly handled. They could have taken on more research, but limited funds reduced the number of graduate students that could be financed. Taking a cue, I explained our support of qualified programs as a possible source of assistance. At the end, I tried to find the old metallographic laboratory, but it had been so greatly altered I did not recognize it.

A group of us had lunch at the Löwenbraü on Kurfürstendamm. A major topic of conversation was the arrival of a baby daughter to Professor Hanemann, by a second marriage. I had known his first daughters as youngsters, forty-seven years previously.

Coming back to Frankfurt, I missed meeting Dr. Max Hansen, technical director of Metallbank, and author of the well-known "Binary Alloys." I did visit his laboratory, and I found it so well equipped and staffed that I could picture myself in a large government or private research institute. I had a special reason for making his acquaintance: He was the only other person, to my knowledge, who had published on the ternary alloys of copper, tin, and zinc. I wished to compare notes with him, as the other member of the Kalchoid Club. Actually, he had simply used my results and diagram for his paper, but he might have his own ideas on that upper heat effect or phase change that I discovered. My phase diagram has been inadvertently ascribed to him, but not *by* him. I regret that we never met.

I was pleased that my mission permitted at least brief visits to Bavaria. We were especially fortunate in being able to see the Passion Play at Oberammergau, and to have comfortable quarters at a new hotel. Mike Van Vleck was with us, and enjoyed the performance, although he had to stand. Returning to the hotel, we found him conversing with an Air Force colonel and his lady, so I joined and introduced myself. He was Col. Sam Francis. I asked, "Not all-American fullback at Nebraska?" whereupon he owned up. We had a few words about Minnesota-Nebraska games and he regretted that his teams had never beaten ours. Things have changed quite a bit of late.

Austria and Switzerland lay just over the border, but were practically off limits for me. I did visit those countries, however, on a visiting-fireman basis.

The Verein Deutscher Metallkunde — the nonferrous counterpart of the iron and steel institute, Verein Deutscher Eisenhüttenleute — was meeting in Vienna, and I was invited to attend. The major subject was the physics of grain boundaries, to which Professor Paul Beck of the University of Illinois contributed a paper in German instead of in English, for which he received special thanks. Professor Novotny, of the University of Vienna, was studying sonic fatigue, which had become important to the aircraft industry. To me, his paper pointed up some of the difficulties in understanding the subject.

I took advantage of being in Vienna to visit the Schoeller-Bleckmann Steel Co., one of Austria's major steel producers.

I have always enjoyed being in Vienna, because of its various attractions, but there is one that remains a fond memory — even more so than those wonderful white Lipizzaner horses — and that is a visit to the city park and Kursalon for hot chocolate, pastries, and a concert of Strauss waltzes.

Leoben is a center for the Austrian mining and metallurgical industries, and has a university and a research institute for the foundry industry. At the research institute, I was especially interested in their method of determining the properties of gray iron by an ultrasonic method. That was new to me, but we have since heard of it, and its use on ductile iron, in the U. S. After a side trip to Gebr. Boehler in Kapfenberg, a quality steel company, and a session with Dr. E. Krainer, we drove past Donnawitz, a federal steel company, and the famous "Iron Mountain" of highly pure iron ore. I did not stop at Donnawitz, because I had previously visited Dr. Hauttmann in the Linz plant, where Dürer's L/D process (our basic oxygen process) had been developed for production. We were then off to Herrenchiem-see and Frankfurt.

Returning to our home base always engendered a feeling of interesting experiences to come, either within or without the Army. For example, we had Bach evenings at the little Johanniskirche in eastern Frankfurt, and "Kiss Me Kate" by local talent. The leading lady, an army officer's wife, was so talented and attractive that she could have stepped into the role for a Broadway performance. Or I could mention luncheon at our casino and a discourse on Kaiser Wilhelm II, with medievalist Professor Walter Kienast from the university. There were also visits to the delightful Kronberg Castle, Bad Homburg (the empress's favorite spa), a concert at Bad Nauheim, and the early Roman camp that is nearby. These diversions, however, were bonuses, and not a part of my mission.

As the occasion permitted, I took in several conventions and symposia to become posted on the state-of-the art of the subjects being discussed. One of those meetings was held in Essen, on nondestructive testing, and it brought out the growing importance of the ultrasonic method. Another was on hardness testing, in Dortmund, and emphasized the need to have international standards for equipment and methods. I thought we should have been better represented, in view of the importance of hardness testing and the development of world trade. It is the principal mechanical test that is universally used.

We did have a notable representative, Mr. V. E. Lysaght, or "Mr. Hardness Tester." After the meeting, he approached me with a request. After the war, he had shipped five hundred Rockwell hardness testers to Russia, and they had been assigned to the lady hardness tester who had presented a paper at this meeting. He wished to learn from her what experience the Russians had had with them. The catch was his lack of Russian and her lack of English. He had found a member of the Russian contingent who spoke German, so would I give him the questions in German, and then translate the answers as they came back from the lady and were given to me in German? The mission was a technical success, but the results were negative. The Rockwell machines had been distributed to various plants, and she had lost contact with them. It was for that reason, she stated with fine diplomacy, that she could not compare the American machines with the Russian equivalents.

The Institute of Metals held its 1959 fall meeting in Stockholm, and there I first observed the rapporteur system in action. A competent man digests a group of papers on a common subject and presents the essential points. The discussions proceed as usual. I though we could benefit by adopting it. The Fall Lecture was very ably given by Professor Rudberg, without notes and in excellent English.

At that meeting, I became aware that the man next to me was attracting my attention. It was Professor Axel Hultgren, the long-time leader of Swedish metallurgy. It was he who overcame hydrogen embrittlement of rails and hardenable steels by cooling them slowly. His method was standard practice before vacuum degassing. I have mentioned that in 1952, Edyth and I had dinner with him when we were in Stockholm.

The English conduct these meetings in fine style, and include in the social events banquets and ambassadorial receptions. I was pleased to see again the G. Shaw-Scotts, whom I had met in Ghent in 1913.

Another meeting was the twenty-sixth International Foundry Congress in Madrid. We were pleased to find that Jim and May Pearce were in the Hilton Hotel with us, and we saw a lot of them. Later, we visited them in Birmingham, shortly after Jim had retired as director of the British Cast Iron Research Association. At that time, he told us about the lost art of making thin-wall, delicately designed ornamental castings of gray iron. It was called "Berlin Iron," and examples can be seen in museums. The foundry was destroyed in 1838 by revolutionaries and so the art was lost.

Dr. Pearce had become director of BCIRA in 1924, when it was a small and floundering organization, and had proceeded to build it into a large and flourishing association before he retired, in 1958. He became a widely known and a highly acclaimed research director. Additional contributions to government, industry, and education added to his activities and stature.

Back to Madrid: Our Spanish hosts really went all out to make the congress worthwhile and interesting. I rated a number of the papers as excellent, but I thought one by Dr. B. Matuschka of Schoeller-Bleckmannsstahlwerke, Austria, had exceptional merit. That could be, just possibly, because it reported on an idea I had had for testing the quality of a heat of steel as-cast, and before it was converted into wrought form. He showed that the ductility of a cast coupon, or test bar, from a poor heat is only a small fraction of the ductility of the standard longitudinal test of the product wrought from it. In those heats, the cast coupons had otherwise unsuspected segregates present at grain boundaries, which, although not detected in longitudinal tests of the wrought form, could cause premature service failures. In spite of the obvious virtue of this procedure, its use would doubtless be limited to special cases where service reliability is featured.

In regard to those segregates that form during solidification, Allianz Versicherung (equivalent to Hartford Insurance in the United States) had a program, at a laboratory in Berlin-Mariendorf, on the ultrasonic detection of dangerous segregates in weld metal, and on the condition of the heat-affected zones of weldments. I visited the laboratory with Dr. O. Neuhof, technical director in their Berlin office. There I learned that the problem is difficult to master, although their large display of service failures showed why they were studying it. The costs incurred by those failures served to demonstrate its economic importance to both insured and insurer.

Dr. Neuhof told me at lunch of the dramatic effect of Hitler's oratory, especially on youngsters. Everyone likes a champion, and Hitler presented himself as a champion of the German people. One had to hear him to understand his persuasiveness. There is much to be learned about Hitler and his Third Reich from scientists and technical people, who were mostly objective observers (although victims) of history in the making. They also make it clear that the abnormal conditions in Germany from 1914 to 1950 were enormously disruptive of any normal way of life.

I was naturally interested in the fate of the Carl Zeiss Works, since I had visited their partly destroyed plant in Jena, in 1945. So, while driving in the vicinity, I visited the new plant at Ober-Kochen, in West Germany. Practically all the important people had come over from East Germany. I was quite impressed by their new line of equipment for controlling closer tolerances and finer surface finishes. However, Zeiss equipment is now well known and need not be touched on here. Of more direct interest to me was their new electron-beam welding, which has become very useful for special applications.

Our last trip was both scenic and fruitful. I had reserved vacation time and had desirable business visits in mind that would take us into Bavaria, Switzerland, and even into France. Our first stop was a favorite, Rothenburg, where we visited Herr Pflüger and his wrought-iron shop. From there, we drove to another medieval city, Dinkelsbühl, on the Romantic Way, and on to Garmisch-Partenkirchen and up to the summit of the Zugspitze. We then drove to Zurich by way of tiny Liechtenstein and stopped at a motel on the shore of a lake. That location gave us a good view of Mount Pilatus.

In Zurich, during a vist to the Institute of Technology and the Federal Testing Institute, I was rewarded with the tale of a gigantic fatigue test. An old and large concrete bridge was being vibrated to destruction. I was told the amplitude had reached about one foot, and traffic was still permitted on it. I decided not to use it.

I regretted not seeing Bob Mehl when I was in Zurich. He had recently retired as dean of the graduate school at Carnegie Tech, and was on a six-year assignment for U. S. Steel, to keep posted on European advances in metallurgy. Also, I may add, to attend meetings and to deliver invited lectures.

At Lucerne, we had a trip around the lake to William Tell country, and then on to Interlaken, where, after a trip up the Jungfrau, we went to an evening concert of Swiss music, in the park with the floral clock. Then it was on to Lausanne and the Beau Rivage, our favorite hotel. While there, we visited the Castle of Chillon. From Lausanne, I visited Battelle in Geneva, while Edyth enjoyed a boat ride on the lake.

By now it was time to head north to Freiburg im Breisgau. The next day, we crossed the Rhine to visit the University of Strasbourg. We had lunch at a little restaurant on the square in front of the famous cathedral, the one whose second spire was never completed. I pointed out to Edyth where Goethe had stood on its flat top, proving to himself he did not shrink from height, as he looked down on the street below.

I next had an appointment with Fritz Förster, who had been connected with the ALSOS mission. After returning from France, he set himself up in business as Institut Dr. Förster to produce devices for nondestructive testing. As we completed our conference, Fritz noticed that there was a lady in my car outside and assumed, correctly, that she was Mrs. Hoyt. He then suggested lunch at a restaurant high on the mountainside. The view was beautiful and peaceful, and again reminded me of Goethe's verse. I commented on the poet's great powers for forceful and beautiful statement, and illustrated it with the refrain of the harpist in the "Apprenticeship of Wilhelm Meister." That inspired Fritz to recite his favorite passage from "Faust." After lunch we drove to his California-style home on the mountainside and met Mrs. Förster and their offspring. Then we were off to Frankfurt and 9 Ernst Schwendlerstrasse to spend our few remaining days on the mission.

In fifteen days we had covered 1254 miles, and enjoyed every minute and each mile. Travel by auto was very similar to what it is in the United States and was quite convenient since motels were available. Similar, that is, until you have to drive through towns. Then it can be, and usually is, an adventure to get from where you are to where you wish you were.

My last official visit came when we were waiting in Bremerhaven to embark on the good ship *Upshur*, to return home. I motored to Bremen, to visit the laboratory of the German Institute for Heat Treating. After viewing the facilities I had a question for Dr. O. Schaber, the director. In my consulting practice, I had run across a mysterious salt bath for surface hardening steel that operated economically at a low enough temperature to avoid undesirable warping and dimensional changes. A large amount of this salt was stored in the client's heat-treat shop, in case deliveries were held up. That may sound odd, and actually, the situation was odd.

My client purchased the salt from a mysterious source, unknown except for an address. Furthermore, attempts to identify the active constituents were said to be completely negative. My client was concerned about this uncertain source, and I was asked to check on the purveyor the next time I was in New York. Not long thereafter, I found the supplier's address, down by the waterfront in an old Civil War type of brick building, but nothing else — no office, no salt, no one at all. I was puzzled, but in a little while the janitor came along and when I asked him where the office was, he just smiled and replied, "Office? All they have here is a mailbox. I think they have a warehouse over in New Jersey, but I don't know where."

So I asked Dr. Schaber about that remarkable and mysterious salt, and I learned that it was of German origin, that he was well informed about it, and also that it was not a secret formulation. The doctor also told me that lightly scaled steel quenches faster than scale-free steel. His comment left me puzzled, as I recalled the superhard surface I had produced on Armco Iron after heating it in hydrogen and quenching it quickly in oil. The clean surface absorbed carbon fast enough to harden it. This shows the effectiveness of a clean surface, although it is not a direct comparison.

All good things come to an end, and that was true of my mission. Just before my departure, Colonel Freeman left for the United States, and that called for a farewell party at the casino for the staff and their families. Although officers throw parties on the slightest provocation, like the one for a general who was being "transferred" to the house next door, this was something special — hence the casino. We even had after-dinner talks and, since I was leaving soon, I took the occasion to express my gratitude for the cooperation and assistance I had received from the European Research Office and the Army in general. Our wonderful Sergeant Gordon was present, so I decided to include him. I had given him a hard time — handling my finances, vouchers, billeting, transportation by train, air and private auto, filing classified documents. Upon my return to the office a short time before, with a particularly complicated voucher to complete, I had told him how grateful I was for all the assistance he had given me. He just looked up and said he wished either that he would be transferred or that I would be. He was very well-liked, and I was glad he had a twinkle in his eye when he said it. As I recounted that incident, and assured the sergeant that his fond wish was about to be fulfilled, my understanding audience greeted it with appreciative laughter.

A question — should I attempt here some general comments on my mission or should I pass up the opportunity? About this time, I saw the cartoon of Dagwood watering his lawn while his neighbor was enjoying a peaceful snooze just a few feet away. Should he or shouldn't he? As Dagwood turns the hose on his neighbor, he says he would rather be sorry for doing than for not doing it. That settled it for me, so here goes.

My experience in and with our Army was pleasurable. In addition, Army personnel performed the many civilian functions with care and efficiency. I often wished that more civilians could have a similar association. My rating of DEF-2 had little effect on the way I was treated, except in specific cases, such as parking my car in places that were reserved for generals, and being assigned to special quarters in our armed forces hotels.

Coming to the scientific and technical import of my mission, I found many important developments, although, admittedly, I had to survey the field in some detail to find the most promising leads. Also admittedly, some of them turned out to be catching-up work instead of being new contributions.

At the time of my mission, we were in the midst of a transition from the older, or conventional, metallurgy to a new and more sophisticated metallurgy. Whereas formerly we had made out well with iron, nickel, copper, aluminum, tin, lead, and zinc, plus alloying elements, the new technology made it necessary to include a large number of other elements. Engineers, physicists, and others were designing new parts, structures, and "hardware," and they were studying the service conditions to which the materials were to be exposed. The end products (missiles, space ships, supersonic planes, nuclear power plants) were well publicized, but metallurgists were in the background, attempting to meet new and difficult requirements. In this way, they provided materials as they set limits to performance. We needed to exploit the new metallurgy to the best of our ability, in view of the new technology. The Cold War called for the coordination of our own research with that of the NATO and other free countries. That was approximately the setting for my mission as I saw it.

I believe it is true that most countries have people of outstanding ability whose contributions should be followed closely. Consequently, I held it necessary to find those working in my field and to report on current work, especially on the programs that had not been published. This was especially true for metallurgy and the materials sciences, because we were inadequately informed on developments in Europe. Language difficulties might account for this, in part, or reliance on publications that have frequently cooled off by the time they appear in print. However that may be, I uncovered more valuable leads than had been implied as possible before I left on my mission.

The scientific and technical gap between the United States and Europe has been discussed repeatedly on both sides of the Atlantic. I don't doubt its existence, but even so, we should not take European work lightly. The gap can also work in reverse, evidence of which I found in certain important instances. That holds as well for the methodology of research and development as for technology. I came away with the feeling that European investigators were more consistently committed to the basic approach, with better control over pertinent variables. Consequently, they are less likely to ascribe results to the deliberately included variables while being unaware of the effects of unsuspected variables. Work with superpure steels, as contrasted with the commercial counterparts, is an example. The latter are not so well "pedigreed" as the former.

To return to our support of research in Europe: Most of the contracts were in chemistry, physics, electronics, astronautics, and the biological and medical sciences. It came as a mild shock to learn that only minor attention was given to metallurgy and the materials sciences. Although our few contracts were in good hands, I had become acquainted with other able investigators who could have contributed substantially to our materials program. I wondered why we were overlooking that potential of brains, experience, and research facilities. According to a review of contracts, only 4.5% of the funds supported metallurgy.

I have long felt that it is the end product that gets attention, and that the involved processes of material selection, procurement, fabrication, and inspection for economical and reliable service are stepchildren of technology. However, if these problems were studied by top-drawer people, the ensuing information would be available to the much more numerous personnel who would apply and use it. A Langmuir in a research laboratory, or an Einstein in theoretical physics, can spin off enough problems to keep many others busy. We need but recall that the number of people who have made basic contributions to physics is drastically limited, yet the effect is like that of the pebble that is tossed into the pond. Their work has spread in the hands of others to produce the new physics, and the many useful applications thereof.

Lest I be misunderstood, let me say that we had people who visited laboratories in Europe and reported on work in progress. I cannot say, personally, what use was made of the reports, although, if they were properly distributed and studied, they should have helped to indicate some worthwhile contributions. However, they cannot substitute adequately for more or less permanent representatives abroad who are competent and experienced. This could be an excellent use of selected individuals who have retired, and who are otherwise qualified for assignment. In this connection I think of Bob Mehl, who spent six years in Europe for U.S. Steel and continued to return to the United States periodically. Our government should be no less aware of the need of this sort of contribution, and this is especially true for metallurgy and the materials sciences. The terms of reference should include the function of developing proposals for submission to one of our European research offices to handle. Mine did not, but I felt so keenly that a man on the spot should act, at least partly, in that capacity, that I went ahead with Bristol-Aerojet on the proposal on superpure steel.

I trust that this mission gives a good example of an exciting and useful contribution that can befall the lot of a metallurgist.

Chapter 10

Consulting Years, 1953-77

Experiences in consulting with typical cases.

RETIREMENT, if you live long enough, is as inevitable as taxes. It brings the old routine to a halt and a new way of life begins — you are on your own. If you are still good at figures, you know when that day will come and can plan your future. I was, I knew, and I did. In 1953, I switched from technical advisor to metallurgical consultant.

When I entered consulting, I wished that I had had a book on the subject that cited experiences and techniques. What follows here does not pretend to meet that requirement, but it should throw some light on problems and situations that may confront a metallurgist. One should not think that the word of an expert is always accepted at face value, even though he is armed with a briefcase and is away from home. I admit that I have had negative as well as positive responses. I believe both are educational.

In most instances, my work as a consultant was similar to that as technical advisor at Battelle. Although that eased the transition, important differences did develop. For example, I have, more recently, become concerned about product liability and the possible consequences of advice that I am duty bound to give. In somewhat similar situations, I have found myself between the Scylla of advising a premium steel at too high a cost, and the Charybdis of accepting a cheaper steel with a greater chance of failure.

Following are a few case histories that serve to illustrate problems and show how solutions were developed — or, in limited cases, negated. In a couple of instances, my advice was either rejected or ignored; arbitrary control was exerted by people who refused to face facts, or did not understand their significance. Those instances underlined a lack of metallurgical knowledge on the part of some decision makers.

A Case of Invisible Cracks (or: "Steel Is Not Just Steel"). The U. S. Army Corps of Engineers had built the Garrison Dam at Riverdale, North Dakota, on the Missouri River, and had installed four penstocks for power development. Upon installing two more penstocks for an addition to the power plant, they ran into trouble. This experience shows that dangerous hazards can arise even though standard engineering and inspection practice has been followed.

The first four penstocks had been installed satisfactorily, so the fifth and sixth were to be handled in the same way. The new steel was purchased to the same ASTM specification, but a different supplier had the low bid this time. The first step was the qualification of the submerged-arc procedure for welding the longitudinal seams of the cans, or penstock sections. Some steel was left over from the earlier contract and it was used for qualification, instead of test plates cut from the new supply. The joint was made in two passes, one from each side. The qualification tests were satisfactory, and work started.

After a series of cans had been welded and placed in the tunnel, a circular plug sample revealed a deep and sizable crack in the weld deposit. Although the seams had been 100% radiographed, a careful inspection of the X-ray film at that location gave no evidence of the crack. Additional sampling revealed more cracks and raised questions of the cause of the cracks and the danger of leaving them in. That was when I was asked to investigate the problem.

Such cracks may extend into the weld metal from dirt streaks in the plates, and that effect has been advanced by some as the cause of the cracks. Those nonmetallic impurities may increase the tendency to cracking, but my experience has shown that they are neither the sole nor ever a necessary cause; the cracks also occur adjacent to sound steel and definitely away from streaks of the foreign inclusions. Those I examined always formed at the junction of two scallops of the weld deposit, regardless of the presence or absence of inclusions.

Upon checking the source of the two steel shipments, I learned that they had been supplied by two different mills. That taught again that steel is not just steel. Obviously, suspicion was cast on the new steel.

My investigation centered on the cracks — would they endanger the penstock, and if so, how were we to locate all of them for repairs? This was before we had quantitative information on fracture toughness and on the influence of defect size. Furthermore, inspection and evaluation, as we now know it, was in its early stages of development, and radiography was the principal method. This sets the scene for what followed.

I first discussed the service conditions with Mr. Nelson, chief of the design section at Riverdale. The nominal hoop stress across the longitudinal weld seams was not excessive, yet was expected to reach the yield stress a number of times over a ten-year period, under abnormal, yet anticipated, conditions. The casing temperature would be in the low 30s.

I was also told by the engineers that a penstock failure would wipe out the power plant. A penstock failure in the East had occurred during the acceptance test, and caused several engineers and helpers to lose their lives. We were all in no doubt that we had to take those cracks seriously, which meant that we must know if they were a hazard.

Charpy notched-bar tests showed that the steel and weld metal did not have crack-stopping ability at low water temperatures, which meant that a running crack would lead to a disastrous failure. The next step was that of determining if the cracks would grow or extend under abnormal service conditions.

I devised a test that was new at the time, but that I thought might work. I had a full-sized test specimen prepared from a section of the penstock that contained a typical weld crack. Its size was established by an ultrasonic method for reference. For this, the pip height on the oscilloscope was set at one inch. The test specimen was then stressed to its yield point when at 32 °F. The pip height was again measured under the same conditions and found to be 1¼ in. After a second loading, the pip height had increased to 1¾ in. These tests demonstrated to our satisfaction that the crack would grow under predicted service conditions. Ergo, those cracks could not be accepted.

The next step was that of locating all the cracks in about 15,000 feet of weld in the cans that were already in place. Radiography was ruled out, because we knew that those cracks would not show up on X-ray films. The dye-penetrant and similar methods for surface cracks would also be useless for those deep-seated cracks.

This brings us to the ultrasonic method, which I proposed. The contractor engaged a testing firm to conduct the inspection, and we learned a lesson. No crack was found where the inspector thought he had located one, while cracks were found even when the inspection had revealed none.

We couldn't give up, so I discussed our situation with Perry Rieppel, who was then in charge of welding at Battelle. Perry told me to see Bob McMaster at Ohio State University, well known for his work in nondestructive testing. Bob came to my rescue by recommending an inspection service.

The lesson was that a qualified ultrasonic-testing operator can locate subsurface cracks. The new inspector not only located cracks up to four feet long, but also identified the quadrant of the weld in which they occurred.

The repairs were made, and the penstocks went through the acceptance tests satisfactorily.

The very substantial repair costs created considerable controversy among the steel supplier, the contractor, and the corps. In brief, the supplier insisted that he was blameless because his steel met the purchase specification and he was not responsible for the welding. The contractor pointed out that the corps had approved the X-ray films for release of the cans. Accordingly, he wanted to be paid for the extra work. The corps insisted that they were buying a sound structure and not a defective one. I felt that the good-workmanship clause in the purchase contract should cover dangerous defects. The case went ultimately to the Board of Contract Appeals for adjudication. The board decided in favor of the corps.

The board paid great attention to the crack defects. It stated in part, ". . . defects, not revealed by the radiographic examination, which are found by such other examinations are sufficient grounds to reject all welding, if in fact the defects affect the integrity of the structure." My testimony on the danger of the cracks to the penstocks and on the need to repair them was convincing, especially because the steel and the weld metal of the penstock would permit a fracture to run.

The failure of the cracks to register on the X-ray film puzzled the board, as did the lack of discrimination of the ASME Boiler Code. It commented as follows: "It is difficult to believe that cracks or defects extending several inches along a joint, and in some instances as much as four feet, should not show up on the radiograph film or that the ASME code is so written as to be interpreted to allow such defects simply because they are not evidenced on the film." That is rather strong language!

Incidentally, the plaintiff in this hearing had an expert on nondestructive testing as a witness. I was highly curious to learn what he would say about my test, because it was new to the art. Instead of attempting to discredit it, he agreed that it was a good one. It was obviously the cardinal point of the case.

The penstocks in place were repaired by manual welding in the tunnel. For the cans that still had to be made, the contractor came up with the happy solution of welding with two passes on each side, instead of the initial single pass. That eliminated the cracks, although the steel and dirt streaks remained the same.

It is a good question — where to place the responsibility for the costs of inspection and repairs, the increased costs of having to use four instead of two passes, and the delay in completing the contract and in producing electric power for revenue? Innocently enough, the corps and the contractor erred by using steel from the first contract for procedure qualification, instead of the steel that was delivered for the job. I advise my clients to use the steel as delivered. If the source changes during the contract, I tell them to requalify even if the second mill belongs to the same steel company.

A consultant is duty bound to help his client, although that may require advice that the client is wrong, or has no case. I have had cases of that kind, and have backed out of them gracefully. I have also had another kind, unhappily, when I have developed favorable evidence for my client only to see him lose out. An example may illustrate this point.

A Case of Reentrant Angles. An architect-engineer designed a star-shaped water reservoir with reentrant angles, which was to be beautiful as well as functional. It must have been a beautiful structure, although I never saw it whole.

During the acceptance test, as the reservoir was being filled with water, the joints failed and the sides were washed down the hillside. One huge slab narrowly missed crashing into a house. Whan I examined the remains, I found the welded joints of the framing at the reentrant angles to be painfully defective, and clearly responsible for the failure.

The board of review held the architect and his design responsible, rather than the contractor and his faulty welding — in spite of test data showing that sound joints would have held. I had written a report that held the defective joints to be responsible, but for reasons unknown to me, I was not called to testify.

Personally, I dislike reentrant angles in welded construction. They create bad notches and they add extra burdens on welding and inspection. That does not imply that they should never be used, yet the design should be studied for the possible use of a casting that would be welded to structural members under favorable conditions.

A Case of a Crash Program. I was involved with ACF Industries in the production of Polaris-missile chambers, for use by our nuclear-powered submarines. That was when we were trying to do something that had never been done before. It was a crash program, replete with problems of steel, forming, welding, inspection, heat treatment, and testing, to say nothing of headaches. Development work went on concurrently with production in the shops. This was also a case of faulty procurement.

The steel was a modified AISI 4340. It had already been made, and was delivered from a warehouse in Ohio to Milton, Pennsylvania, as light-gage sheet. The various parts, with nozzles, were welded together to make the casing as a huge cylindrical shell with both ends closed. The casing was then shipped to Minneapolis, where it was quenched and tempered to 200,000-psi yield strength. Sounds simple.

Forming dished heads with this steel had proved to be a difficult operation. The best that had been accomplished by others was forming pie-cut shapes and welding the segments together. When Rudy Furrer, then vice-president of ACF, asked me to look into it, I discussed the problem with Mr. H. LeVan, chief engineer of the Milton plant and an old hand at forming steel. Milton had a large press, so we believed we could produce single-piece heads by forming them at about 1200 °F. With skilled help at the press, we demonstrated that our method was practical. That put us in business.

We ran into problems with the steel almost immediately, which pointed up the seriousness of procurement. I couldn't blame the supplier, because the steel met the purchase requirements, especially for composition and strength, and he would doubtless have supplied a better quality had one been specified. We were not alone, at that time, in having problems with special-quality steels, and AIME sponsored a special meeting in Pittsburgh in 1958 on "Quality Requirements of Super-Duty Steels" — and "How to Meet Them" could have been a subtitle. The meeting could be of no help to us in the case of the steel on hand, yet that meeting doubtless spurred producers in their efforts to supply satisfactory steels for missiles and other special applications.

It is sometimes surprising what crops up to stop or slow down production. A good example of a gremlin of this stripe occurred at a simple routine operation. After welding the longitudinal seam, the cylinder was sized and rounded up to conform to the head closures. The heat-affected zone alongside the weld stretched more than the parent metal, and reduced that section more than could be tolerated. Another gremlin was retained austenite, which bobbed up once in a while.

This work was not only new to us who were directing it, but also new to the shop personnel. The attitude of the individual worker was especially important. I am reminded of a little tale that must have originated in the Middle Ages.

A cathedral was being erected and a man approached a group of stonemasons who were working on a wall. He asked the first what he was doing. "I'm laying these stones," was the reply. Then he asked the second what he was doing. He replied, "I'm building this wall." When he put the same question to the third mason, he received the answer, "I'm building a great cathedral."

Fairing the weld bead into the parent metal to produce a streamline stress flow across the joint was a very important operation. To be sure that this operation was properly done, I worked closely with the man who was selected to do it. He was careful and skillful, and truly making a missile and not just dressing a weld bead. Eliminating that dangerous notch was one step in production that needed no further attention — or so I thought.

Upon arriving at Milton on my visits, I used to go into the shop first, for a survey of work in progress. One morning, I found a new man on that operation, and was shocked to see that he was leaving the welded joint in a dangerous condition. My work in training the first man had gone down the drain. The general manager told me that union rules required switching the men from one job to another, with or without adequate training for the new job. With all the problems we had, I felt that this was one we could do without. We learned later that a completed chamber was lost in hydrostatic testing because of a sharp or sudden transition between weld bead and parent metal. The new man was simply grinding a weld bead, instead of making Polaris missiles for our submarines. This point was nicely emphasized by Adm. Hyman G. Rickover when he stated that "every person involved must constantly bear in mind that he personally is responsible for the entire ultimate result." I soon generated a responsible attitude in the new man.

Before shipment, we had to proof test the casings hydrostatically to a hoop stress that was but a little below the yield strength. The factor of safety for beginning plastic deformation was only a little over one to one, as compared to the more usual two to one. Even tiny defects could initiate a failure, as one did. Heat treating to a reduced strength to increase the notch toughness would provide a temporary remedy, but would reduce the thrust, and therefore the range of the missile, by an unacceptable amount.

We discussed the frequently posed question of whether the hydrostatic-test pressure should be applied three times, as was required, or only once. This question has developed differences of opinion, with some favoring a single test and some three tests. If there are no defects of dangerous size present, it would make no difference. On the other hand, a small defect can increase a little in size during the first test. Subsequent tests can make it grow and initiate a failure. Conceivably, the same could happen when the missile is fired, causing premature failure. Though we were required to test three times, we felt that it was the sounder procedure.

One of our hydrostatic tests strengthened that opinion. This chamber took the full test pressure twice, but ruptured when pressure was still being increased on the third test. We felt that that missile would have been unsafe to fire had we tested only once. To give an idea of the size of defect involved, one of them initiated a failure that was smaller than the white of my little fingernail.

When a casing failed in hydrostatic testing, it was closely examined at Milton, and at the ACF research laboratory at Berwick. The origin that initiated the failure was easily located, but the true cause was not as easy to diagnose; it might have been the steel, the welding, the heat treatment, the repairs, or some false step in nondestructive testing or fabrication. Poor or nonuniform steel was always in our minds, because it could be the direct cause of, or else aggravate, a borderline condition.

Obviously, inspection for defects was important. Although we had very satisfactory X-ray and dye-penetrant inspections, neither was sufficient. I wondered why we had not been provided with a reliable inspection procedure. I believe the answer was to be found in the state of the art at that time, as was brought out at a special meeting on nondestructive testing sponsored by the government. We were trying to do something before we knew how to do it.

A bright spot in this program at Milton was the cooperation of Dr. Irwin of the Naval Research Laboratory. By that time, he had begun his studies of fracture toughness, a field in which he was about to become the leader. I had visited him at NRL and was well aware that we needed the type of assistance he could give. Each time we had a failure, he came to Milton to examine the evidence. Between him and the ACF research laboratory in Berwick, we had as good a diagnosis as was available in those days. I enjoyed his visits, and not only for his lucid discussions of fracture mechanics.

We soon knew that the initiating defects were small, and that we were missing at least some of them with radiographic and dye-penetrant inspection. Ultrasonic inspection was being used to some extent at steel mills and by industry; it was our only hope for locating subsurface defects, although we would have to learn how to apply it at the same time we were getting out production.

I had only a short time to work with a UT inspector, and then with indifferent results. Locating tiny defects in light-gage sheet imposed a severe restriction. Unhappily, I had to discontinue this association with Polaris missiles when I left for Germany on my Defense Department mission.

A Case of Fortuitous Inspection. One day in May, Mrs. Hoyt and I were enjoying the attractions of Yosemite Park when a telegram came from Walla Walla, Washington. We were later to take in the Seattle Fair and then go on to Walla Walla, where Edyth was to lecture, and we were to stay with friends. Thinking that the telegram must be for her, I handed it over without reading it. It came right back with the comment that she didn't understand it, and, besides, it was for me. Would I get in touch with Otto Lunn, chief of the design section of the U. S. Army Corps of Engineers of the Walla Walla district, in connection with a problem they wished me to look into?

A little later, I was in Otto's office discussing a huge vertical-lift lock gate at the John Day Dam on the Columbia River. The immediate problem was a large crack that they had accidentally discovered in a special high-strength, quenched-and-tempered steel side plate. Thus commenced another interesting experience on the metallurgy of dam structures.

The John Day Dam was most impressive on account of its huge size, the long reach of its "millpond," and its tremendous hydroelectric power capacity. The water was to rise so high along the river banks that highways, railroad tracks, and a town or two were being moved to higher ground. All that was impressed upon me as Alvin Ross, project engineer, introduced me to the project.

The crack in the side plate was a toe crack at a full-penetration T-joint, which had penetrated 15% through the thickness and was 1⅝ in. long. The steel had many stringer inclusions (weakness in the Z, or short transverse, direction) along which the crack had run stepwise until it had penetrated in depth. I was surprised that it existed, and of course it would have to come out.

Additional discussions with the engineers made it clear that the failure of this gate would have evil consequences, such as the loss of the river for transportation. This caused me to consider the whole of the gate, not just the plates, and I suspected the framing that holds the parts of the gate together. If one bad defect had not been found by the standard inspection, others might also be undetected. Accordingly, I had them spot check some of the weld joints by ultrasonic testing and found, indeed, that the joints had defects. We also tested specimens of the framing steel and learned that it had very little resistance to catastrophic brittle fracture at low winter temperatures. The corps had specified premium steels for both side plates and framing, but had not included a provision for normalizing the latter.

Inasmuch as I did not wish to have the gate rejected, I proposed that all of the welded joints of the framing be inspected ultrasonically and that defects be repaired. Before proceeding, we discussed all angles of the problem, inasmuch as the repairs could be extensive and costly. In particular, I pointed out that with notch-brittle steel in the framing, the necessary safeguard was the elimination of fracture-initiating defects. Since the steel was ductile at the notch-brittle temperature, the stress would be streamline and would be harmless.

Our decision to inspect and repair reflected the change in attitude during recent years on such matters. If welded joints were unsound, and knowing that the steel lacked adequate notch toughness, no responsible authority would accept a structure in that condition for important and critical service.

During one of our sessions, we had the pleasure of having Leon Bibber join us. He was an old hand at welding, having been with the Navy, and later with U. S. Steel as their top man on that fast-growing technology. Leon had retired by then and was in shipbuilding — his own. His was a gigantic hobby, a welded oceangoing cruiser — if he ever finished it. His friends were betting that he would never cease working on it, he was having so much fun. He vowed to me, however, with a slight show of emotion, that he would take it down the Ohio and Mississippi Rivers, out into the Gulf of Mexico, and on to the islands of the Caribbean.

I had asked the corps to invite Leon to come in on welding, but we found him loath to leave Pittsburgh and the Bibber Shipyard. With the promise of a $1 consulting fee and transportation, he gave in. I believe there was a hint of something about the fair in Seattle.

Leon's exposition of the welding problems of the gate was timely, and must have helped the designers to handle T-1 steel and others of that same type. It is a high-strength alloy structural steel that is quenched and tempered to 100,000-psi yield strength in thick sections. It must be correctly made and handled to take safe advantage of its weight saving. Leon had grown up with it, and had taught its welding technology to many users. He took a dim view of certain reentrant angles of the gate, and recommended a steel casting for that design. The casting would have generous radii instead of sharp angles, and would eliminate uncertain welding and inspection.

The actual inspection of the gate was in competent hands, and disclosed an impressive footage of cracked or defective joints. After repairs, a final inspection showed that the joints were sound, and the gate was put in service.

A Case of "To Heat or Not To Heat." I had another case at Walla Walla. A similar vertical-lift lock gate had been installed in the Ice Harbor Dam on the Snake River, just before it flows into the Columbia River. The experience at the John Day Dam convinced the corps that it, too, should be studied. The possibility of having the combination of notch-brittle steel and defective welded joints could no longer be accepted. Winter temperatures as low as −20 °F could make the gate vulnerable to service hazards.

Since the safety of the gate was at stake, the problem involved the design, stresses, the presence or absence of defects, and the notch toughness of the steel. If the present condition of the gate seemed to require it, I had in mind heating the gate when the ambient temperature fell below the danger temperature. My plan included the detection and repair of defects, and the determination of the notch toughness of the steel. Either the steel might be vulnerable to a running fracture (loss of the gate) or it might stop a fracture that had started (a simple repair).

A review of the structure with the design engineers located the important stress-carrying members. They were sampled, and transition-temperature curves were obtained to determine the notch toughness and the brittle-fracture temperature. The gate was also inspected ultrasonically, after which repairs were made and the gate reinspected.

This rather thorough investigation was followed by a meeting at Walla Walla with the designers, inspectors, and operating personnel, and all phases of the problem were discussed. The question was — to heat or not to heat? After correlating the steel quality with the stresses on the members, I could say, "Do not heat." I am sure we all felt better with the assurance the investigation gave. I can only hope that both lock gates withstand the rigors of many more winters to come.

Here were two important structures on which a lot of time and money had been spent after fabrication; yet, superficially, on delivery they had been in good condition for service. Was the extra repair cost called for? Why all the fuss? A low winter temperature introduces no hazard, because those steels retain their ductility at much lower temperatures; the structures were made by established contractors, and the steel was supplied by mills accepted as bidders; furthermore, many outdoor structures had been made under no more favorable conditions, and they were still functioning satisfactorily. However, when defects are known to be present, and the steel is known to be of subnormal quality in view of their occurrence, should one reject, or should one take a chance that everything will turn out all right?

I gave a lot of thought to this situation, although my function was to develop the facts and not to make the decision. When I was working on this project, the evils and treachery of notch-brittle steel were well known, as were the facilities for producing sound weldments. In other words, was there any excuse for delivering an unsound structure to the user? And, if so, should the corps accept the gate as delivered, knowing that it was defective and vulnerable to low temperatures? In brief, acts of God were no longer an acceptable excuse, either for failure to inspect, or for an ultimate failure of the structure. In another light, the damage that would result from the loss of the gate would be so great that the repair cost would be insurance against the cost of replacement and liability claims.

A Case of Remove or Leave as Is. A consultant also becomes involved in situations requiring him to advise that no action is needed. Although a recommendation to insert an operation not included in the work schedule, to lessen the chance of a premature failure, would be an easy way out, the added work could impose an unnecessary expense and delay construction.

The Bechtel Corp. of San Francisco was installing a large boiler in southern California, and I was asked to advise on the desirability of stress relieving the support girders for the boiler tubes. They carry the weight of the tubes.

The girders were made of ASTM A7 steel, which got me off to a bad start because of its variable quality. Although I was somewhat concerned about its notch brittleness, the only critical moment would come when the girders were supporting the heavy boiler tubes during their installation. I was informed that it was scheduled for the middle of December. At that time of year, the temperature can fall to 20 °F, which is below the transition temperature for brittle fracture of A7. (I wondered why it was my fate to have it come in winter, instead of during the hot summer months, when the girders would be at a safe temperature.)

I considered warming the girders for erection, but after securing information on inspection, and other pertinent data, I decided it should not be necessary to tear the structure apart, stress relieve it in Los Angeles, and reinstall it. That completed my assignment.

I drove past the power plant later, on my way to a job for Kaiser Steel in Fontana, and noted that it was in operation. The girders had held when the tubes were hoisted into place. The visual evidence made me feel better.

A Case of Off-Standard Notched-Bar Tests. In my experience, different steel mills supply steel of different quality to the same specification. Common structural steels such as A7 or A36, which have no provision for Charpy tests, will vary considerably in notch toughness. Also, a mill that produces a low quality in one type may produce a high quality in other types. I emphasize this point because the mill that made the "poor" steel in this case has also produced the best quality I have ever tested of structural steel of another type. To secure good steel from some mills, a premium price would have to be paid, but that would not be necessary with a source whose normal practice produced the better quality. This case is an example.

The Pullman-Standard Car Manufacturing Co. was experiencing premature brittle failures of their welded bolsters on freight cars. These bolsters must withstand rough service conditions in low winter temperatures. Although the research laboratory at Hammond, Indiana, had improved design and welding practice, the bolsters were still not satisfactory. That was when I was called in. I soon found I was engaged in a problem of brittle fracture that went beyond mere testing, and involved the philosophy of the treatment of brittle fracture.

My first order of business was an examination of the voluminous test results of full-scale components. These were so-called bump tests, in which a stationary freight car containing the test bolster is impacted by a second freight car that rolls down an incline at a specified velocity. The impact velocity is increased in graduated steps until the bolster fails. The test bolster was maintained at about −25 °F.

The object of this study was to discover if steel quality had contributed significantly to the results. Bearing in mind that other variables had been involved in the bump tests, my chances were not too bright, but I did spot a variation from bad to better. With this clue, I hoped to develop a correlation between laboratory notched-bar tests and the test performance.

Although an acceptable correlation with service performance was rare, our work on ship steel, and our tests of turbine forgings at Schenectady, had shown it to be possible to develop one.

This lead was followed by standard Charpy V-notch tests from +80 to −40 °F. The 15-foot-pound transition temperatures of thirteen selected bolsters came as follows: one at zero, two at 20, four at 40, and six at 80 °F. This was a large variation in notch toughness, and no bolster could be rated as safe at subzero temperatures.

According to the philosophy of the day, Pullman-Standard should use a premium steel that would run at least 15 foot-pounds at −25 °F (or even lower, to be on the safe side). That would be a normalized, fine-grain steel. It would make good bolsters, but its cost would be so high that I suspected the front office would say I was suggesting that they get out of the freight-car business. I recalled that Bill Kepler had told me, "Anybody could do this job for ten dollars; I must do it for five."

We needed a modified notched bar that would differentiate between acceptable and nonacceptable steels at −25 °F, bearing in mind that the test bolsters did differentiate. The V notch was obviously too drastic, because all the steels were notch brittle when tested with that notch. In situations of this kind, it has been wrongfully suggested that the V-notch bar be used, but at a higher testing temperature. However, changing the temperature changes the very properties that control the notch behavior, and different steels respond differently. My concern was with the properties at −25 °F and the behavior under somewhat less severe conditions than those of the V notch.

The standard keyhole notch with a 1-mm radius was still too sharp, because it, too, produced brittle fractures of all the test steels at −25 °F and gave values of about 5 foot-pounds. We then tried U notches of 2-mm and 3-mm radii, but all the steels now broke, in ductile fashion, with high impact values. Those notches were too generous, and we next tried 1.5 mm.

The U notch with 1.5-mm radius gave us a Charpy test by which steels could be checked for bolster performance with a high probability of success. A truly satisfactory steel, to be mentioned shortly, ran about 70 foot-pounds whereas the steels that had failed in service or in the bump test ran only 20 foot-pounds or less. Reducing the notch radius to a little less than 1.5 mm probably would have discriminated even better.

By this time, I believed that the problem was mainly one of procurement. I learned that the steel that failed came from the same mill that had given excessive rejections in pipe or casing manufacture at A. O. Smith. The steel was a semikilled steel, and from my studies at A. O. Smith and

Battelle I knew that it was relatively cheap and, when made right, served satisfactorily in many applications.

There were two possibilities, although a requirement that the steel meet the modified Charpy test in procurement had to be ruled out as unacceptable by the supplier. That left one possibility — securing steel from the mill that produced the good skelp for A. O. Smith.

I was sufficiently confident of my appraisal of the problem and its solution to arrange a trial order of steel from that mill. The purchase order was the standard used by Pullman-Standard for bolsters. Our laboratory tests with the 1.5-mm notch at −25 °F ran about 70 foot-pounds, which indicated that the new steel should perform satisfactorily, or at least better than the steel from the regular supplier did.

The new steel was thereupon used for a test bolster, and subjected to the bump test at −20 °F. By way of comparing the old steels with the new test steel, bolsters of the former failed in brittle fashion at a striking velocity of about four to six miles per hour. The bolster with the new steel withstood bumps to above ten miles per hour. Instead of fracturing, the test bolster simply deformed slightly, but it could still function as a bolster, and it had absorbed well over three times as much energy.

That was a formidable example of "know your supplier." The improvement in performance was so drastic and impressive that one might expect a quick shift in suppliers — or, at the very least, an application of pressure on the old supplier. It also suggests that design and fabrication engineers have a sympathetic grasp of problems of brittle fracture.

A Case of Mismanaged Procurement. This case involved the 13% chromium stainless steel. Although that is a fine steel, it does have problems from manufacture to end use. Most of it goes into light-gage sheet for intermediate corrosion resistance. It is also used in wrought and cast form in thick sections where corrosion and wear resistance and moderately high strength are needed. Combining steelmaking, composition, and heat treatment to meet the requirements for thick sections, and also writing a purchase specification, can be difficult and complicated.

This steel had failed badly in a similar application, and had caused considerable damage and financial loss. Both spalling and corrosion were involved. A subcontractor was completing an order for a new structure, but had had it rejected as not being in compliance with the specification. That was the setting when I was called in to determine if the new product was really satisfactory for the end use, which I found it was.

The reason for the rejection was difficult to understand, and seemed, at the start, to be based on an unrealistic decision. The special requirement of the specification that had not been met was *not* one that directly controlled the service performance. That was the cardinal point that was missed by management.

The original failures had been investigated by several parties in considerable detail, but the metallurgical examination was too limited. The various reports offered no definitive statement on cause or cure, for either spalling or corrosion.

This uncertain situation had created considerable confusion, inducing the vice-president of purchasing to deal directly with a major producer. The latter proposed a new specification, which restricted the as-quenched hardness and, in effect, imposed a tighter control on carbon, but did not control chromium as well as I believed desirable. The implausible assumption was that only heats that met the special carbon requirement would not spall or wear, yet would be adequately corrosion resistant. Tacitly, the higher carbon content of my client's product was the justification for rejection.

My assignment was to determine if the steel was or was not acceptable for the application. If so, my client would request reconsideration of the rejection, because considerable money was at stake.

I studied the original failures, and the various reports I have mentioned, and finally ferreted out the reasons for both spalling and corrosion. I then examined and tested the rejected steel in the light of all the information I had assembled. For this study, I had the benefit of reports on this steel by U. S. Steel.

My studies included checks for spalling, and also corrosion tests, and were much more thorough than the acceptance tests required by the user's new specification. The results showed that the rejected steel met the requirements for the end use. It was of no avail, because the steel was finally rejected. That was in spite of its being much better-pedigreed, and more likely to stand up in service, than a replacement shipment would.

This was a management decision that ignored competent technical appraisal of the specification, the steel, and the service. It was based on the misconception that the carbon content must be severely restricted or else the steel would fail in service; it also disregarded possible adverse effects of faulty heat treatment. I am not one to ignore sound requirements for composition, or to make an arbitrary rejection for deviations that are of no significance. This was also a case of a frustrated consultant, because I hated to have my client incur a large financial loss unnecessarily.

A Case of When Is a "Perfect" Weld Perfect? Cases of "Why did it fail?" come frequently to the attention of a metallurgical consultant. They are too numerous to mention, but one that I worked on was of special interest to me. This was a case of When is a "perfect" weld perfect?

In the mid-1960s, a chlorine cylinder, as it was about to be loaded on a Navy freighter in Oakland, fell off a hand truck and broke open a welded seam. Chlorine gas escaped, by ill fortune close to an intake for fresh air to ventilate the ship. Unhappily, many of the crew were gassed, causing a heavy financial burden on the Navy. My client was the Admiralty Section of the U. S. Department of Justice. Was it the Navy's fault for allowing the cylinder to fall, or was it the responsibility of the company that sold the cylinder to the Navy? In other words, what is a "perfect" weld?

The cylinder had been made in 1928, by roll welding the two heads into the cylinder body. One of those welds had split open. According to the Interstate Commerce Commission specification under which the cylinder was made, those welds were required to be "perfect." The former owner argued that one should not allow a chlorine cylinder to drop, so the quality should not be judged by what happens under such unusual and avoidable circumstances. Besides, the cylinder had been in service for more than twenty-five years, which should prove that the welds were perfect for the intended use.

In my report, I gave metallographic evidence that the weld was not perfect, according to any definition of the term with which I was familiar. I went further, and stated that accidents of the kind involved must be reckoned with, and therefore the weld should be correspondingly strong, and not just strong enough to contain the pressure. The weld failed, and therefore was less than perfect. I went still further, and gave my opinion that the cylinder should be capable of surviving a *much greater drop* without springing a leak.

About a year later, and before the case went to trial, we had another accident in Oakland. A truck went out of control on the elevated Nimitz Freeway and dumped five chlorine cylinders down to the earth below. The fall was about ten times that of the Navy cylinder, yet none failed. If one or more had, the escape of chlorine gas could have had disastrous consequences. One cylinder, in particular, had sustained much larger dents than the Navy cylinder.

On studying the cylinder with the large dents, I learned that it had been made by the same company, to the same ICC specification, by the same process, and in 1927, or one year before the cylinder in suit. Its welds had held and must have been strong enough to withstand a much greater drop. Reporting on this situation, I did not hesitate to call attention to this very neat confirmation of my prior attention. This, then, was also a case of a happy consultant.

*　　*　　*　　*　　*

An entirely different assignment came from a request of George B. McMeans, then vice-president of Kaiser Steel, to investigate the desirability of establishing a research and development department at the Fontana plant. The question was not so much how to as whether to. It was not to be an ivory-tower operation, but one that would benefit their operations and their market. Ever since my steel-mill experiences of World War I days, I have been fascinated by both products and operations, so I entered this assignment enthusiastically. Fine cooperation from the technical and operating staffs at Fontana indicated to me that they felt the need of a research and development department.

My first visit to Fontana convinced me that a research and development department should tend to Fontana's knitting, as its principal contribution, but that it should also be the center of information on both basic and industry-wide types of investigations, as the brains of the technical staff. I felt there would be enough work to keep a small yet diversified staff gainfully occupied.

I had no difficulty in finding specific problems that offered good prospects of worthwhile contributions. Accordingly, I had in mind a formal attack on immediate problems, which experience has clearly shown to be successful. It would be a game of skill rather than a game of chance, as someone has classified it.

I confess that one project in my mind was the problem of semikilled steel. My experience at A. O. Smith with skelp, at Battelle with Otiscoloy and ship steel, and at Pullman-Standard with bolster steel had demonstrated the quality and cost advantages of well-made semikilled steel. It had a good surface and ordinary mechanical properties, a good structure with no high-phosphorus bands, good weldability and notch toughness, and uniformity from heat to heat. We also had many data that demonstrated the wide variability in notch toughness and quality of commercial structural steels, which informed customers regarded with a jaundiced eye.

I reasoned that as customers became better acquainted with quality, and its advantages in fabrication and use, they would prefer a source with a reputation for high quality and uniformity. They would also want to purchase the product without paying a premium price. At A. O. Smith, that was back of our switch of orders of skelp to the one source, and it paid off handsomely. It also gave Otis Steel a cost advantage for Otiscoloy in a market that was dominated by more expensive fully killed steels. The old-timer would regard such a program as gilding the lily, but I had faith in quality coming to the surface, where it would be visible.

Research and development has a task force of special talents and experience. Its people are professional in their approach to their work. Furthermore, their compensation, advancement, and status in their fields depend upon the success of their work. No other department can provide the in-depth assistance of an R&D department when it is in the hands of qualified people who love the excitement of making contributions, and who must produce or else.

Some steel executives had already sensed the need to control their operations more firmly than in the old days. They also realized that the steel business had become complex, and that many details must be better understood if it were to meet the requirements of a developing technology. Production staffs have neither the time nor the facilities to provide that function. Logically enough, steel companies had established research departments to implement new programs, of which U. S. Steel was a prime example.

Undoubtedly, the work during and after World War II on structural steels, hardenability, weldability, the boron steels, and the National Emergency steels demonstrated the possibilities inherent in steel metallurgy. It also taught the economy and utility of determining in the laboratory, prior to production, composition ranges, treatments, and manufacturing variables. The production people have enough problems of their own, but they should be posted on the basics as a guide to good practice, and the target at which to aim.

Our development of Otiscoloy at Battelle for production at Otis Steel is just one example. ASTM now has more than 120 steel specifications and the SAE-AISI steels add many more, meeting most of the requirements of an advanced technology. These steels don't just happen these days; a lot of study precedes the standardization and manufacturing practice of each one.

Those were some of my thoughts when considering the type and the size of the research and development department I intended to propose. Steel technology had developed far enough then to convince me that the establishment of such a department would be a logical step to take. Incidentally, this extended to the steel salesman on the firing line. He should have more detailed information on his steels for sales in a developing technical market. I have run into the lack of this information, to the disadvantage of the supplier.

Looking back, many a laboratory contribution of one day has entered the technology of the next. Although one's experience is necessarily limited, I have found, however, that steel suppliers that have research and development backing have more essential information than those that do not. As consumer and user become more sophisticated in keeping abreast of the new technology, this differential is bound to be of advantage to the better-informed supplier.

Kaiser did not produce many of the steels I mentioned, yet its market included grades whose compositions, properties, and characteristics should be closely controlled for satisfactory performance in fabrication and end use. Furthermore, we already knew that impurity elements could be either harmful or beneficial, and that, when the steel is made right, alloying elements can improve steels seemingly beyond what might be expected from the small amounts that are added. I visualized close cooperation between research and development and production to make the steel right.

Although I don't mention production costs here, when I discussed them with staff members they seemed to feel that close studies of plant operations would have a good chance of reducing them. We figure that a small saving here, and another there, would help substantially to pay for research and development.

I felt strongly that Kaiser Steel should have a research and development department when I submitted my report. That completed my assignment. Since then, a number of prominent steel companies have followed the lead of Armco and U. S. Steel.

One final comment on my consulting experience. I was on an assignment by Borg-Warner to discuss with Walther Mathesius a situation that we thought he was best capable of handling. It was our last meeting. I recalled our first, in Charlottenburg in 1911, and asked if he had visualized the brilliant career he had just completed. After a moment's reflection, as though taken by surprise, he replied frankly that he hadn't.

Walther had started with U. S. Steel in a lowly capacity, but his technical talents were soon recognized and he advanced up the ladder through blast-furnace superintendent and general superintendent of the South Chicago mill, then of all U. S. Steel mills in the Chicago district, to vice-president in Pittsburgh and, finally, to president of Geneva Steel, at Provo, Utah. It was his vision and planning that led to U. S. Steel's decision to build Geneva, and he was the logical person to take charge. I was told by one who worked closely with him that Walther was an excellent manager who kept a close watch on steel quality and costs. From my visit at Provo, and from the viewpoint of a consumer, I felt that he had established good practice.

My consulting experience ran from research and development for a number of companies, through engineering design, construction, inspection, procurement, and failure analysis, for various individuals, companies, and organizations. Procurement stood out more and more as a vital, although neglected, engineering and economic function. By that I mean it is not given the attention it deserves. It is a field of its own, and goes well beyond working stresses, yield strengths, and price paid, and requires an understanding of the material and its production, as well as the way it will function in processing and service. I like to discuss procurement while design is still on the drawing board, because that is when I can be of most value to a client. I have had my share of cases of unintelligent procurement that has added extra costs, along with the uncertainties of service performance. The more I think of it, the more I believe procurement should be included in a course in metallurgy. As a student I was taught nothing about it, and as a teacher in academe I knew too little to cover it adequately. It should be taught in technical courses today, because practicing metallurgists and engineers should be posted on its principles. Securing adequate quality, yet paying no more than is justified, begins with procurement. Without attempting a primer on the subject, a few examples will bring out its significance.

A prominent case of poor procurement and its remedy is that of merchant-ship plate of World War II. Welded fabrication replaced riveting out of necessity, yet the same steel specification (unpedigreed A7) was used. A sidelight illustrates a blind spot in procurement. During the early experience with brittle ships, a citizen informer initiated a suit for $100-million against a steel company for shipping steel that failed slightly in tensile strength to meet the specification. That suit was brought under a Civil War statute. The argument stated that the steel was not as strong as it was supposed to be — no wonder the ships broke in two!

No fraud was involved here, only ignorance, and the suit was dropped in 1951. The blind spot was in placing significance on tensile strength and ignoring notch toughness. However, that was not the only confusion surrounding the problem of brittle ships and steel procurement.

A committee was formed on ship steel to look into the situation, and it met once in New York. Three of us advocated examining the steel quality of the failed ships, especially for notch toughness. We got precisely nowhere.

One resigned because he would not work with a group that was so blind to the effects of notches as a cause of the premature failures. The second was an Englishman who also gave up, saying that an English blacksmith knew more about notches than the committee did. I, being more accustomed to frustration, stayed on to see what would happen. Fortunately, that committee was dissolved; it was not until a Committee on Ship Steel was formed that we helped effectively on steel. Its sponsorship of a substantial program at Battelle on the effects of manufacturing variables on the quality of ship steel developed the information that steel producers should have to produce material of *adequate* quality at a reasonable cost.

Steel procurement, based on sound principles for making notch-tough steel, was ultimately recognized as the most effective factor in combating brittle ships. Knowing how to buy steel — procurement — at last got our welded-ship program off on a sound basis. At the same time, welding and inspection were also improved.

Lloyd's of London introduced a new angle in steel procurement that is an improvement on simply preparing a standard specification for all suppliers. Their inspectors studied manufacturing practice, either to approve modifications or to require them, before accrediting suppliers. That is the practical equivalent of prequalification, a procurement device that has great merit when the low bid must be accepted.

Another important element of procurement arises in connection with the purchase of fine-grain steel. On its face, it appears to be a high-quality steel, but that is not so unless it is made right and is also normalized. I had an experience with a premium fine-grain steel that was in the hot rolled condition but with no notched-bar test included for acceptance. Charpy tests showed that it was notch brittle at the low service temperature. As a bit of missionary work, I had specimens normalized. This time, the Charpy values were satisfactory. It is a good question — why pay for a premium steel if it is notch brittle, when the application requires it to be notch tough? It gives no more assurance against brittle fracture than a cheaper steel would. It would also carry a higher stress, making it even more vulnerable. ASTM specifications that do not, but should, require the check of a reliable notched-bar test are inadequate in this respect, and the unwary purchaser pays for quality he may not get.

A surprising case of a steel of the T-1 type gives an example. The low carbon content, judicious use of alloying elements, plus a full quench and temper, presumably ensures low-temperature notch toughness. But this is not necessarily true, especially if it is purchased to an ASTM specification that has no Charpy-test requirement. When I ran a transition-temperature curve on this particular steel, I had to test above 200 °F to get acceptable values. Ordinarily, these steels remain notch tough down to low subzero temperatures.

A quality feature even more difficult to control by standard specifications is what I call the wrought iron effect, or weakness in the thickness direction of a rolled member. When a wrought iron bar is nicked and then bent, it starts to break at the notch, but as soon as the fracture encounters the longitudinal slag stringers it turns abruptly and runs along them. The notch is eliminated, and from then on the bar bends like a plain bar. That is why wrought iron enjoys a reputation for being tough.

Improperly made steel can have a somewhat similar structure that is missed, however, by conventional mechanical tests, including Charpy tests. This weakness is likely to occur in fine-grain steel that some say is "dirty steel." To minimize this effect, the steel must be made right, and the correct amount of aluminum must be added to avoid the interdendritic sulfides that produce lamellar inclusions. Their effect can be demonstrated with the Charpy bar by making the notch parallel to the plate surface, instead of the specified perpendicular orientation. This test gives abnormally high impact values.

If the condition is truly bad, ultrasonic inspection detects it because the sound beam is dispersed and does not penetrate to the opposite surface of the part. The T-joint test is a formal test of this condition that Leon Bibber publicized when he was with the Navy, and which we studied at Battelle. Still another way is that of making a tension test of a coupon that is cast when the heat is poured. As Matuschka showed at the Madrid meeting of the International Foundrymen's Congress, the percentage elongation is abnormally low when sulfide envelopes separate the dendrites. Of course, low sulfur is beneficial.

Broadly speaking, procurement involves materials of many different properties and characteristics such as corrosion, stress corrosion, creep behavior at elevated temperatures, resistance to scaling or surface deterioration, cryogenic behavior, delayed cracking that is erroneously called static fatigue, weldability, forming, deep drawability, hardenability, wear, cutting efficiency, stress/weight ratio, expansion coefficient, magnetic properties, conductivity and brittle fracture. And others, in addition to strength.

The T-1 steel of U. S. Steel and similar steels of other producers offer an exercise in procurement. Good use is made of these steels, but they have problems that are inherent in the product. ASTM has two specifications that may be used for these steels, or else "T-1 or equivalent" steel can be ordered. Those steels should be notch tough, although, unless a notched-bar test is specified, there is no guarantee of that quality. I am always reminded of that "T-1 or equivalent" steel that was notch brittle even at high summer temperatures, whereas brochures and literature cite good notched-bar values at low subzero temperatures. The unsuspecting can trip up badly here, when there is no requirement for guaranteed notched-bar values.

"Know your supplier" is good advice in procurement. This is true even for steels of presumably high quality, but it becomes critical when a semikilled steel, if well made, would be adequate. Mills do not ordinarily accept orders for this type of steel if a Charpy test is included, but good steel *can* be secured if you know your supplier.

Why all this emphasis on quality and on Charpy tests? The adoption of welding in fabrication and construction is a major reason, combined with advances in the engineering arts. As I pointed out in my Adams Lecture before AWS, welding produces a monolithic structure which makes it vulnerable to a running fracutre. More sophisticated designs require higher and more uniform quality. To keep pace with technological advances, new specifications have been added, and the process is in a state of flux. The A7 Specification of ASTM, as an example, had an old and venerable life, but it has now passed into the limbo of history. This is in spite of the relatively low incidence of service failures due to steel, or the fact that early structures of presumably poor steel, such as the Brooklyn and Eads bridges, are still going strong.

As a comment, an interesting account of Eads' bridge over the Mississippi River at St. Louis is given in "Catfish and Crystal," by Ernest Kirchsten. Andrew Carnegie had a contract to supply the steel, but he canceled it, claiming that Eads had adopted an unspecified quality of steel. The substitute steel is not described, but Eads had it made by the Butcher Steel Co., of Philadelphia. He installed a special testing machine at the mill for acceptance tests. After overcoming many obstacles in building the bridge, General Sherman drove the last spike, and the bridge was ready for use.

Eads devised an unusual nondestructive test for acceptance. To demonstrate the soundness and safety of his bridge, he simply rounded up fourteen locomotives and ran them out on the bridge. Kirchsten refers to Eads' structure as a "magnificent bridge which still serves and adorns St. Louis" (1960).

An excellent review of these problems appeared in a 1963 paper by H. Thielsch for ASME, "Quality Control and Service Performance." He states: "The level of quality specified should be commensurate with the service conditions. The tests specified should be sufficient to insure satisfactory performance." The key phrases are level of quality, tests specified, and satisfactory performance.

I was interested in his comments on specifications. Thielsch wrote that various agencies and companies "increasingly consider the technical society and trade association standards inadequate." (That reads like a mild Rickover!) It also agrees with my own experience, and that of others who have expressed opinions.

This situation requires a little closer scrutiny. Thielsch's inadequacy may be the omission of desirable restrictions (blind spots), or too-wide spreads in allowable compositions or properties.

Speaking of codes and specifications, although they are mostly sound, I have found them to have blind spots. The Board of Contract Appeals of the U. S. Army Corps of Engineers pointed out the glaring code deficiency of allowing four-foot-long cracks if, or when, not detected by radiographic inspection. (I have heard some insist that a weldment was sound because it had passed 100% X-ray inspection.) The ultrasonic method for detecting deep-seated cracks was known at the time, yet was not specified. This situation is unacceptable when a weldment is simply required to meet a certain code. A somewhat similar situation exists when steel is procured by simply writing in a specification number. When hazardous service is involved, which I assume was so here, I used to wonder at the failure to keep pace with metallurgical knowledge and inspection practice.

A good example of a blind spot in a specification is the omission of a notched-bar test when notch toughness is expected and needed. This is true for steels "with improved transition properties," structural steel for bridges and ships, pressure vessels, fine-grain steels, and even for the quenched-and-tempered steels. The innocent user of such specifications will feel that he will receive steel with adequate resistance to brittle fracture, but that is not necessarily so. Although it is true that these steels are used, a risk of premature failure is run unless nondestructive testing is used to guard against dangerous defects and design notches.

There is a natural spread in composition and properties of production heats that must be permitted to the producer if he is to keep his manufacturing cost and sales price down. I found evidence of this spread when canvassing producers for data for the ASME handbook, "Metals Properties." It is also well documented by the distribution plots in Volume 1 of the ASM "Metals Handbook." When a tighter specification is used, or heats are selected to meet it, the producer must accept the risk, and the consumer must pay the price, of higher rejections, or of diversions to other orders. However, one should not modify a loose specification unless fabrication costs justify it or service reliability requires it. Knowing when to do so is a part of good procurement.

I have found that manganese is mishandled in some specifications. It is an important additive or alloying element in many structural steels, yet some specifications give only a maximum value. That looks good on paper, yet if no minimum value is included the steel must be accepted even though manganese is lower than is desired for the application. If only a single value is specified instead of a range, I would prefer it to be a minimum; the producer will not add much more.

Additional provisions for quality and tighter restrictions increase manufacturing costs, which must be paid for by the user. A pertinent question, however, is — why pay for quality and not get it? Procuring steel today is no longer the simple process it was in the past. It requires more knowledge on the part of the consumer if he is to get what he needs without paying for something he does not use. Good enough is best.

One point that consumer and user must accept is the possible higher cost of incoming material. As things become larger or smaller, more complex and critical in performance, and the consequences of failure more drastic, better, and usually more expensive, materials are needed.

It is precisely in procurement that an experienced metallurgist can help both consumer and user to secure the needed steel or other metal without paying more than is necessary, by taking fabrication and inspection costs as well as material costs into account while improving the probability of satisfactory service performance and reducing the chances of service failure and liability costs.

I believe I have made it clear that dilemmas arise as to whether to underbuy for costs or to pay more for security, and as to whether to use a specification as written or to modify it. These situations call for decisions that are based on knowledge and experience. In addition, new knowledge of metallurgy and new and more exacting technology have changed procurement philosophy — early in some instances and late in others — and as we ask more of metals, we are less lenient of failures as acts of God. The start is procurement.

Before passing on to experiences with management, which is involved in all decision-making fields, several subjects that bear on metallurgy in consulting should be considered, however briefly. They are patents, specifications, and product liability.

The patent system is beneficial in its over-all effect, yet ownership of a patent frequently has unhappy consequences. The patent grant is intended to benefit the public by improving technology. To achieve this, the government encourages and promotes invention by protecting the interests of the inventor by granting a seventeen-year monopoly. However, the need to secure validity of a patent by adjudication at the hands of trial lawyers and civil judges means that the inventor can lose his monopoly prematurely. I have cited the statement of an executive that his company had never won a patent case as plaintiff, and had never lost one as defendant. The patents were all legally granted by our government.

In some instances, the inventor has a choice between securing a patent and facing a possible lawsuit, and keeping his invention as secret practice. If he sells a product on the market, he doesn't have much of a choice. Actually, only important patents are likely to show up in a law court, not useless ones. I recall a patant that reversed the usual steps of melting and refining steel — the deoxidizer was added first and iron ore last. I doubt if it was infringed or adjudicated.

As examples of what can happen, I mentioned the A. O. Smith cellulose-covered welding electrode case and the two Carboloy cases, involved patent cases covering epoch-making inventions. In both instances, large and valuable industries were established as a direct result. The public benefited enormously, but patent validity was denied and the ingenuity of the inventors only partly rewarded.

This is no place to discuss legal argument, but my contacts with patent and civil cases remind me of the Lord's question to Job — "Who is this that darkeneth counsel by words without knowledge?" (Job 38:2).

This experience suggests that we should have a technical judge sitting with the civil judge on cases that require decisions on technical points. The technical judge would issue the controlling decision on technical points, and his decision would be binding on the civil judge. As an example, the Smith patent and our Carboloy patents would have been correctly ruled valid and infringed.

In *Metalworking News* for September 27, 1971, in a report on the status of the basic-oxygen patent suits, I read an interesting confirmation of the need of a technical judge for such cases. It came in a ruling of the Supreme Court on the right of a patent holder to relitigate a court decision of validity if "there is proof that the judge declared the patent invalid because he did not understand the technical intricacies of the subject matter."

In closing my comments on patents, I suggest "Open Letters," the President's Lecture to the Institute of Metals, in 1969, by H. W. G. Hignett. It is excellent reading.

Turning to specifications, they are house rules that control purchases of materials or goods. I have already mentioned that some specifications have blind spots, and that others should at times be modified or amplified, the better to meet the requirements of fabrication and end use. Blind spots can be especially disagreeable, and can give rise to controversies between the supplier and the consumer. As this is written, I have a case that involves steel "made to fine-grain practice." The steel is so laden with foreign inclusions or stringers of "dirt" that it is vulnerable in fabrication and in use, and to stresses that attack the steel in the Z, or thickness (short transverse), direction. The consumer is placed in a dilemma. He expects to get good steel that involves no extra costs in fabrication and inspection, yet if he includes a clause in his purchase specification that limits the inclusions to an acceptable level, he may price himself out of the market.

The advent of much-improved controls in steelmaking, and a better understanding of the processes, should make it possible to maintain a more uniform production and higher quality, hence fewer missed heats in production. I believe that some specifications are too loose for many applications, but I don't doubt that if they are tightened to ensure adequate quality, the customer would have to pay too high a premium. Even steel of subnormal quality can be used if proper safeguards in design, fabrication, and nondestructive testing are utilized. Use of the term "quality" is all-inclusive. For example, although carbon is an important alloying element in steel, if it is too high it is in effect an impurity.

I have been reading lately about product liability; apparently, the number of those cases is increasing. I have wondered if a consultant can become involved as a defendant in a damage suit, and if he must have an attorney review each report he writes. I have had to determine the cause of failure by methods of failure analysis, but as a consultant and expert witness, I had no responsibility for the damage. But what would be my responsibility if I advise what to do beforehand? Loose specifications, or specifications that are modified, could produce some interesting arguments. Court cases that fall in this category should use a technical judge.

Management is the echelon at which procurement and production are brought together for comprehensive coordination with policy on markets, finance, personnel. The metallurgist has his important role of assembling facts and presenting them to management, in understandable and convincing fashion. I think of Glen Riegel in this connection, when, as chief metallurgist of Caterpillar Tractor, he had his company pay more for steel by insisting on tighter procurement, including notch toughness, but was able to show that the extra material cost was worth it in terms of smoother production and enhanced reliability in service.

This function of the metallurgist should be brought to the attention of students as a part of their training, and to broaden their understanding of the role and opportunity of their profession. As Finnaston stated, we should divest ourselves of any "mouse-like" qualities and view our work as an integral and important function of the company's operation. I have given examples, and other metallurgists must have improved their functions in the same manner upon broadening their outlook. Of course, the recent tendency to advance materials people into management makes communication easier.

Speaking of management, the story of Ernst Abbé has fascinated me ever since I lived in Jena with the family of an employee of the Carl Zeiss Optical Works. Abbé was the model of a great manager. His series of apochromatic objectives is one of the great achievements of science and technology. Although we now take his optics for granted, we must be forever grateful for his genius in making high-power microscopy a reality. He started as a scientist in optics, but later he dominated Carl Zeiss in management.

Prior to his start, in 1870, the optical microscope was limited to low magnifications. Attempts to reach high magnifications failed because the images were blurred, lacking in detail, and had color fringes. To break new ground, Abbé decided that he must develop the theory of image formation, including resolving power, if he were to improve microscopic examination. His work led to the identification of spherical and chromatic aberrations and the chromatic difference of spherical aberration as the reasons for imperfect images and color fringes.

There was one huge deficiency, however. No optical glasses were available with which to implement theory. With the financial assistance of the Prussian government, he developed a whole new system of optical glasses with Schott and his associates. With them he was able to make his apochromatic lenses, which he announced in the classic paper of 1886 before the Jena Society for Medical and Natural Sciences.

For reasons best known to him, although they have been ascribed to his reticence, Abbé did not patent his inventions. The only advantage to Carl

Zeiss was the virtues of the mineral fluorite, as a lens material, which he played down.

When I was in Jena, Abbé was revered as the principal architect of Carl Zeiss, and for his humanitarian treatment of his employees. With complete disregard of personal emolument, he turned the ownership of the company over to the workers and established the Zeiss Foundation for their benefit during and after employment. His motivation stemmed from the sad plight of the industrial workers of the 1840s. In terms of a biographer, "he left full freedom of thought and action to each man, requiring only of them devotion to their particular tasks." That seems to sum up his management philosophy.

To return to experience with management, mine has varied from satisfactory to unsatisfactory, in proportion to management's comprehension of metallurgical principles and practices.

Nondestructive testing is not a metallurgical function, yet the metallurgist should have a guiding voice in its application and in the evaluation of the results. It is a new profession that has come a long way since Lester's early radiographic inspection of welds and castings. A mighty example is that of Saturn V and the Apollo moonships. According to R. W. Neuschaefer of quality and assurance at the Marshall Flight Center, X-ray and neutron radiographic, magnetic-particle, eddy-current, liquid-penetrant, thermal, and light-ray inspections were used. Those inspections located defects and permitted necessary repairs to be made. The success of those flights attests to the importance of nondestructive testing.

How much inspection should be used, or what is sound inspection philosophy? The decision on this, and on methods, must rest on judgment and experience for each case, unless a code or other requirement must be met. Another point is the decision on which defects are to be removed and which are judged to be innocuous. Although the use of fracture-toughness testing provides a sound procedure for determining the maximum size of a harmless defect, that technique is seldom used for production. In many instances, the defect is so large that it must be removed. In others, the rejection limits that have been established for radiographic inspection are used. When the fracture toughness of the material is not known, the decision must be arbitrary, unless Bill Pellini's technique can be used. Codes and standards insert a somewhat implausible clause that all cracks must be removed. This raises questions of when is the defect (subsurface) a crack, and how small a crack can be detected by the inspection method?

When a defect in a structure is discovered accidentally, my practice is to have a more complete inspection performed. Several times, this has revealed a condition that the engineers could not tolerate and required extensive repairs for the safety of the structure. I was pleased to note a